Fire Mountains of the Islands
A History of Volcanic Eruptions and Disaster Management

in Papua New Guinea and the Solomon Islands

Fire Mountains of the Islands
A History of Volcanic Eruptions and Disaster Management
in Papua New Guinea and the Solomon Islands
R. Wally Johnson

Australian National University

E PRESS

Published by ANU E Press
The Australian National University
Canberra ACT 0200, Australia
Email: anuepress@anu.edu.au
This title is also available online at http://epress.anu.edu.au

---

National Library of Australia Cataloguing-in-Publication entry

Author: Johnson, R. W. (Robert Wallace)

Title: Fire mountains of the islands [electronic resource] : a history of volcanic eruptions and disaster management in Papua New Guinea and the Solomon Islands / R. Wally Johnson.

ISBN: 9781922144225 (pbk.) 9781922144232 (eBook)

Notes: Includes bibliographical references and index.

Subjects: Volcanic eruptions--Papua New Guinea.
Volcanic eruptions--Solomon Islands.
Emergency management--Papua New Guinea.
Emergency management--Solomon Islands.

Dewey Number: 363.3495095

All rights reserved. No part of this publication may be reproduced, stored in a retrieval system or transmitted in any form or by any means, electronic, mechanical, photocopying or otherwise, without the prior permission of the publisher.

Cover image: John Siune. 'Dispela helekopta kisim Praim Minista bilong PNG igo lukim volkenu pairap long Rabaul'. 1996. 85 x 60 cm. Acrylic on paper mounted on board. R.W. Johnson collection. Intellectual property rights are held by the artist.

Cover design and layout by ANU E Press

Printed by Griffin Press

This edition © 2013 ANU E Press

Contents

Tables	ix
Illustrations	xi
Foreword	xvii
Acknowledgements and Sources	xxi
Volcano Names and Totals	xxiii

1. Burning Islands and Dampier's Voyage: 1700 — 1
Track of the *Roebuck* 1
Near Oceania, Melanesia and Melanesians 9
Early Ideas about Volcanic Activity 14
Preview 17

2. Volcano Sightings by European Navigators: 1528–1870 — 21
Saavedra, Retes and Mendaña 21
Schouten, Le Maire and Tasman 24
Carteret, Hunter, D'Entrecasteaux and Parker Wilson 26
European and Melanesian Viewpoints 33

3. European Intruders and the 1878 Rabaul Eruption: 1870–1883 — 39
Blanche Bay and the Tolai 39
Miklouho-Maclay 43
Traders, Missionaries and a Gentleman Explorer 46
1878 Eruption at Rabaul 50
Powell's Voyage and a Possible Eruption 'Pulse' 54
Volcanological Events Elsewhere 59

4. Volcanic Events of the German Era: 1884–1914 — 63
Colonial Partitioning 63
Ritter Island Disaster 65
Hahl and Sapper 71
Time Cluster of Eruptions 77

5. Australian Colonists and the Volcanoes of Mainland New Guinea: 1849–1938 — 85

First Impressions 85
British New Guinea and Victory Volcano 87
Evan R. Stanley in Papua 90
Australians in the Territory of New Guinea after 1920 93
Australians in the Territory of Papua before 1938 97

6. Calderas, Ignimbrites and the 1937 Eruption at Rabaul: 1914–1940 — 103

Garrison Life and Volcanoes 103
Australian Expedition along the Bismarck Volcanic Arc 106
Calderas and Ignimbrites 109
Eruption at Rabaul in 1937 112
Subsequent Investigations at Rabaul 119

7. Eruptions during the Pacific War and Postwar Recovery: 1941–1950 — 129

Fisher and Renewed Activity from Tavurvur 129
Kizawa and the Sulphur Creek Observatory 131
Eruptions at Goropu Volcano, Papua 136
Hiroshima, Surges and Postwar Recovery 140
Changing the Volcanological Leadership 143

8. Disaster at Lamington: 1951–1952 — 149

Higaturu and the Orokaiva 149
Build-up to Catastrophe 151
Relief and Recovery 156
Seeking Explanation and Meaning 167
Aftermath 171

9. Tony Taylor and an Eruption Time Cluster: 1951–1966 — 179

Eruptions of 1951–1957 179
Experiments in Prediction 182
Long Island Evacuation of 1953–1954 183

Bam Tragedy of 1954–1955 187
Evacuation of Manam and the 1956–1966 Eruptions. 193
Tuluman 1953–1957 and the Obsidian Miners of Lou 199

10. Plate Tectonics and False Alarms: 1960–1972   207

Advances in Science and Technology 207
Gas Emissions from Two Highlands Volcanoes 212
Volcanic-Disaster Preparations at Wau Township in 1967 214
Evacuation from Dawson Strait in 1969. 217
Origin of the Dawson Strait Earthquakes and a Note on Volcanic
False Alarms. 221
Tectonic Earthquakes and the End of the Taylor Era 224

11. Cooke-Ravian and a Volcanic Resurgence: 1971–1979   231

New Eruption Time Cluster 231
Ulawun and the Threat of Cone Collapse 235
Long Island Disaster and Tibito Tephra. 239
Yomba and Cook: Two 'Mystery' Volcanoes. 245
Fatal Eruption on Karkar in 1979 246

12. Eruption Alert at Rabaul Caldera: 1971–1994   255

Crisis Build-up and Stage-2 Alert. 255
Scientific Responses to the Caldera Unrest. 263
Worldwide Volcanic Crises and Developments in Risk Awareness 267
Costs, Benefits and Crisis Decline: 1985–1994 273
Volcanic Alert on Simbo Island 277

13. Eruptions at Rabaul: 1994–1999   283

First Three Weeks. 283
Physical Damage 289
Post Mortem and New Directions for RVO 293
Ongoing Eruptions and New Insights 298
Restoring the North-eastern Gazelle Peninsula. 303

14. Eruptions of the Early Twenty-first Century: 1998–2008   311

International Developments and Modern Near Oceania 311

Ulawun: A Decade Volcano 313

Ulawun 2000: A Short-lived, Powerful Eruption 316

Pago 2002–2003: An Unexpected Eruption 319

Threat of a Caldera-Forming Eruption at Pago-Witori 322

Manam 2004–2005: Abandoning a Volcanic Island? 326

Rabaul 2006 and other Volcanic Crises in 1999–2007 330

15. Reassessing Volcanic Risk in the North-eastern Gazelle Peninsula: 2000–2012   341

Restructuring a Society 341

Volcanic Hazards 344

Early Warnings and a New Model for Rabaul Volcano 348

Determining Risk Today 351

16. Historical Analysis and Volcanic Disaster-Risk Reduction   359

Patterns in the Historical Record 359

Evacuations, Early Warnings and False Alarms 363

Artefacts and Oral Traditions 367

International Disaster-Risk Reduction 372

Observatories and Volcanic Disaster-Risk Reduction 373

Strengthening At-Risk Communities in Near Oceania 376

An Epilogue   381

Appendix: Acronyms and Glossaries   383

Index   387

Tables

Table 1. Volcanoes in Eruption in Near Oceania from 1875 to 1878	58
Table 2. Volcanoes in Eruption in Near Oceania from 1884 to 1899	78
Table 3. Volcanoes in Eruption in Near Oceania from 1951 to 1957	180
Table 4. Volcanoes in Eruption in Near Oceania from 1972 to 1975	234
Table 5. Stages of Volcanic Alert at Rabaul	258
Table 6. VEI Values for Major Eruptions	268
Table 7. Dates of Major Eruptions from Witori (W–K) and Dakataua (Dk) Volcanoes	324
Table 8. Nine Bismarck Volcanic Arc Volcanoes in Eruption or Restless between August 2002 and October 2006	334
Table 9. Thirteen Evacuations in Papua New Guinea	363

Illustrations

Volcano Names and Totals

Volcanoes of Near Oceania	xxiii

Chapter 1

Figure 1. Map by William Dampier showing track of the Roebuck in 1700	2
Figure 2. Ulawun volcano on map and sketch by William Dampier in 1700	4
Figure 3. Ritter Island as seen by William Dampier in 1700	5
Figure 4. Long and Crown islands as seen by William Dampier in 1700	7
Figure 5. Portrait of William Dampier	8
Figure 6. New Oceania showing selected volcanoes and modern bathymetry	11
Figure 7. Giant beneath Etna volcano in 18th century engraving	15

Chapter 2

Figure 8. Detail of New Guinea area from the 17th Century chart *Insulae Molvccae*	23
Figure 9. Manam Island as seen by Abel Tasman in 1643	25
Figure 10. Rabaul volcanoes as seen by Philip Carteret in 1767	27
Figure 11. Rabaul volcanoes as seen by John Hunter in 1791	28
Figure 12. Bagana volcano as seen by John Parker Wilson in 1842	31

Chapter 3

Figure 13. Rabaul Harbour as mapped by Simpson and Greet in 1872	41
Figure 14. Fergusson Island geothermal area as seen by John Moresby in 1874	43
Figure 15. Miklouho-Maclay in staged photograph probably in late 1870s	44
Figure 16. Manam Island as painted by Nikolai Miklouho-Maclay in 1877	46
Figure 17. Rabaul Harbour in 1875 as shown in map by G.E.G. von Schleinitz	48
Figure 18. Portrait of Wilfred Powell	49
Figure 19. The larger of the Beehives in Rabaul Harbour in 1883	53
Figure 20. Lolobau Island as seen by Wilfred Powell in 1878	55
Figure 21. Ulawun and Bamus volcanoes as seen by Wilfred Powell in 1878	56
Figure 22. Volcanoes of the Willaumez Peninsula area as mapped by Wilfred Powell in 1878	57

Chapter 4

Figure 23. Manam Island as seen by Otto Finsch in 1884	66
Figure 24. Ritter Island as seen by G.E.G. von Schleinitz in about 1887	68

Figure 25. Aerial view of modern Ritter Island — 70
Figure 26. Rabaul Harbour after the 1878 eruption as shown in map published in 1888 — 72
Figure 27. Portraits of Karl Sapper and Albert Hahl — 73
Figure 28. Bamus volcano as seen during voyage by L. Couppe in 1894 — 79

Chapter 5

Figure 29. Detail from 1875 map of journey in highlands of New Guinea by J.A. Dawson — 87
Figure 30. Volcanic features of Mount Victory — 89
Figure 31. Evan R. Stanley and family in 1919 — 91
Figure 32. Detail from 1924 geological map of Papua — 92
Figure 33. Modern view of Giluwe volcano — 96
Figure 34. Profiles of Bosavi volcano — 98
Figure 35. Volcanoes of the Fly-Highlands province — 99

Chapter 6

Figure 36. Australian troops at entrance to Rabaul Harbour in about 1918 — 104
Figure 37. Australian troops at Tavurvur, Rabaul, during the First World War — 105
Figure 38. Pago volcano in eruption in 1918 — 107
Figure 39. View of Dakataua volcano in 1921 — 109
Figure 40. Krakatau-type caldera formation — 110
Figure 41. Glen Coe-type caldera formation — 111
Figure 42. Rabaul area shortly after the 1937 eruption — 114
Figure 43. Vulcan eruption in 1937 on front page of Daily Telegraph — 115
Figure 44. Lightning in Vulcan eruption cloud in 1937 — 116
Figure 45. Vulcan and Tavurvur in reduced eruption in 1937 — 119
Figure 46. C.E. Stehn visiting the *Durour* in 1937 — 121
Figure 47. N.H. Fisher at Vulcan in 1937 — 123
Figure 48. Volcanological observatory building at Rabaul in about 1940–1941 — 124

Chapter 7

Figure 49. Tavurvur in eruption in 1941 — 130
Figure 50. Takashi Kizawa at the Sulphur Creek Observatory, Rabaul — 133
Figure 51. Bomber attack at Rabaul Harbour in 1943 — 135
Figure 52. Devastated area at Goropu volcano in 1943 — 138
Figure 53. Goropu volcano as seen on modern topographic map — 139

Illustrations

Figure 54. Base surge at Long Island in 1955	141
Figure 55. Omori seismograph at Rabaul in the early 1950s	145

Chapter 8

Figure 56. Aerial view of Lamington eruption cloud on 21 January 1951	155
Figure 57. Area of volcanic destruction at Lamington in 1951	158
Figure 58. Victims on the road to Higaturu in 1951	160
Figure 59. Destruction from pyroclastic surge near Higaturu in 1951	161
Figure 60. Requiem mass on a jeep on road to Higaturu in 1951	162
Figure 61. Tony Taylor at Popondetta Airstrip in February 1951	163
Figure 62. Aerial view of Mount Lamington from the north in February 1951	164
Figure 63. Visit to active lava dome and crater on Lamington in February 1951	165
Figure 64. Pyroclastic flow at Sangara Plantation in March 1951	166
Figure 65. Cross section through an active pyroclastic flow	168
Figure 66. Ceremonial opening of the memorial cemetery at Popondetta in November 1952	173
Figure 67. Leslie Topue being awarded his British Empire Medal in November 1952	174

Chapter 9

Figure 68. Taylor descending into crater at Langila volcano in 1952	181
Figure 69. Long Island topography and settlements	185
Figure 70. Surtseyan eruption from Lake Wisdom, Long Island, in 1953	186
Figure 71. Bam Island topography	189
Figure 72. Bam Island from the south-west in 1970	190
Figure 73. Bam islanders evacuating by boat in 1954	190
Figure 74. Radial valleys and summit craters of Manam Island	194
Figure 75. Village damage on Manam Island in 1958	195
Figure 76. Pyroclastic flow in North East Valley of Manam Island in 1960	197
Figure 77. South West Valley of Manam Island in 1963	198
Figure 78. Admiralty Islands including Lou and Tuluman islands	199
Figure 79. Tuluman volcano in eruption in 1957	201

Chapter 10:

Figure 80. Small surtseyan eruption at Kavachi volcano in 1977	208
Figure 81. Subduction and formation of a volcanic arc	209

Figure 82. Leslie Topue and Ben Talai in the RVO recording room at Rabaul — 211

Figure 83. Landslide at Koranga Crater in 1967 — 215

Figure 84. Volcanoes and settlements of Dawson Strait — 218

Figure 85. Boundaries of tectonic plates in Papua New Guinea and the Solomon Islands — 223

Figure 86. Madang and Josephstaall earthquake aftershock zones — 225

Figure 87. Portrait of G.A.M. Taylor — 226

Chapter 11

Figure 88. Portrait of R.J.S. Cooke — 233

Figure 89. Ulawun volcano in 1967 — 236

Figure 90. Ulawun volcano topography and settlements — 237

Figure 91. Incandescent lava flows at Ulawun in 1978 — 238

Figure 92. Matapun Beds on Long Island — 242

Figure 93. Long Island and inferred original thicknesses of Tibito Tephra — 243

Figure 94. Karkar volcano topography and the seismograph/Kinim link used in the 1970s — 247

Figure 95. Devastated campsite on Karkar volcano in March 1979 — 249

Figure 96. Chris McKee at Karkar observation camp in 1979 — 250

Chapter 12

Figure 97. Blanche Bay and Rabaul town in the 1970s — 257

Figure 98. Aerial photograph mosaic of north-eastern Blanche Bay — 259

Figure 99. Monthly earthquakes totals at Rabaul for 1971–1988 — 261

Figure 100. Two different patterns for the 'seismic annulus' at Rabaul — 265

Figure 101. Cross-section through magma reservoirs beneath Rabaul Caldera — 266

Figure 102. Estimated airfall-ash thicknesses for a large eruption at Rabaul — 272

Figure 103. Ceremony commemorating the 50th anniversary of the 1937 Rabaul eruption — 275

Chapter 13

Figure 104. Ash clouds from Tavurvur on 19 September 1994 — 285

Figure 105. Eruption from inclined vent at Vulcan on 19 September 1994 — 286

Figure 106. Westward progress of Vulcan eruption cloud on 19 September 1994 — 287

Figure 107. View from space of Rabaul eruption clouds on 19 September 1994 — 288

Illustrations

Figure 108. Damage in part of Rabaul town — 290
Figure 109. Relative ash-damage map for Rabaul town — 291
Figure 110. Portrait of Ben Talai in about 1999 — 297
Figure 111. Strombolian eruption at Tavurvur volcano on 14 March 1997 — 299
Figure 112. Monthly earthquakes totals at Rabaul for 1968–1994 — 300
Figure 113. Epicentres of 'north-east' earthquakes for 1994–1998 — 301
Figure 114. Seismic velocities beneath the Rabaul area — 302
Figure 115. Population movements from Rabaul to the Warangoi Valley area — 304

Chapter 14

Figure 116. Volcanoes of central-north New Britain — 314
Figure 117. Volcanic-cone collapse and tsunami generation — 315
Figure 118. Pyroclastic flow at Ulawun volcano in 1985 — 317
Figure 119. Lava issuing from Pago volcano during 2002–2003 — 320
Figure 120. Contoured thicknesses of lava flow erupted at Pago volcano in 2002–2003 — 321
Figure 121. Obsidian artefacts from Talasea-Mopir area of New Britain — 325
Figure 122. Pyroclastic flows at Manam volcano in 1996 — 327
Figure 123. Manam volcano imaged from space in November 2004 — 329
Figure 124. Tavurvur volcano in explosive eruption in October 2006 — 333

Chapter 15

Figure 125. Economic zones and new settlement belt in northern East New Britain Province — 343
Figure 126. Ash-fall on Rabaul including the Cathedral of St Francis Xavier — 345
Figure 127. Three volcano systems of Rabaul volcano — 350
Figure 128. Two magma reservoirs beneath the Rabaul area — 351
Figure 129. Night-time view of Tavurvur in eruption and town lights of Rabaul and Kokopo — 353

Chapter 16

Figure 130. Jonathan Kuduon at Bokure village, Manam Island — 377

An Epilogue

Internal volcano architecture by C.K. Wungi — 381

Foreword

The famous British geologist, Arthur Holmes, was greatly admired by my geology teacher at Gateshead Grammar School on Tyneside in the north-east of England. This was not just because Holmes was author of the definitive textbook *Principles of Physical Geology* (1944) that we used in class in the late 1950s, but also because Holmes was a local lad, a Tynesider, and therefore a Geordie. Holmes' geological researches — including determining the age of the Earth using the principles of radioactivity — propelled him to international fame, if not geological immortality.

Holmes as a boy lived at 19 Primrose Hill, Low Fell — the same part of Gateshead where I was born and raised. I don't recall being particularly impressed by Holmes' achievements at school as my clear ambition was to become a veterinary surgeon rather than a geologist. I became more familiar with the importance of Holmes' contribution to the evolution of geological ideas, however, after the revolutionary theory of plate tectonics emerged in 1967 and after I arrived in Papua New Guinea as a volcanic geologist in 1969. Holmes said that heat is generated deep within the Earth by radioactivity, which in turn generates large-scale convection currents that bring hot rocks towards the Earth's surface. These rocks melt to form magma as a result of their upward, convective transport into a lower pressure environment closer to the Earth's surface. Some of the magmas so formed may be erupted from volcanoes. This simple idea of 'decompression melting' and magma eruption can be traced back to the mid-nineteenth century.

The 1960s were a formative decade in other ways too. 1960 was when I started, rather unenthusiastically, an undergraduate course in geology at the magisterially named Imperial College of Science and Technology, part of the University of London. And in 1960, too, British Prime Minister Harold MacMillan gave his 'Winds of Change' speech in Africa, acknowledging the end of imperialism — at least of the British sort. A further step in my commitment to volcanic geology followed a field season of undergraduate mapping in eastern Iceland in 1962, supervised by a motivational lecturer in mineralogy at Imperial, G.P.L. Walker — a brilliant observer in the field, startlingly insightful in his geological interpretations, and infectious in articulation of his geological knowledge. There was no escape from such geological mentoring. I therefore started fieldwork for a PhD in Africa in 1963 on the geology and petrology of a young volcano in the Eastern Rift Valley of the British colony of Kenya, where I was hosted by local Masai people and supported by generous-minded geologists of the Kenya Geological Survey. Kenya obtained its independence from Britain that year and I saw the Union Jack lowered and the new Kenyan flag raised in a ceremony in Nairobi, while encompassed by the shrill warbling of enthusiastic Kikuyu people. Similar decolonisation ceremonies would take place in the 1970s in Melanesian countries in the south-west Pacific, including Papua New Guinea.

My career in 1966 seemed set, albeit vaguely, on becoming a teacher at a university somewhere, probably in North America and, indeed, I next undertook a period of postdoctoral work at the University of California at Berkeley. A strong school in volcanology existed at Berkeley in the 1960s, its traditions having been formed by the research of, amongst others, Welshman Howell Williams and the intellectually impressive Belgium-born volcanologist John Verhoogen, who had worked on volcanoes in the Belgian Congo of central Africa. Williams had contributed significantly to an understanding of the formation of calderas, or large volcanic craters, based on his fieldwork in the Cascades volcanic chain of the western United States and elsewhere, and had published in 1941 a well-known and benchmark paper 'Calderas and their Origin'. Nevertheless, and despite the presence of these high achievers, an academic career in volcanic geology lost its appeal for me irretrievably while at Berkeley and I decided to join a government geological survey like the one I had so identified with while in Kenya.

An interview in London in 1968 for a position with the national geological survey in Australia turned out to be more fruitful than I had expected. I was summoned to Australia House on the Strand to be assessed by Dr N.H. Fisher, then chief geologist of the Australian Government's Bureau of Mineral Resources, Geology and Geophysics, known ubiquitously as 'the BMR', and based in Canberra. 'Doc' Fisher was a tall, dark-suited, somewhat gruff and aloof public servant carrying a superficially imperious demeanor, not unlike that of Verhoogen. Fisher's underlying interest in volcanoes — was it a passion? — soon became apparent and there followed a lively, back-and-forth conversation between the two of us about volcanoes and Australian Government volcanologists — Tony Taylor, Colin Branch, and the Imperial-trained David Blake. Fisher himself, as a young man, had published a well-known report, amongst others, on the 1937 volcanic eruption at Rabaul on New Britain Island in the Territory of Papua and New Guinea to the north of Australia in north-western Melanesia.

The interview ended well for me, even though the only other member of the selection panel — a career-diplomat type — had been somewhat sidelined from the specialised volcanological conversation. I would be joining a BMR field party in mapping the mainly volcanic geology of New Britain, which included some spectacular but unmapped volcanoes and caldera structures, including at Rabaul. Our first field base-camp in January 1969 was at Pomio on the south coast of New Britain, hemmed in by the magnificent but brooding Nakanai Mountains, and facing Jacquinot Bay and, beyond, Palmalmal Plantation across the blue water. Fireflies would blink in unison at night in bushes alongside the earthen footpaths at Pomio, and I saw for the first time how the wakes of boats could sparkle with tropical phosphorescence. The coral reefs and their banded sea snakes were breathtaking.

Foreword

The first geological traverse I made in New Britain in 1969 was led by geologist Peter Macnab through the Nakanais — its aim, to map some of the older volcanic rocks of the island. We took off from Pomio by helicopter, landing less than half an hour later on a boulder bed in the lower reaches of the Ip River to the east. I quickly realised on the flight over that I, as a rock-seeking volcanic geologist, was in serious trouble. Tropical rain forest covered every piece of country in sight, including the fiercely rugged Nakanai Mountains. There were no rocky crags to be seen anywhere and I was later to discover that even though rocks were exposed in steep creek beds in these moss-forest covered mountains, most of them were inaccessible in the higher reaches of the streams anyway. Furthermore, the rain and mist were incessant. Neither were there any tracks, maps, indigenous people, and certainly no GPS gadgetry, to assist in traversing this deserted and dreadful piece of forested country. Navigating was in fact a nightmare, unless — like Macnab — you could use instinct, together with black-and-white aerial photographs from the Second World War to see where you were going by adjusting your eyes and viewing the water-saturated photographs in stereoscopic vision in the incessant mist, cold, and pouring rain. My pocket stereoscope simply steamed up.

A three-day traverse had been planned, but we took five days to reach the pre-arranged helicopter pick-up point. We and our Melanesian team of rucksack carriers — no vehicular roads in this devil country — returned to Pomio exhausted, bitten by voracious insects, evil beetles and so forth, and scratched and poisoned by the most vicious of vines, nettles, and roots imaginable. I was informed with true Australian drollery by colleagues back at the Pomio base camp that the five days represented just a 'shakedown' traverse. Pomio is synonymous with high rainfall, mosquitoes, and malaria. I fell ill and, after a few days of high fever, began hallucinating, possibly through overdosing on anti-malarial tablets. I was air-lifted to Rabaul and admitted to Nonga Hospital.

This was not the best beginning to volcanological studies for the Australian Government in what is now Papua New Guinea, but the work and my state of mind did improve. I remained with BMR, now Geoscience Australia, for 36 years and grew to appreciate the volcanoes, volcanic rocks, and people of Papua New Guinea — a truly extraordinary and inspiring part of the world — enough, that is, to want to write this book.

The book is dedicated to Jill who, curiously, has never shared my volcanological obsessions during our more than 40 years of marriage, but she came along anyway.

R.W. Johnson

Canberra

Acknowledgements and Sources

This book could not have been written without the support of the Australian Government since 1968 while I was employed at Geoscience Australia (GA) and while working for the Australian Agency for International Development (AusAID) and its predecessor. Neither of these agencies, however, can be held responsible for any errors of fact or poor judgments of events. Any such flaws are entirely my own doing.

Hundreds of people have contributed in different ways to this history, but here I acknowledge by name the strong personal influence of eight of them: N.H. Fisher, G.A.M. Taylor, D.H. Blake, R.J.S. Cooke, C.O. McKee, N.A. Threlfall, R.J. Blong, and K.J. Granger. Omitting all of the other names is both pragmatic and unforgiveable. Nevertheless, I must extend my appreciation to colleagues and friends at the Rabaul Volcanological Observatory (RVO) with whom I have worked over many years, including the current staff led by Ima Itikarai. This work, since the 1994 eruption at Rabaul, has been undertaken jointly by RVO and GA staff, and made possible by means of project work supported by the Australian Government through AusAID. I acknowledge here the much-valued friendship over those 18 years of GA project manager Shane Nancarrow. Shane has remained consistently loyal and tenaciously committed to our cause of supporting the technical work of RVO.

The following, in no particular order, have generously given up their time to review draft chapters, or to contribute materially to them: the late Hank Nelson, Elena Govor, Neville Threlfall, the late Herman Patia, Chris Ballard, Des Martin, Ima Itikarai, Shane Nancarrow, the late Robert Blakie, Gianni D'Addario, Steve Saunders, Klaus Neumann, Lucille Piper, Hilary Howes, David Blake, Hugh Davies, Monica Russell, Maclaren Hiari, Robin Hide, John Horne, the late Norman Fisher, the late Margaret Spencer, John Latter, Pat Durdin, David Marsh, Chris McKee, Betty Forster, Russell Blong, Colin Pain, Bob Tilling, Alanna Simpson, Brad Scott, Levi Mano, Beddie Jubilee, Gerry McGrade, Vince Neall, Pip Earl and, not least, Albert Speer. Bryant Allen and Chris Ballard heroically reviewed the entire manuscript.

Hank Nelson and Chris Ballard, both Pacific historians at the Australia National University (ANU) in Canberra, have been consistently supportive of my attempts to write as if a historian, as has ANU as a whole in allowing me to be a visiting fellow in the College of Asia and the Pacific. Pacific historians past and present at ANU have set for me an impossibly high standard in writing stylish English and telling stories.

An attempt has been made to quote only peer-reviewed publications, but I make no apologies for also referring to contributions from the 'grey' literature, much of which is of high quality. I also acknowledge with gratitude access to and assistance from the respective staff of, the 'Doc Fisher' Library — named after N.H. Fisher — at Geoscience Australia, as well as of the National Library of Australia, the Menzies Library at ANU, the National Archives of Australia, and the Australian War Memorial. These agencies are all in Canberra. I must include also the National Archives of Papua New Guinea in Port Moresby (and its impressive staff), what remains of the library at the Rabaul Volcanological Observatory, the libraries of the University of Papua New Guinea and the National Research Institute, both in Port Moresby, and the Mitchell Library in Sydney.

Selecting photographs has been a challenge, firstly because there is such a wide range to choose from and, secondly, because in some cases, and despite determined investigative efforts, I have not been able to identify the original photographers. I here offer my sincere apologies to these anonymous contributors and thank those whom I found.

Many of the photographs are from a large collection kept by GA, each one of which has an individual serial number and some of which are copies of originals. The photographers for many of these images are former GA officers. GA also has a collection of negative copies for photographs taken by other people, and obtained by other GA staff, including myself, over the years. The locations of the original negatives for some of these photographs are unknown, as indeed are the names of some of the photographers. Staff at Bica Photographics in Canberra helped greatly in improving the quality of some of the digital images and showed a great deal of personal interest in the volcano-related photographs.

Finally, I wish to acknowledge the production team at ANU E Press, led by Duncan Beard, who have been a constant source of friendly, professional, support and advice. I am grateful also to Silvio Mezzomo of Geoscience Australia, Karina Pelling and Jennifer Sheehan of the Digital Design team in the ANU College of the Pacific, as well as freelancers Ian Scales and Peter Johnson, for finalising line diagrams, most of which were designed by them specifically for this book. I also thank the ANU Publication Subsidy Committee, which awarded a small grant that offset most of the cost of copyediting the final version of the book. Justine Molony undertook the task of copyediting with aplomb, patience and professionalism.

Volcano Names and Totals

Volcano names used in this book conform generally to those listed under the 'Melanesia & Australia (05)' category in the volcano database managed by the Global Volcanism Program of the Smithsonian Institution, Washington D.C. Details from the database were published most recently by Siebert et al. (2010) who listed 64 Holocene volcanic centres for the area shown in the map below. The names of these volcanic centres were based initially on the work of Fisher (1957) and many of them have synonyms. Details on the volcanoes and their activity are updated from time to time on the Global Volcanism Program website.

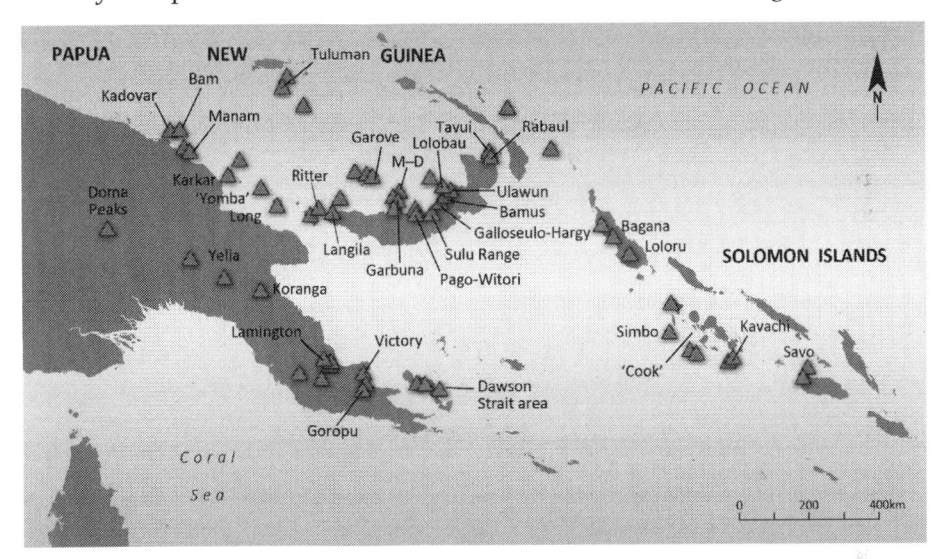

The triangles represent volcanoes that have had known or inferred Holocene eruptions, or those with possible but uncertain Holocene eruptions, together with three active geothermal fields — which are not named on this map — in areas where there is no known Holocene volcanism. M-D is Makalaia-Dakataua.

Source: Adapted from maps by Simkin & Siebert (1994, p. 58) and Siebert et al. (2010, p. 75).

'Cook' in the Solomon Islands has been reported in the literature to be an active volcano and is listed by Siebert et al. (2010) who, however, correctly label it as 'Not a Volcano'. Five other volcanoes, mainly classifying as 'Uncertain', are four possible submarine eruptive centres, as well as 'Yomba', a volcano whose existence is based on local legend. Furthermore, Musa River is a geothermally active area 40 kilometres south of Lamington volcano, but no Holocene eruptive centres have been found there (Fisher, 1957). The total number of Holocene volcanoes, therefore, reduces to 57 if these seven less-certain localities are excluded from the list of 64 volcanoes given by Seibert et al. (2010). Note, however, that many of these 57 volcanoes are only inferred to be Holocene, as historical evidence for eruptions or Holocene geochronological data are absent for them. Some of these may well be truly extinct.

Volcanoes referred to most commonly in this history are shown by name in the accompanying map. The first-named volcano in hyphenated double names, such as Pago–Witori, refers to a younger cone — such as 'Pago' — which is contained within the caldera of an older and second-named volcano, as in 'Witori'. Other maps showing volcano localities in greater detail are found throughout the main text of this book.

References

Fisher, N.H., 1957. *Catalogue of the Active Volcanoes of the World including Solfatara Field*. Part 5, *Melanesia*. International Volcanological Association, Napoli.

Global Volcanism Program website: http://www.volcano.si.edu/index.cfm

Siebert, L., T. Simkin & P. Kimberly, 2010. *Volcanoes of the World*. 3rd edn. Smithsonian Institution, Washington D.C., University of California, Berkeley.

Simkin, T. & L. Siebert, 1994. *Volcanoes of the World*. 2nd edn. Smithsonian Institution, Washington D.C., Geoscience Press, Tucson.

1. Burning Islands and Dampier's Voyage: 1700

The Island all Night vomited Fire and Smoak very amazingly; and at every Belch we heard a dreadful noise like Thunder, and saw a Flame of Fire after it, the most terrifying that ever I saw.

William Dampier (1906)

Track of the *Roebuck*

William Dampier, former buccaneer, rounds the north-western Bird's Head Peninsula of New Guinea Island in a British Royal Navy 'fifth-rater', the *Roebuck* at the beginning of a new century in 1700. He will subsequently record the presence of five 'burning' islands or mountains — volcanoes — in the New Guinea region, a not inconsiderable achievement. Indeed, this is something to envy if they were all 'burning' in full eruption. Dampier will not see the Bird's 'tail' in the south-east where there are active volcanoes, including one later called Lamington that would produce a major disaster in 1951. Nor will he enter the waters of the Solomon Islands where a Spanish explorer, Alvaro de Mendaña, may have seen Savo Island in volcanic activity more than a century previously. But four of the 'burning' islands and mountains to be observed by Dampier are those of a 1,000-kilometre-long chain that contains most of the active volcanoes of this region and which, today, are known collectively as the Bismarck Volcanic Arc.

Volcano discovery is not, of course, the purpose of the *Roebuck*'s voyage. Dampier has been given a more strategic aim: to provide his masters in Britain with information about the eastern side of New Guinea, and about the eastern seaboard of Australia, then called 'New Holland'. The *Roebuck* is on a voyage of exploration — something that, nevertheless, carries a certain amount of scientific interest. Spanish, Portuguese and Dutch explorers had all been active in the region, and Britain needed to catch up in discovering new openings for commerce and trade. Dampier was selected to lead the *Roebuck* expedition because of the experience he had gained in circumnavigating the globe, in incisively observing and recording his encounters with the natural world and, subsequently, in publishing in 1697 a bestseller, *A New Voyage Round the World*. This description of Dampier is up-beat, but the reality may have been different. Dampier is 'in command of a cheap expedition, with a rotten ship and an inferior crew, and without a single officer of any moral quality to supply his captain's deficiencies'.[1]

---

1   J. A. Williamson, in Dampier (1939), p. xxxi.

Fire Mountains of the Islands

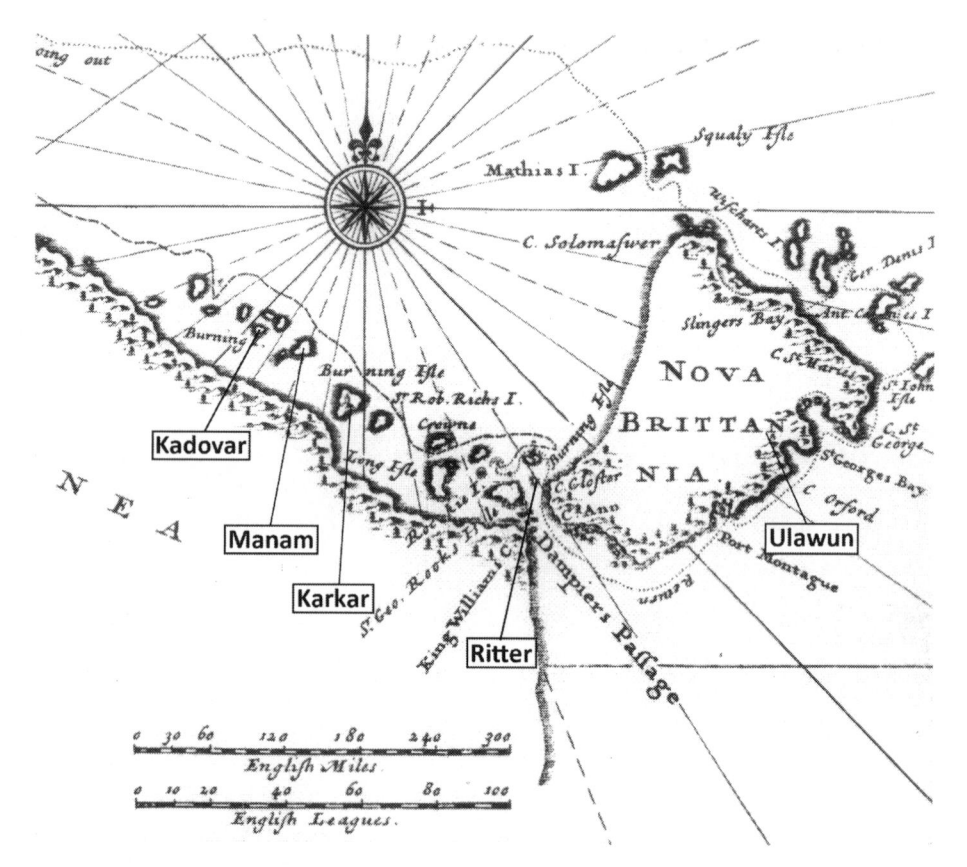

Figure 1. The clockwise route of Dampier's 1700 voyage is plotted as a dotted line in this detail from a chart in *Voyage to New Holland*. 'Nova Britannia' is shown much larger than the combined actual sizes of New Britain and New Ireland because both the strait between the two islands, and the northwestern coast of Nova Britannia, were unknown to Dampier. The names in boxes are those used today for some of the volcanoes seen by Dampier. Bam is the island immediately to the east-north-east of Kadovar.

Source: Adapted from Dampier (1939, fold-map between pp. 208 & 209).

The *Roebuck* is swept by favorable winds to the east, crosses the equator well to the north of both New Guinea and what would later be called the Admiralty Islands, then back across it and down to the eastern side of what is now known as New Ireland, but which Dampier calls 'the Main'. He cruises alongside and between four island groups offshore from, and in a chain parallel to, New Ireland, evidently not recognising them as volcanoes, although, had he landed at Lihir — site today of a huge goldmining enterprise in volcanically heated rocks — or at Feni, where there are geysers and hot ground — he might have deduced that these islands too were 'burning'. Dampier does not give names to the islands as they have been named already by the Dutch explorers, Willem

Schouten and Jacob Le Maire, on a previous voyage of discovery, but he will give names to many other geographical features during his voyage, ensuring that saints, British royalty, Royal Navy admirals, and special patrons are all acknowledged.

He names Slingers Bay, on New Ireland, after a confrontation — a 'first contact' — with several hundred Melanesians, many in canoes, who attempt to entice the *Roebuck* closer to shore. Dampier is curious about them too, offering beads, knives, and glass. He chooses prudence and starts sailing out further, but the Melanesians begin '… to fling Stones at us as fast as they could, being provided with Engines [i.e. slings] for that purpose; (wherefore I named this place Slinger's Bay)'. The *Roebuck* carries 12 guns, 14 less than it is capable of carrying, and Dampier has already armed his crew with 'all our small Arms, and made several put on Cartouch Boxes [for cartridges] to prevent Treachery … But at the Firing of one Gun they were all amaz'd, drew off and flung no more Stones … some of them were killed or wounded'.[2]

Dampier rounds a cape at the southern end of New Ireland, naming it Cape St George — for Dampier is an Englishman, although not of aristocratic origins — and then crosses westwards and names St Georges Bay, not realising that the bay is actually a strait separating New Ireland from New Britain. He mentions seeing high but cloudy land to the south-east, which is probably Bougainville Island where there are active volcanoes, and he may even have seen the volcanic peaks of the Rabaul area to the north-west, but he decides to set a westward course most likely because of difficult north-west headwinds at that time of year. Dampier then records the first volcano on his voyage, across low land between the Gazelle Peninsula and Nakanai Mountains in New Britain: '… we saw a Burning Mountain in the Country. It was round, high, and peaked at the top (as most Vulcano's are), and sent forth a great Quantity of Smoak'.[3] This almost certainly is Ulawun volcano on the north coast of New Britain, one of the most consistently active volcanoes in the region and the highest in the Bismarck Volcanic Arc.

The *Roebuck* continues its voyage westwards along the south-facing coastline, and Dampier at one point respectfully names Cape Orford 'in Honour of my noble Patron', Lord Orford, Admiral of the Fleet. Orford is 'a gentlemen of a sanguine complexion, inclining to fat; of a middle stature' according to a footnote quotation included by another of his editors — with some relish one assumes — in Dampier's book of the voyage.[4] Dampier then makes the most significant geographical discovery of his voyage: he finds a strait between the

---
2   Dampier (1906), p. 526.
3   Dampier (1906), p. 533.
4   J. Masefield, providing an unattributed quotation in Dampier (1906), p. 533.

New Guinea mainland and the landmass whose southern coastline he has been tracking. The separate landmass he calls 'Nova Brittannia', still believing New Ireland to be part of it, and the strait eventually receives his own name, Dampier Passage or Dampier Strait.

A crucial decision has to be made around this time. Does he take a high-risk option and continue south-eastwards in his leaky vessel and with substandard crew into unchartered waters and find the east coast of Australia, or does he cut his losses, take advantage of the strait he has just discovered and head back westwards to the Bird's Head? The lower risk option is chosen and the east coast of Australia is not reached by the British until 70 years later when James Cook — evidently a superior captain to Dampier and in a better ship — takes the honours.

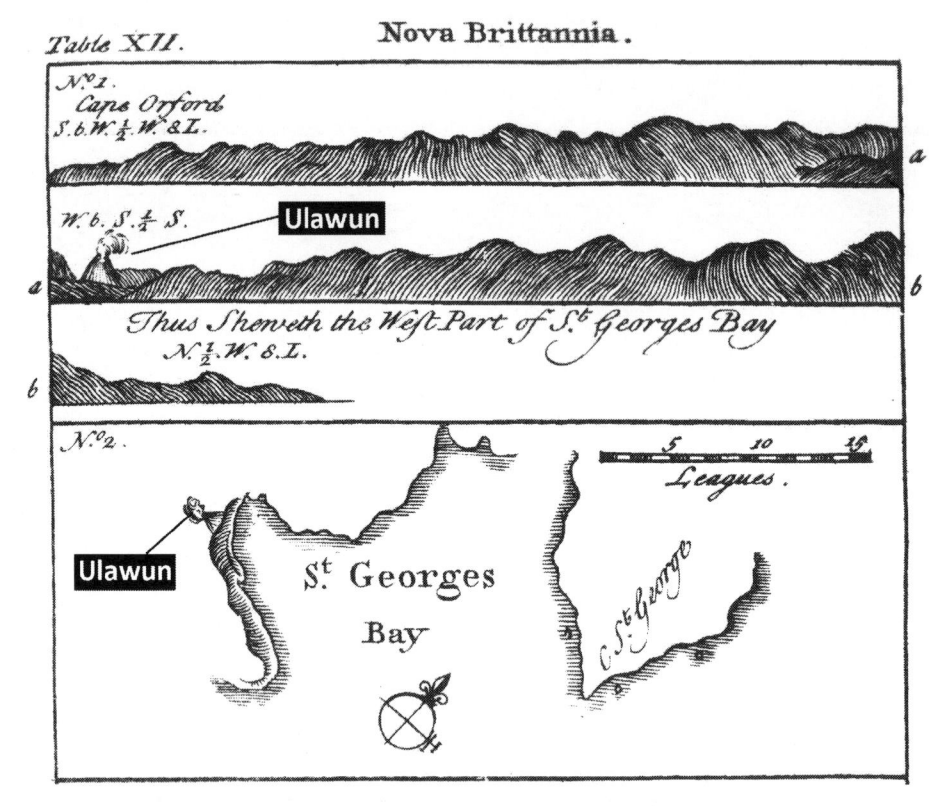

Figure 2. Ulawun is shown by Dampier in both a profile sketch and a map of St Georges 'Bay'.

Source: Adapted from Dampier (1939; sketches 1 & 2, Table 12, between pp. 206 & 207).

Dampier makes another significant volcanological discovery — that of an island volcano in full eruption sitting within the strait west of Nova Brittannia. The volcano eventually is named Ritter by German colonists late in the nineteenth century. This is how Dampier described the eruptive activity on the night of 24 March 1700:

> At 10 a Clock I saw a great Fire bearing North-West by West, blazing up in a Pillar, sometimes very high for 3 or 4 Minutes, then falling quite down for an equal Space of Time; sometimes hardly visible, till it blazed up again.[5]

And, on the next night, the intervals between the volcano's 'belches':

> were about half a Minute; some more, others less: Neither were these Pulses or Eruptions alike: for some were but faint Convulsions, in Comparison of the more vigorous; yet even the weakest vented a great deal of Fire; but the largest made a roaring Noise, and sent up a large Flame 20 or 30 Yards high; and then might be seen a great Stream of Fire running down to the Foot of the Island, even to the Shore. From the Furrows made by this descending Fire, we could in the Day Time see great Smoaks arise, which probably were made by the sulphureous Matter thrown out of the Funnel at the Top, which tumbling down to the bottom, and there lying in a Heap, burn'd till either consumed or extinguished; and as long as it burn'd and kept its Heat, so long the Smoak ascended from it; which we perceived to increase or decrease, according to the Quantity of Matter discharged from the Funnel.[6]

Figure 3. Steep-sided Ritter Island is shown in eruption in 1700 in this illustration from Dampier's book.

Source: Dampier (1939; Sketch 3, Table 13, between pp. 216 & 217).

---

5 Dampier (1906), p. 541.
6 Dampier (1906), p. 542.

This dramatic description is illustrative of Dampier's skill as an author, why he was widely appreciated in Britain as a travel writer as much as for being an explorer and, indeed, why in part he was chosen to lead the *Roebuck* voyage in the first place. Dampier is obviously impressed by the eruption, as well he might be for night-time ejection of incandescent material from volcanoes is always an imposing sight. More particularly, however, Dampier is here describing what may be ***pyroclastic flows***[7] — that is, hot avalanche-like 'floods' of volcanic blocks, ash and gas. Pyroclastic flows can originate in many ways, but in this case the flow was evidently formed by collapses of fountains of incandescent lava fragments that had been flung up from the vent, which then fell back onto the steep slopes of the island, avalanching down to the sea, and leaving a furrowed, and still hot, 'block and ash' deposit. Such flows have since been seen at other volcanoes of the Bismarck Volcanic Arc, including Ulawun that Dampier had viewed only a few days before.

More volcanological discoveries follow as the *Roebuck* makes its way westwards through the volcanic islands along the north coast of New Guinea, a coast that had been traversed earlier by both Spanish and Dutch voyagers. Dampier makes profile drawings of the islands and the mainland, but these are neither well labelled nor closely linked with the text of his book or his map. This is because 'Dampier was at sea when the second part of his book was printed, and was dead when the second edition appeared in 1729'.[8]

Historians, geographers, and volcanologists have since made their own interpretations of what Dampier saw, and not all of them are in agreement. There is little doubt, however, about the recognition of Long and Crown Islands, which Dampier names, and between which he sails, the former described as a 'long island with a high Hill at each end' and both appearing 'very pleasant, having Spots of green Savannahs mixt among the Wood-land: The Trees appeared very green and flourishing and some of them looked white and full of Blossoms …'.[9] His perception also is that both islands have few people on them. Volcanologists have been intrigued by these evidently simple observations because of their implications for the date of a major eruption that devastated the island some time within the last few hundred years.

Long Island is shown on Dampier's map as a narrow strip of low land linking the two peaks, and that certainly is the impression when the island is viewed from the east or west. But this view, today, is illusory because the low land is actually the rim of a 13-kilometre-wide ***caldera***, or large crater, and the island is roughly hexagonal in outline. The caldera also has a lake that was not

---

7   Unavoidable volcanological terms used for the first time are shown in bold italics and are listed with acronyms in the Appendix.
8   J.A. Williamson, in Dampier (1939), p. viii.
9   This, and the remaining quotations in this section, are all from Dampier (1906), pp. 545–48.

discovered by Europeans until the late 1930s when it was named Lake Wisdom by Australians, after an administrator of the Territory of New Guinea. Calderas of this size represent collapse of the roofs of large, shallow, **magma** reservoirs, and their formation is accompanied by powerful and voluminous explosions of pumice, ash and gas. Melanesians now living on Long Island have stories about the most recent catastrophic eruption on the island. This was a volcanic disaster of widespread impact which, were it were to happen today, would have significant consequences to life and property in the region. A fuller story can be told about this catastrophic eruption at Long Island.

Figure 4. These two island profiles from Dampier's book are of Long Island (left) and Crown Island, but are here switched around so that the coordinates are consistent from left to right.

Source: Adapted from Dampier (1939; Sketch 4, Table 13, between pp. 216 & 217).

> On Tuesday the 2d April, about 8 in the Morning, we discovered a high peeked Island to the Westward, which seem'd to smoak at its Top. The next Day we past by the North-side of the Burning Island, and saw a Smoak again at its Top; but the Vent lying on the South-side of the Peek, we could not observe it distinctly, nor see the Fire.

Dampier is here describing the impressive and towering stratovolcano of Manam, one of the most active volcanoes in the region. The name 'Burning Isle' on Dampier's map presumably refers to Manam, even though the label is shown closer to Karkar. Two volcanologists were killed on Karkar in 1979, and the people of Manam suffered a major eruption in 2004–2005 that covered the entire island leading to its evacuation. Pyroclastic flows, such as those witnessed by Dampier at Ritter, have been observed many times at Manam since the 1950s.

The identity of the next 'burning' island along the chain that Dampier mentions has been the subject of even more uncertainty. Dampier writes: 'We also saw another Isle sending forth a great Smoak at once; but it soon vanished, and we saw it no more'. The island is one of a cluster known today as the Schouten Islands, amongst which Bam has been the only known historically active volcano — in the mid-1950s when the islanders were evacuated, with tragic results. Perhaps, therefore, Bam is Dampier's 'other' island. This view, however, was not supported by R.J.S. 'Rob' Cooke who in the 1970s was head of the Rabaul Volcanological Observatory. Cooke, one of the volcanologists killed at Karkar in 1979, promoted the view persuasively, if not adamantly, to many of us as a result of an analysis of Dampier's records, that the island must be Kadovar. The label

Fire Mountains of the Islands

'Burning I.' on the map in this case is directly on Kadovar and definitely not on Bam. A prominent area of hot ground appeared on Kadovar in 1976, leading to fears of an eruption, but none has followed — at least, not so far.

Figure 5. William Dampier's portrait was painted by Thomas Murray in 1697–1698, just before his voyage to Melanesia, and after Dampier had achieved favourable recognition from the British establishment as naturalist, navigator, explorer and writer — his early buccaneering, if not piratical, days in the Caribbean notwithstanding. Dampier brought the leaky *Roebuck* back to the south Atlantic via the Cape of Good Hope, after his final rounding of the Bird's Head of New Guinea, but the vessel foundered and sank off the volcanic island of Ascension, without loss of life.

Source: The portrait shown here is a copy painted from the Murray original by Edmund Dyer in about 1835. National Library of Australia, Canberra.

Dampier is now about to complete his loop journey through the New Guinea region. He returns westwards to the Bird's Head part of the island and sees on 17 April 'a high Mountain on the Main, that sent forth great Quantities of Smoak from its Top: This Vulcano we did not see on our Voyage out'. There are no young volcanoes in this part of New Guinea, so the implication is that Dampier has mistaken the 'smoak' for weather clouds on a peak that was not a volcano. The question, then, may be asked: what kind of 'smoak' did Dampier see at

Kadovar, Manam, and Ulawun? Or, more specifically, were these volcanoes in actual eruption, such as Dampier saw undeniably at Ritter? A volcano can be described as 'active' in a general sense if it is one that has been seen, and noted to be, in eruption previously, or one that has the potential to produce another eruption. White vapour emerging from an 'active' volcano, however, which is more prominent after heavy rain or during cooler times of the day anyway, does not mean that the volcano is in actual eruption — that is, emitting volcanic ash or extruding a lava flow. Ash-laden clouds from volcanoes in full eruption are densely dark grey to brown — colours that Dampier does not mention — and white vapour can emerge from active volcanoes when they not in eruption but only passively degassing or 'drying out' *between* eruptions.

The conclusion, therefore, is that Dampier may have seen only one volcano, Ritter, in full eruption during his voyage; the 'smoak' of three others was not necessarily ash-laden, so the volcanoes may not have been in eruption; and, the fifth volcano was a misidentification.

Near Oceania, Melanesia and Melanesians

Dampier did not write that the people he encountered at Slingers Bay were Melanesians because the name 'Melanesia' was not used until after 1832 when the French navigator Jules Sebastien César Dumont D'Urville introduced it.[10] Jorge de Meneses, Portuguese governor-elect of the Moluccas, in 1526 had described the inhabitants of the New Guinea region as 'black people with frizzled hair', and Melanesia therefore means 'black islands'. 'Melanesian', far from meaning a distinctive racial type, simply refers to the people of several different origins who live on the following islands:

- New Guinea, the largest island, together with some smaller islands nearby immediately to the west
- the Bismarck Archipelago, including the largest islands of New Britain, New Ireland, and Manus
- the Solomon Islands chain, including Bougainville Island
- the Santa Cruz Islands in the far east of the modern-day Solomon Islands
- Vanuatu, Fiji, New Caledonia and the Loyalty Islands.

Islands of the first three of these categories are generally larger and more closely spaced, and therefore more likely to be inter-visible from sea level, than are the islands to the south-east and out to Polynesia. These three sets of islands represent an eastward extension of island South-East Asia. They have been

---

10   Dumont D'Urville (1832).

called 'Near Oceania', which is separated from 'Remote Oceania' by an important biogeographical boundary running between the Santa Cruz Islands in the east and the Solomon Islands chain, including Savo, in the west. This boundary in prehistoric and geological times has restricted the movement of flora, insects, and fauna, including people.[11]

The islands of Near Oceania formed as a result of two near synchronous, geologically recent, tectonic collisions. One is the northwards collision of the Australian continent with an arc of islands, causing a major uplift of what is now New Guinea Island. The other is a westwards collision of the huge, submarine, Ontong Java Plateau, with what is now the Solomon Islands chain. Near Oceania today is literally squashed between these two colliding masses. The region closely matches a geologically diverse area of active tectonics and volcanism — a unified tangle of interconnecting tectonic plate boundaries and volcanic provinces of remarkable complexity in comparison with the areas around it and, indeed, with anywhere else in the world. The name 'Near Oceania' is not used widely outside of academic circles, but its brevity is of considerable convenience here because it encompasses precisely the area of volcanoes of interest. The name covers, in terms of modern political boundaries, the two independent Melanesian states of Papua New Guinea and most of the Solomon Islands, together with the province of Papua in eastern Indonesia.

Dampier would not have known that Melanesians represent an aggregation of different periods of immigration by different peoples. The first journeys of humankind out of East Africa evidently included people whose descendants travelled along the Indian and Indonesian coastlines to East Asia, Taiwan, New Guinea, and Australia. These descendants have been in Near Oceania occupying volcanically active areas for at least 35,000 years,[12] meaning that their experience with volcanic eruptions covers not only the most recent 11,700 years of the **Holocene**, but at least another 23,000 years of the preceding **Pleistocene**.

The human inhabitants of Near Oceania have left no calendars, hieroglyphics, alphabets, or writings. Their culture is based on a paradigm that links past and present with place and belonging, rather than favouring the present over a separate, historical past. Nevertheless, non-documentary knowledge about significant events in the past, such as volcanic eruptions, is passed down from one generation to the next by means of oral history. Knowledge in Melanesian society is transferred by an extraordinary range of languages and dialects. Near Oceania is 'the most linguistically diverse part of the planet. Here, in an area

---

11 Pawley & Green (1973) and Green (1991). Note that the Santa Cruz Islands — including the active volcano Tinakula — are in Remote Oceania in the far east of the maritime territory of the nation-state of the Solomon Islands. They are part of a separate tectonic setting to the islands of Near Oceania and relate more closely to the islands of Vanuatu.
12 Torrence et al. (2004) and Ballard (2010).

1. Burning Islands and Dampier's Voyage: 1700

that has less than one per cent of the world's land mass, we find almost 20 per cent of the world's languages — roughly 1,100 mutually unintelligible tongues which average only 2,000–3,000 speakers each'.[13] These languages are divisible into two groups. The first is an older, diverse, 'Papuan' or Australo-Melanesian group that dominates New Guinea Island, but which is found in some of the smaller islands too. The second is a younger Oceanic Austronesian group. These people appear to relate, at least in part, to the incursion about 3,500 years ago of seafarers, who may have left behind distinctive, dentate-stamped 'Lapita' pottery along shorelines and coastal areas, including the volcanically active zones of Near Oceania, and who may be the ancestors of Polynesians.[14] A boundary between the Austronesian and Papuan language groups even divides the island people of Karkar volcano into two parts. The most modern migrations to Near Oceania include not only 'reverse' immigrations of Polynesians, but also, most recently, incursions of European and Asian people.

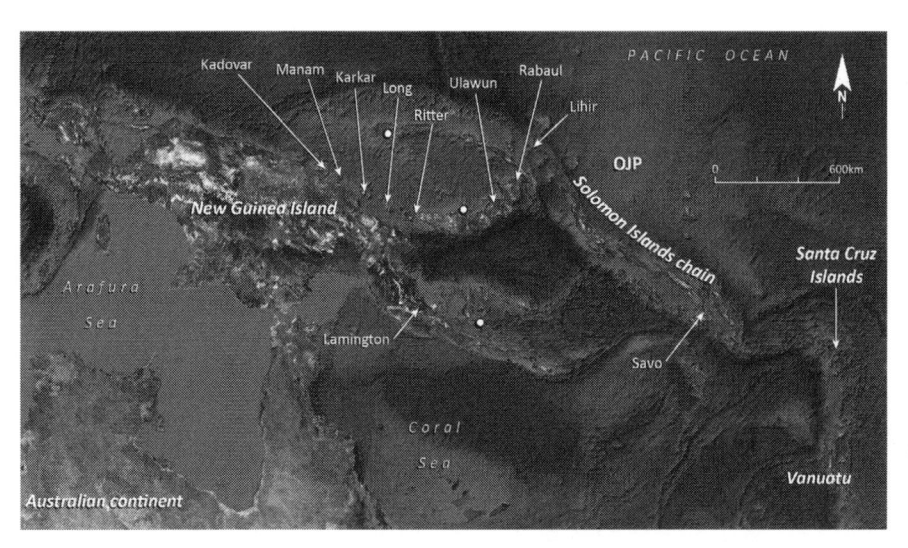

Figure 6. The contorted pattern of sea floor depths shown by the different shades of blue reflects the complex modern tectonics of the New Oceania region, which is shown here together with Australia in the south-west and — in Remote Oceania — modern-day Vanuatu in the south-east. Near Oceania includes the modern-day Indonesian province of Papua in western New Guinea. Note, especially, the deep, and bent, submarine trench seen immediately south of Ulawun on New Britain Island. The three main sources of obsidian used traditionally for trade and exchange are shown by the white-filled circles. These are, from north to south, Lou-Pam-Hahie, Talasea-Mopir and Fergusson-Sanaroa. OJP is the submarine Ontong Java Plateau.

Source: Google Map base.

---

13 Pawley (2005), p. xii.
14 See, for example, Kirch (1997) and Spriggs (1997).

Dampier would not have known that the Melanesians of his time are both maritime and montane peoples, as they are today. They live in villages and hamlets on island coasts and tiny islands, on forested inland ridges and valley slopes, on the edges and flood plains of rivers including major riverine systems like the Sepik, Ramu and Fly, on lake shores, and in cold-weather highland regions where intermontane valley floors are used for successful experiments in agriculture — indeed, amongst the earliest anywhere in the world. Melanesians also live on volcanoes. Ridges, where they exist, are favoured for settlement for defensive purposes. There are no towns or cities, kings or kingdoms, princes or fiefdoms, no centralised systems of government. Homes are made from 'bush' materials. Food is obtained from the sea, rivers and forests.

Melanesian men gained status and became 'big men' and leaders through success in intertribal battles and by acquiring wives, gardens, pigs and the heads of victims, rather than inheriting positions of leadership. There are rich cultural traditions of myth, dance and body decoration, as well as in sculpture in places like the Sepik. Kinship ties and lines of descent form the unity of community relationships. Land is not owned, bought, and sold in the European way of freehold or other forms of individual titles, but rather is passed on through a complex system of customary land tenure based on rights and obligations at tribal, clan, family and individual levels.

Trade-and-exchange arrangements underpin intercommunity dealings — perhaps most famously the 'Kula ring' in the Trobriand Islands, documented subsequently by pioneer anthropologist Bronislaw Malinowski.[15] More relevant here, however, is the distribution throughout Melanesia and beyond of quality *obsidian* from different sites in three main volcanic areas in Near Oceania: the Admiralty Islands, including the Lou and Pam Islands; the Talasea-Mopir area of New Britain; and Fergusson Island and adjacent islands in the D'Entrecasteaux Islands.[16] The concept of 'exchange' and reciprocity is rooted even more deeply in the world view of some Melanesian groups, and is thought to have determined attitudes to meaning and explanation amongst the Orokaiva, for example, following the catastrophic volcanic eruption at Lamington volcano in 1951.[17]

Belief systems strongly integrate the natural world with the ghosts of ancestors and through acknowledgement of the reality of a wider spirit world. Spirits are sources of knowledge, danger and protection. They influence human lives and need to be shown deference, but may be controlled in some circumstances, or at least assuaged, through ritual. Sorcery is practiced. Spirits live in many places,

---

15   Malinowski (1922).
16   Summerhayes (2009). The literature on obsidian and prehistory in Near Oceania is extensive, but early studies include those of Key (1968), Ambrose (1976), Smith et al. (1977), Specht (1981) and Torrence (1992).
17   Schwimmer (1973).

including the craters of volcanoes. An especially striking example of the volcanic spirit world is from Manam volcano where the terrifying female spirit Zaria lives and who is regarded as the volcano itself, its eruptions an expression of her moods:

> She is described as a wild-looking creature who spews fire from her armpits and vagina. When she walks about she wears an incandescent skirt aglow with flickering flames … . When humans provoke her or she becomes angry, Zaria emerges from her cavernous home in the crater's depths and roams the slopes of the volcano, leaving a trail of fire and burning lava in her wake.[18]

Zaria is the origin of fire, which represents an important traditional aspect of transformative female power. She is a force of destruction and renewal, represents both death and life and, even more broadly, 'is a symbol of the cyclical process of destruction, transformation, and renewal that characterises human life and the natural world.'[19]

European settlement and colonisation would cause the disintegration and disappearance of some Melanesian culture, and other parts of it would be altered and redefined in the form of cargo cultism, particularly during the twentieth century. Cargoism represents the belief that European wealth in the form of material goods or 'cargo' will arrive if people become cult members and perform prescribed rituals. Cults have a range of expression, many of them bizarre to the Europeans, but invariably they are focused on the belief that the cargo will arrive supernaturally by ship, aeroplane, helicopter, or by some other means. Comparisons have been made by one anthropologist between Britain in the seventeenth century and cargo cultism in the southern Madang Province of twentieth century New Guinea. He wrote that cargoism is the 'devastatingly reasoned' outcome of a belief that true knowledge is attributable to divine revelation, rather than to secular and empirical knowledge.[20] This situation was not too dissimilar to beliefs in the Britain of Dampier's time, when secular science eventually replaced many magical explanations of 'reality' that were previously associated with, and endorsed by, Christianity. Volcanoes at this time, for example, were a reminder to Christians in Europe of the fires of Hell burning below, a place where sinners are incarcerated.

---

18   Lutkehaus (1995), pp. 5–6.
19   Lutkehaus (1995), p. 8.
20   Lawrence (1982), p. 66.

Early Ideas about Volcanic Activity

'Smoak' is only one word in Dampier's vocabulary that intimates what people at the beginning of the eighteenth century thought about the cause of volcanic eruptions. Dampier also uses, as seen in the above quotations, 'burning', 'blazing up', 'fire', 'flame', and 'funnel' — words that are indicative of combustion. Volcanology today includes terms such as **pyroclastic** — meaning fire-broken — as well as *ash*, as part of its established vocabulary. The title of this book is also fixed in the language of combustion.

A combustion theory of volcanism emerged from the Mediterranean region where the development of Greek and Roman civilisations involved accounting for the origin of natural events, including both the earthquakes and volcanic eruptions of the region.[21] The theory evolved from an even earlier concept proposed by Anaxagoras, a Greek natural philosopher of the fifth century BC, who said that eruptions were caused by great winds stored inside the Earth. Aristotle (384–322 BC), said to be the father of natural history, also believed that earthquakes and volcanic eruptions were formed by subterranean winds — underground 'weather' — forcing their way out to the surface. He drew comparisons with human flatulence: 'For we must suppose that the wind in the earth has effects similar to those of wind in our bodies whose force when it is pent up inside can cause tremors and throbbings'.[22]

Other 'ancients', and even naturalists well into the eighteenth century, added that the winds ignited underground flammable substances, such as sulphur, or pyritous and combustible stones, even oils and fats from buried animals. This volcanism was thought to have been caused by the Earth's internal combustion. Dampier's account of burning islands reflects acceptance of this origin by combustion — a theory, however, that has long since disappeared. Chemists of the seventeenth century pointed out that huge amounts of air would be required to continue the burning and that any combustion likely would be choked off deep in the essentially airless interior of the Earth. The chemists themselves favoured volcanism being caused by heat-producing chemical reactions that did not require fuelling by air. This 'chemical theory' of volcanism is also outdated, but it lasted even into the twentieth century.[23]

---

21 See, for example, Sigurdsson (2000).
22 Quoted by Sigurdsson (1999), p. 37.
23 Curiously, however, the chemical theory of volcanism does emerge again in accounting for a volcanic crisis on the New Guinea mainland at Wau in 1967.

1. Burning Islands and Dampier's Voyage: 1700

Figure 7. The ancient Greeks, like the Melanesians, had a rich mythology concerning volcanoes and their eruptive activity. A mythical giant Enkelados, also known as Typhon or Typhaeus, is seen here lying on his back beneath Etna volcano, north-eastern Sicily, in this copperplate by Bernard Picart from 1731. Enkelados in Greek mythology was one of the giants who battled the gods on Mount Olympus, and who became paralysed when a spear struck him in battle. The gods buried him beneath Etna, but he was not quite dead. Subsequent eruptions at Etna were believed to be the breathing of Enkelados, and there were earthquakes every time he stirred. The aerial figure in the upper right is Zeus hurling lightning bolts at the monster buried beneath the volcano.

Source: Sigurdsson (2000; frontispiece of *Encyclopedia of Volcanoes*). Haraldur Sigurdsson also provided the detail for this caption.

Rocks exposed at the Earth's surface had been regarded for centuries as products of a former world of supernatural violence and gigantic convulsions, of mysterious cataclysms and floods of biblical proportions, including Noah's Deluge, and of supernatural extinctions and creations of life. Such catastrophism regarding the world's origin still prevailed in Europe at the time of Dampier's voyage. Yet, one British historian of science has, with good reason, portrayed the course of the seventeenth century as 'one of the great episodes in human experience, which ought to be placed ... amongst the epic adventures that have helped to make the human race what it is ... [Since] the rise of Christianity, there is no landmark in history that is worthy to be compared with this'.[24] The power of the Church and State, and the 'magic' embedded in Christianity, were both challenged by this Scientific Revolution as the seventeenth and eighteenth centuries progressed. Witchcraft was still punishable by law in seventeenth century Europe when Dampier was born. The seventeenth century in Europe also included the development of politically separate 'nation states', a concept that would later, in colonial times, split Near Oceania into different countries separated by the most arbitrary of territorial boundaries.

Development of science in the seventeenth century was underpinned by an understanding of the influence of gravitational attraction between the Sun and planets and their relative motions, as theorised by Isaac Newton (1642–1727). This and other theories led to establishment of the foundation principles of mathematics and physics when Newton, the founding father of science, published in 1687 his monumental *Principia Mathematica*, arguably the fundamental publication in the history of science. Development of today's many scientific sub-disciplines and specialisations would follow, but geology — and therefore a clearer understanding of the origin of volcanoes — would not begin to emerge strongly until late in the eighteenth century. Solid-earth tides, as opposed to ocean tides, generated by the gravitational attractions of Earth, Moon and Sun, as proposed by Newton, would later be regarded by some volcanologists as a possible mechanism for the triggering of volcanic eruptions in Near Oceania.

Newton was the personification of the Enlightenment during what has been called the 'long eighteenth century'. This intellectual and idealistic movement, which started late in the seventeenth century and ended in the early nineteenth century, was an empirical methodology guided by the light of reason and logic, and claimed as a new way of thinking by its proponents, the *philosophes* — scientists, philosophers, and writers. Its methodology was not in fact entirely new, but the expectation was that new knowledge of the natural world and derived universal and absolute truths would provide liberation from ignorance and superstition leading to progress, freedom and the happiness of mankind. Whether such ambitious ideals have been reached is debatable, but the Enlightenment nevertheless did stimulate the global exploration of the natural world. Dampier, in this context, may be regarded as the first Enlightenment voyager in Near Oceania, and his observations and records as pioneering contributions to the nascent history of volcanological studies in the region.

---

24 Butterfield (1962), pp. 179, 190.

Preview

This book is about how an understanding of the volcanoes of Near Oceania has gradually unfolded. The historical coordinates and navigational waypoints of the enterprise are a series of significant volcanic disasters, but different histories are interweaved between them. First is the discovery, exploration, settlement and colonial history of Europeans, together with postcolonial events — that is, a political and military history. Then there is the history of volcanology, particularly the ways in which key discoveries and interpretations of major eruptions elsewhere in the world have impacted on an understanding of volcanic eruptions in Near Oceania — a history of volcano science. A much longer and largely undocumented history of Melanesians in the region, is revealed to some extent by the results of volcanological archaeology and anthropology, including oral traditions, myths and stories — a prehistory of the modern independent states of Near Oceania. Care will be taken, however, in the interests of focus and length, not to expand into the larger histories of the development of geological theories in general and the related discovery of mineral resources in volcanoes, or to divert into the equally fascinating evolution of petrological ideas of how magmas form deep within the Earth and how these relate to tectonic structure in this remarkably complex region. Similarly, description and discussion of the modern technologies and instrumentation that can be used to monitor volcanoes will have to be avoided, as will delving into the engrossing and extensive sociological literature on risk perceptions and community vulnerabilities to natural hazards.

Stories are presented of volcanic crises and disasters, of lessons learnt and mistakes made, and of the key players who have made advances in volcano understanding. There are questions of community vulnerability to, and risk from volcanic eruptions; of the difference between hazard and risk; of the value of traditional knowledge and oral history; of the application of concepts and suitably sustainable technologies for volcanic disaster risk reduction (DRR) in Third World countries; of recognition that volcanic eruptions are only one kind of natural hazard and that, as such, they fit into the broader theatre of international natural hazard disaster risk reduction. This book is not a volcanological text or systematic source book, directory, or gazetteer of volcano-related information written in specialist scientific language, but rather a history — imperfect as all histories are — with loose ends, dead ends and gaps, filtered knowledge and uncertainties, and opportunities still to be explored. The history, still, has the potential to be rich and potentially important for ongoing volcanic disaster risk reduction work in two of the contemporary nation states of Near Oceania — Papua New Guinea and the Solomon Islands.

References

Ambrose, W.R., 1976. 'Obsidian and its Prehistoric Distribution in Melanesia', in N. Barnard (ed.), *Ancient Chinese Bronzes and Southeast Asian Metal and Other Archaeological Artifacts*. National Gallery of Victoria, Melbourne, pp. 351–78.

Ballard, C., 2010. 'Synthetic Histories: Possible Futures for Papuan Pasts', *Reviews in Anthropology*, 39, 232–57.

Butterfield, H., 1962. *The Origins of Modern Science 1300–1800*. 2nd ed. Bell and Sons, London.

Dampier, W., 1906. *A Continuation of a Voyage to New Holland, &c. in the Year 1699*, in *Dampier's Voyages*, ed. J. Masefield, 5, pp. 451–573. E. Grant Richards, London.

——, 1939. *A Voyage to New Holland*, ed. J.A. Williamson. The Argonaut Press, London & Hereford.

Dumont D'Urville J.S.C., 1832. *Voyage de la Corvette L'Astrolabe execute par Ordre du Roi, pendant les Annees 1826–1827–1828–1829, sous le Commandement de M.J. Dumont Durville, Capitaine de Vaisseau. Histoire du Voyage*. Tastu, Paris.

Green, R., 1991. 'Near and Remote Oceania — Disestablishing "Melanesia" in Culture History', in A. Pawley et al. (eds), *Man and a Half: Essays in Pacific Anthropology and Ethnobiology in Honour of Ralph Bulmer*. Polynesian Society Memoir, 48, pp. 491–502.

Key, C., 1968. 'Trace Element Identification of the Source of Obsidian in an Archaeological Site in New Guinea', *Nature*, 219, pp. 523–34.

Kirch, P.V., 1997. *The Lapita Peoples: Ancestors of the Oceanic World*. Blackwell, London.

Lawrence, P., 1982. 'Madang and Beyond', in R.J. May & H. Nelson (eds), *Melanesia: Beyond Diversity*. Research School of Pacific Studies, The Australian National University, Canberra, pp. 57–72.

Lutkehaus, N.C., 1995. *Zaria's Fire: Engendered Moments in Manam Ethnography*. Carolina Academic Press, Durham, North Carolina.

Malinowski, B., 1922. *Argonauts of the Western Pacific: An Account of Native Enterprise and Adventure in the Archipelagoes of Melanesian New Guinea*. Routledge and Kegan Paul, London.

Pawley, A., 2005. Preface, in A. Pawley et al. (eds), *Papuan Pasts: Cultural, Linguistic and Biological Histories of Papuan-Speaking Peoples*. Pacific Linguistics, Research School of Pacific and Asian Studies, The Australian National University, Canberra, pp. x–xvii.

Pawley, A., & R. Green, 1973. 'Dating the Dispersal of the Oceanic Languages', *Oceanic Linguistics*, 12, pp. 1–67.

Schwimmer, E., 1973. *Exchange in the Social Structure of the Orokaiva*. St Martin's Press, New York.

Sigurdsson, H., 1999. *Melting the Earth: The History of Ideas on Volcanic Eruptions*. Oxford University Press, New York.

——, 2000. 'The History of Volcanology', in H. Sigurdsson (ed.), *Encyclopedia of Volcanoes*. Academic Press, San Diego, pp. 15–37.

Smith, I.E.M., G.K. Ward & W.R. Ambrose, 1977. 'Geographic Distribution and the Characterization of Volcanic Glasses in Oceania', *Archeology and Physical Anthropology in Oceania*, 12, pp. 173–201.

Specht, J., 1981. 'Obsidian Sources at Talasea, West New Britain, Papua New Guinea', *Journal of the Polynesian Society*, 90, pp. 337–56.

Spriggs, M., 1997. *The Island Melanesians*. Blackwell Publishers, Oxford.

Summerhayes, G.R., 2009. 'Obsidian Network Patterns in Melanesia — Sources, Characterisation and Distribution', *Bulletin of the Indo-Pacific Prehistory Association*, 29, pp. 109–23.

Torrence, R., 1992. 'What is Lapita About Obsidian? A View from the Talasea Sources', in J-C Galipaud (ed.) *Poterie Lapita et Peuplement*. ORSTOM, Noumea, pp. 111–26.

Torrence, R., V. Neall, T. Doelman, E. Rhodes, C. McKee, H. Davies, R. Bonetti, A. Guglielmetti, A. Manzoni, M. Oddone, J. Parr & C. Wallace, 2004. 'Pleistocene Colonisation of the Bismarck Archipelago: New Evidence from West New Britain', *Archaeology in Oceania*, 39, pp. 101–30.

2. Volcano Sightings by European Navigators: 1528–1870

… a large volume of dense white smoke [was observed] to issue forth, & continue high in the air, in the vast conical mass, which on examination with a Telescope was plainly seen to emanate from a volcano in active operation! all down the sides were numerous furrows or channels from whence the smoke arose, as though from a recent deposition of molten lava.

John Parker Wilson (1842)

Saavedra, Retes and Mendaña

The Bible contains the following information:

> And King Solomon made a navy of ships … And Hiram sent in the navy his servants, shipmen that had knowledge of the sea, with the servants of Solomon. And they came to Ophir, and fetched thence gold, four hundred and twenty talents, and brought it to King Solomon.[1]

Discovering the location of Ophir and its gold was one of the factors that drove the Spanish to explore the south-west Pacific, despite both the brevity of the Bible's reference to Ophir and the absence of any hints to its actual location. The belief in Ophir was interwoven with tales of a suspected southern antipodean continent — *Terra Australis Incognita* — as well as of gold-bearing islands that the Spanish conquistadores had heard about from Incan historians. Further back still were classical references to the golden and silver islands of *Khryse* and *Argyre*. The Spanish embarked against a background of belief in the existence of Ophir — 'a loose amalgam of biblical and classical tradition, scholarly and cartographical deduction and conjecture, report and rumour, fact and fiction'.[2]

Gold is intimately connected with the volcanological history of Near Oceania. This is not only because volcanoes were discovered, coincidentally, in attempts to locate the metal during this Spanish 'discovery' phase of the sixteenth century, but geologically as well, for many of the economic gold and copper deposits of the region are found in the roots of geologically recent volcanoes and in the alluvial deposits derived from them. King Solomon's sources of gold were

---

[1] King James Bible, Chapter 9, Verses 26–28.
[2] Jack-Hinton (1969), p. 27.

never found in Near Oceania, or elsewhere, but mineral-extraction companies in the twentieth century created their own mines there, such as at Panguna, Ok Tedi and Lihir, after geological exploration in volcanic areas. Furthermore, the king and the Ophir legend are now embedded in the biblical naming of the Solomon Islands.

New Guinea Island in the early sixteenth century was at the edge of Europe's known world, when the powers of Spain and Portugal were jostling for economic ascendancy amongst the spice-rich islands of south-east Asia.[3] Their aspirations were controlled to an extent by treaties and Papal Bulls aimed at defining two world hemispheres whose longitudinal boundary on the other side of the world from Europe — defined by the Treaty of Tordesillas — ran somewhere near the spice-rich Moluccas just north-west of New Guinea. Jorge de Meneses, the Portuguese governor-elect of the Moluccas, encountered *Ilhas dos Papuas* and the Papuans living there, at the north-western end of New Guinea in 1526.

The Spaniard Alvaro de Saavedra Cerón attempted to return to Mexico from the Moluccas in both 1528 and 1529, by sailing eastwards along the north coast of New Guinea then north-eastwards, becoming the first European to encounter Manus, the largest island of what the British would later call the Admiralty Islands. Iñigo Ortiz de Retes in 1545 also attempted to cross back across the Pacific to Spanish America by first sailing eastwards along the north coast of the island that he named *Nueva Guinea*, New Guinea.[4] His vessel, the *San Juan*, was twice forced northwards from New Guinea by difficult winds, but it seems to have reached about 5 °S — perhaps somewhere near Karkar volcano, where the north coast swings southwards — before the voyage was abandoned. The return voyage to Portuguese territory was through the small islands situated well to the west of Manus, where Melanesians attacked the ship wielding spears tipped with 'flint suitable for striking fire'.[5] One interpretation is that this is 'a clear reference to obsidian spearpoints' traded from the Admiralty Islands.[6]

Original documentation on the voyages of both Saavedra and Retes is not detailed, and there is no known record of volcanoes having been seen by them. The label 'Los Volcanes', however, appears in different spellings on several European maps showing the north coast of New Guinea, and which were published later in the sixteenth century and into the seventeenth century — including, for example, on Abraham Ortelius' *Typus Orbis Terrarum*, or Map of the World, the

---

[3] Wright (1945), Sharp (1960) and Whittacker et al. (1975, Documents B1–B9).
[4] See, for example, Sharp (1960).
[5] Sharp (1960), p. 31.
[6] Spriggs (1997), p. 226. Obsidian, however, is unlikely to produce a spark hot enough for fire-lighting purposes. The term 'flint', used by the Spaniards and translated by Sharp (1960) may, therefore, simply be a general reference to stone that was shaped by flaking, which would be familiar to people in the sixteenth century as gun flints and as natural flints, such as chert, used for fire lighting (J. Kennedy, personal communication, 2012).

world's first modern atlas, which was first published in 1570.[7] The early voyages by Saavedra and Retes, then, could be the source of this brief volcanological information. One is tempted to imagine that the youthful volcanic forms of, or even eruptions from, at least some of the islands must have been identified by Saavedra, or Retes, or both, as they passed along the north coast of New Guinea Island during their respective passages.

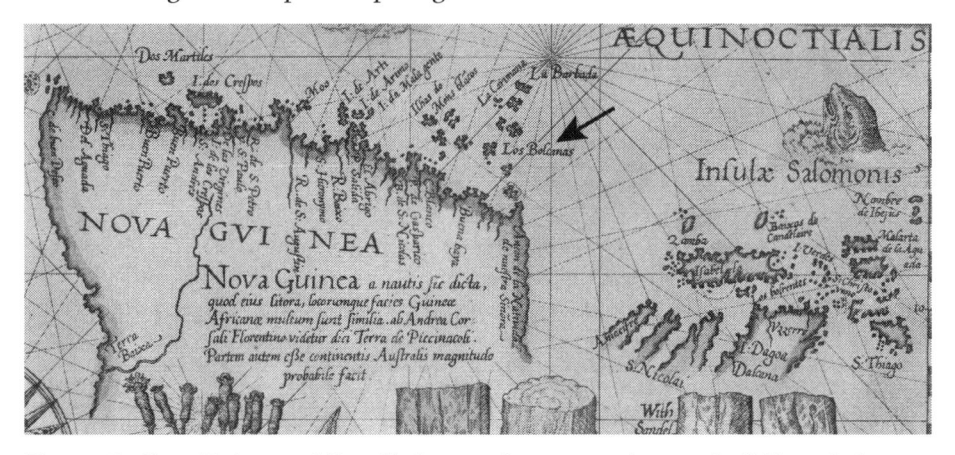

Figure 8. 'Los Bolcanas' lie off the north-eastern 'corner' of New Guinea Island in this detail from the decorative chart *Insulae Molvccae*, which is dated 1617, but is based largely on 16th century sources.

Source: C.J. Visscher (1617), 'Insulae Moluccae celeberrimae sunt ob maximam aromatum copiam quam per totum terrarum orbem mittunt'. Mitchell Library, State Library of New South Wales (Safe/M2 470/1617/1).

Another sixteenth century Spanish record of volcano observation is from 1568 when Alvaro de Mendaña, 'in a spirit of colonialism, commercialism, Catholic proselytism and romantic curiosity'[8] crossed the Pacific and found what would subsequently be called the 'Islands of Solomon'. One of the islands is Savo volcano, about 35 kilometres north-west of present-day Honiara on Guadalcanal, 'which is always throwing out a great deal of smoke', wrote Hernando Gallego, chief pilot of the expedition.[9] Whether this means that Savo was in actual eruption, or was in a passive state and simply sending out a plume of water vapour, is unknown, but there are no references to either falls of ash or glows from the central crater. Gallego also referred, rather peculiarly, to what 'appears to be a road descending from the top to the sea'. The 'road' almost certainly was a gully down which some sort of flowage had recently denuded the vegetation — perhaps a rush of water, or a flow of mud, or a pyroclastic flow. Thus, although Savo does not seem to have been in eruption during Mendaña's visit, it may have been in eruption a short time before, or else Mendaña was there during a relatively 'quiet' phase of a longer eruptive period.[10]

---

7   Jack-Hinton (1969) and, for example, Ehrenberg (2006).
8   Jack-Hinton (1969), p. xv.
9   Tyssen-Amherst & Thompson (1901), p. 30.
10 Petterson et al. (2003).

Fire Mountains of the Islands

Schouten, Le Maire and Tasman

The motivations of the Dutch in exploring the south-west Pacific in the seventeenth century were more specifically oriented to trade and profit than were those of the Spanish. Their pragmatic mercantile ambitions and bulbous cargo ships had already led to the establishment of commercial interests, particularly in relation to the spice trade, in the form of the *Verenigde Oostindische Compagnie* or United East India Company. The Dutch had aspirations to widen their web of influence, and they would eventually colonise what is now Indonesia, including western New Guinea. The Dutch, as the governing power in the Dutch East Indies, would have to deal with disasters from volcanic eruptions, perhaps most famously from Krakatau volcano in 1883. The Spanish, in contrast, had lost a colonising interest in the region. Their power as a major maritime nation had declined by the end of the sixteenth century, although they had undertaken a second voyage to the Islands of Solomon in 1595.

Accounts of two Dutch voyages refer to volcanoes and their activity in Near Oceania. The first of these was undertaken in 1615–1617 by Willem Cornelisz Schouten, 'a man well experienced and very famous in navigation', and by Jacob Le Maire — supercargo, commander of the voyage, and son of Isaack Le Maire 'renowned merchant of Amsterdam ... being very inclined to trade in strange and far distant parts'.[11] Schouten and Le Maire crossed the Pacific from Cape Horn, sailed up the eastern side of New Ireland, rounded its north-western end, then crossed the Bismarck Sea to the New Guinea north coast, where Saavedra and Retes had sailed in the previous century. The Dutchmen saw, on the morning of 7 July 1616, 'a burning island, emitting flames and smoke from the summit, wherefore we gave it the name Vulcanus. The island was well populated and full of coker-nut trees'. The volcano is Manam and this statement represents the first known European report of an unequivocal volcanic eruption in Near Oceania. Schouten was at first inclined to think the volcano might be Api, in the Banda Sea to the west of New Guinea, so similar did the sizes and shapes of Manam and Api appear to him. Large imposing stratovolcanoes such as these do indeed look alike, their great height and imposing symmetry reflective of a global constancy in the geological processes that form them.

The second Dutch voyage of relevance is that of Abel Janszoon Tasman who, in 1643, after mapping parts of the coastlines of Tasmania and New Zealand, entered Near Oceania, following a similar track to that of his countrymen, Schouten and Le Maire, 27 years earlier. Tasman sailed between the volcanic Witu Islands and the north coast of New Britain, and reached the offshore islands of New Guinea where, on the night of 20 April 1643, he saw 'a large flame issue steadily from

---

11 Villiers (1906), p. 166.

2. Volcano Sightings by European Navigators: 1528–1870

the top of the mountain. This is the volcano which Willem Schouten refers to in his Journal'.[12] 'Vulcanis' is Manam, and a drawing of the 'burning island' in Tasman's journal is indeed a good likeness of the island volcano.

Figure 9. An English translation of the caption seen on this drawing of Manam volcano from Tasman's journal reads: 'A view of the burning Island when it bears from you north-west'. The ravine on the left running from the near the top of the island down to the sea, is the south-western 'avalanche' valley on Manam. There are four such valleys on Manam, which deliver sediment into the sea, building up coastal fans, such as the one seen here. The billowy cloud at the summit of the volcano is not necessarily of volcanic origin and may simply be a decorative weather feature.

Source: Tasman (1898, detail of figure opposite the journal-facsimile page containing entries for 13–14 April 1643).

Tasman noted two days later, on 22 April, 'At this time we had the high burning mountain east-south-east and south-east by east from us at 7 miles' distance. At night the flames were very violent',[13] which, again, refers to Manam. However, *two* 'burning islands' are shown on the charts that were compiled by later cartographers of the Tasman voyage, including the well known 'Bonaparte' and 'Eugene' maps, one island corresponding to Manam, the other to Karkar Island. This may be a cartographic error, as Tasman refers to only one volcano in activity. In addition, one of Tasman's recent editors, historian Andrew Sharp, concluded in his 1968 study that the active volcano seen on 20 April 1643, and illustrated, was actually Karkar and that Tasman was mistaken in identifying it as Manam. The evidence for this allegation, however, is not convincing as the volcano in the illustration closely resembles that of Manam. Tasman, in any case, would have been lucky to see any eruptive activity from Karkar during the short time he was near the island. This is because there have been only two reported observations of eruptive periods at Karkar during the past four centuries — one in the late nineteenth century, the other in the 1970s. Manam, therefore, which is much more frequently active, was more likely to have been the only volcano that Tasman saw in eruption in 1643.

---

12   Tasman (1898), p. 46. See also Sharp (1968), pp. 229, 231.
13   Tasman (1898), p. 49.

These early references to volcanoes represent all that is known from the Spanish and Dutch periods of exploration in Near Oceania. All of the descriptions are frustratingly brief and, in some cases, unclear or ambiguous, and none of them contains the descriptive detail of a volcanic eruption that William Dampier subsequently provided for Ritter Island in 1700. The voyage by Mendaña in 1568, when Savo volcano was seen, is perhaps the most convenient start to the 'historical' period of volcanic activity in Near Oceania, but the description by Schouten and Le Maire of Manam in 1616 is the least ambiguous of the few early observations of eruptive activity in the region. A 'historically active' volcano, then, is one that has been in eruption since 1616, or 1568, even though there are large gaps in the documentary record of activity during subsequent centuries.

Carteret, Hunter, D'Entrecasteaux and Parker Wilson

The French and British together took an interest in Near Oceania after, and to an extent because of, Dampier's 1700 voyage. Both nations had similar longer term global interests in the expansion and securing of empires, colonies, and power — interests that at times brought them into direct military conflict in Europe during the French Revolutionary and Napoleonic Wars of 1792–1815. The respective characteristics of their expeditions to Near Oceania in 1700–1870 were similar too, at least to the extent that scientific objectives were articulated together with the broader aim of 'discovery', and that their ships of exploration carried natural scientists. 'Rediscovering' the Solomon Islands and locating them more precisely also featured strongly in the aims of the eighteenth century voyages. These goals were a reflection of the emergence of science during the Enlightenment movement of eighteenth century Europe, when the rational faculties of the human mind were expected to deliver a greater understanding of the world at large.

Philip Carteret undertook a round-the-world voyage for the British Crown in 1766–1769 in the *Swallow*. It was a remarkable effort because of the endurance of its long-suffering crew and because the vessel was a decrepit and heavy man-o'-war — both slow and difficult to handle. The *Swallow* was in the northern Solomon Sea on 9 September 1767 when Carteret found the channel or strait between New Britain and New Ireland that his countryman, Dampier, in 1700, had thought to be a bay. Carteret was assisted in finding the strait by the south-east trade winds prevalent at that time of year, September; whereas Dampier, who sailed there in March, found that the north-western monsoon had set a south-easterly current which stalled any northwards track into the 'bay'.

Carteret gave the name Duke of York's Island to the largest of the cluster of islands in the middle of the St Georges Channel, which would be used as an important base for European settlement more than a century later. More particularly, however, Carteret noted that on the western side of St Georges Channel were 'three remarkable hills close to each other, which I called the Mother and Daughters. The Mother is the middlemost and largest, and behind them we saw a vast column of smoke so that probably one of them is a volcano'.[14] The three hills are prominent, peaked, and inactive volcanoes in the Rabaul area of northeast New Britain. The 'smoke' was evidently from one of the smaller unseen volcanoes that nest inside the protected natural harbour at Rabaul and which may have been either Rabalanakaia or Tavurvur. Carteret was the first European to notice parts of the Rabaul volcanic complex, which, today, is regarded as the highest risk volcanic centre in Near Oceania.

Figure 10. The volcanoes of the Rabaul area are highly exaggerated in this sketch from the Carteret voyage to the area in 1767. They are certainly steep, but do not overhang as shown here.

Source: Hawkesworth (1773, 1, detail from figure facing p. 368). Clifford Collection (RB CLI 3561), National Library of Australia, Canberra.

French aristocrat and voyager, Louis-Antoine de Bougainville, reached the Solomon Islands from the south-west on 28 June 1768 and, in early July, he sailed north-westwards along the eastern coast of what would become recognised as the largest island of the Solomons. His was not a volcanologically significant exploration because, unlike Carteret the year before, Bougainville evidently did not see any eruptions in the region during his voyage, or recognise volcanoes on the island named after him.[15] This, however, is not surprising because the volcanoes there, when viewed from the east, form the high central spine of the island, which during the day is commonly covered by cloud.

The year 1788 is a pivotal one for the volcanological story of Near Oceania. This is not because of any particular volcanic happening, but rather because the British 'First Fleet', under the command of Captain Arthur Phillip, arrived that year at Botany Bay on the east coast of Australia with its cargo of convicts, marines,

---

14   Hawkesworth (1773), p. 376. See also Wallis (1965).
15   Bultitude (1981).

officers, and crew. It is, therefore, the year that marks the beginning of European settlement of the eastern coast of Australia and, gradually, establishment of regular shipping lanes northwards through Near Oceania to China, south-east Asia and India, and to London via Batavia, capital of the Dutch East Indies. Both Dampier Strait and St Georges Channel were used for this purpose although, for safety reasons, many ships would sail the longer routes east of Bougainville Island.[16]

Ships stopped in Near Oceania to take on fresh water and food, the number of 'first contacts' between Melanesians and Europeans thus increasing.[17] Europeans intent on trade and resource exploitation stopped over, too, later including those involved in the infamous 'blackbirding' of Melanesian labour for the sugarcane fields of Queensland. Whalers arrived, working grounds near Bougainville Island for example, and missionaries were attracted to the region by the prospect of spreading the Word of God and saving needy souls amongst the godless 'savages'. Adventurers and naturalists were drawn in search of excitement, uncertainty and discovery of the unknown in a part of the world still distant, mysterious and isolated from Europe. Beachcombers and men stranded by shipwreck found their own refuges amongst the islands. All of these people are potential sources of volcano observations — at least, those of them who were sufficiently literate and sufficiently motivated to record them.

Figure 11. Canoes of the Duke of York Islands decorate this drawing from John Hunter's account which uses the volcanoes of the Rabaul area, somewhat disproportionately, as a backdrop. The large central cone is Kabiu or the Mother volcano. Tovanumbatir, or North Daughter, is to its right, and what appears to be Watom Island is on the extreme right. Tavurvur is shown in eruption, in 1791, between Turagunan or South Daughter, on the left, and the Mother. The caldera wall of Rabaul volcano is shown on the extreme left behind Turagunan.

Source: Hunter (1968 (1793); Plate 11). National Library of Australia, Canberra.

---

16   Whittacker et al. (1975, Documents C1–C11).
17   See, for example, Thomas (2010).

John Hunter was Phillip's second-captain in the First Fleet and in command of the armed tender *Sirius*. He returned to England on the hired Dutch vessel *Waaksamheyd* by way of St Georges Channel and, between 22–25 May 1791, was anchored at the Duke of York Islands. Hunter wrote in his account that 'The hills mentioned by Captain Carteret … by the name of Mother and Daughters are very remarkable. A little way within the south-eastermost Daughter there is a small flat-topped hill, or volcano, which all the time we were within sight of it, emitted vast columns of black smoke'.[18] The volcano is Tavurvur, the black smoke is ash laden and Hunter, therefore, saw the volcano in full eruption.

Comte de La Pérouse and his ships the *Boussole* and *Astrolabe* were at Botany Bay in January 1788, even as the British First Fleet's convicts and crew were being settled at Sydney Cove. Jean-Francois de Galaup de La Pérouse was on a major voyage of discovery for the French, supported by King Louis XVI, and aimed at rivalling the British successes of Captain Cook. La Pérouse left Sydney in March 1788, but he was wrecked in the Santa Cruz Islands, never to be seen again — at least by Europeans. A search mission in 1792–1793, supported by a now post-Revolution government — the Bastille had been taken on 14 July 1789 — was led by Contre-Amiral Antoine-Raymond-Joseph Bruny-d'Entrecasteaux, who undertook two voyages into Near Oceania using the *Recherche* and the *Esperance* — 'search' and 'hope' respectively.[19]

The first voyage was in mid-July 1792 when D'Entrecasteaux sailed up the western side of Bougainville Island, but nothing of volcanological significance was reported, and the first known and recorded claim of seeing volcanic activity on the island was not until four years later. Captain Hogan of the *Marquis Cornwallis* was sailing en route from Port Jackson to Canton, China, when he saw on 6 July 1796 'a great quantity of black, or rather sulphureous smoke of matter emitted from the earth'.[20] The coordinates given by Hogan for the emission, however, correspond to a point near the north-eastern coast of the island where there are no volcanoes. The report may therefore be spurious, but this did not stop a 'Cornwallis volcano' being marked on later charts, even into the twentieth century.[21]

A voyage along the western coast of Bougainville by the *Margaret Oakley*, sometime after October 1834, and possibly in 1835, resulted in the frustratingly brief statement that '… the summit of a mountain was crowned by what appears to be two extinguished volcanic craters'.[22] And the French naval explorer

---

18   Hunter (1968), p. 155.
19   Labillardiere (1800), Beautemps-Beaupré (1807), Rossel (1808) and Duyker & Duyker (2001).
20   Hogan (1801). See also Jack-Hinton (1969) and Bultitude (1981).
21   Guppy (1887) appears to have been the first to cast doubt on the validity of the Hogan report. Bultitude (1981) gave examples of the later false mapping of Cornwallis volcano.
22   Jacobs (1844), p. 221. See also Bultitude (1981).

Jules Sebastien Cesar Dumont D'Urville in December 1838 made an even more intriguing observation that the highest of the island's summits 'expose to view some hues which, from the reflection of the sun, resemble piles of snow in the gullies'.[23] This may refer to the bare solfataric area on the summit of Balbi volcano, although Dumont D'Urville did not refer to any volcanoes as such.

The first unequivocal observation of an active volcano on Bougainville Island by a European was not until 50 years after D'Entrecasteaux's visit, when John Parker Wilson, ship's surgeon on board the British whaler *Gypsy*, described and sketched Bagana. The volcano was near the centre of the island and Parker Wilson saw it from the south-west on 15 March 1842.[24] His description of 'numerous furrows and channels' marking the sides of the volcano is typical of Bagana today, where masses of slowly moving lava spill imperceptibly down its flanks from a central, vapour-producing crater, creating a strongly 'furrowed' appearance to the sides of the mountain, and sending off water vapour at cool or wet times of the day. The description is also the first recorded observation of probable lava extrusion at a volcano in Near Oceania. **Lava flows** are distinct from 'pyroclastic flows'. They represent the down-slope movement of a coherent stream of hot 'sticky' liquid — much more viscous than, say, water — which cools gradually on the outside producing a rough rocky exterior and eventually freezes. This is in contrast to faster moving pyroclastic flows, which are hot, highly mobile avalanches of hot rock, dust, and gas. Lava flows are usually easy to walk away from. Pyroclastic flows are not.

The second voyage by D'Entrecasteaux, in 1793, was somewhat more volcanologically significant than his first one the year before. He sailed through the Louisiade Archipelago off the south-eastern tip of New Guinea, which had been named previously by his countryman Bougainville, then north-westwards through islands that received his own name — the D'Entrecasteaux Islands — and on to Dampier Strait. The islands of both the Louisiade and D'Entrecasteaux groups include youthful volcanoes, but these eruptive centres are small or inconspicuous and would easily have been overlooked. Then, in Dampier Strait on 29 June 1793, a volcano in eruption is spotted. Two expedition accounts of this eruption are available. D'Entrecasteaux himself wrote of seeing:

> an enjoyable spectacle: a sudden eruption of a volcano on the island closest to the coast of New Britain. The flames were not visible, as it was daytime. But masses of thick smoke could be seen coming out of the summit of the mountain; and a flow of lava was dashing to the sea, forming several cascades, from which columns of white smoke could be seen rising at different heights, even on the sea-shore.[25]

---

23 Dumont D'Urville (1843), p. 93. The English translation is from Bultitude (1981, p. 232).
24 Wilson (1839–1843).
25 Duyker & Duyker (2001), p. 258.

2. Volcano Sightings by European Navigators: 1528–1870

Figure 12. Dr Parker Wilson in 1842 drew, in the margin of his diary, this simple sketch of Bagana showing impressionistically the channels formed by the steep-sided, blocky, lava flows that make up the volcano.

Source: Wilson (1839–1843), Royal Geographical Society (RGS-IBG Collections, reference 'ar JWI/1/15031842').

The naturalist Jacques Labillardiere provided another witness account, referring to the

> great quantity of burning substances [being] thrown out of the aperture of the volcano, which lighting upon the eastern declivity of the mountain, rolled down the sides till they fell in the sea, where they immediately produced an ebullition in the water, and raised it into vapors of a shining white colour. At the moment of the eruption, a thick smoke, tinged with different hues, but principally of a copper colour, was thrown out with such violence, as to ascend above the highest clouds.[26]

These two accounts are remarkably similar, both to each other and to the one given by Dampier from his 1700 voyage in that, again, some form of pyroclastic flow was seen racing down the flanks of a steep island into the sea. The Frenchmen, however, did not realise that the volcano they were observing

---

26   Labillardiere (1800), pp. 447–48.

was the same one, later named Ritter, that Dampier had seen nearly a century before. Labillardiere, particularly, thought incorrectly that Dampier's volcano was another nearby but inactive volcanic island, and that they had witnessed eruptive activity from a different and hitherto unknown volcano.

D'Entrecasteaux passed through Dampier Strait and sailed eastwards along the north coast of New Britain, passing the Witus Islands, and noting on 3 July three peaked islands in a chain running north from New Britain. These he named Willaumez, Raoul and Gicquel, after members of the voyage.[27] The 'islands' are part of Willaumez Peninsula, a 60-kilometre-long chain of volcanoes that runs northwards from the north coast of New Britain. Some commentators have concluded that there must have been modern uplift of the islands and emergence of the sea floor between them.[28] A more likely explanation, however, is that D'Entrecasteaux did not realise, because of his low and distant vantage points, that the high points which he thought were islands were actually linked by low land. D'Entrecasteaux also recognised another island to the east, which he named Du Portail after another expedition member, but which is now known to be Lolobau volcano, a true island. Pic Deschamps on the mainland was named too — the feature is Likuruanga or North Son volcano. None of the many peaks and islands given French names during the D'Entrecasteaux expedition were recognised as being volcanoes.

European voyages through Dampier Strait and along the north coast of New Guinea continued throughout the nineteenth century and reports from them contain some mention of volcanoes. Dumont D'Urville, for example, gave the dimensions of Ritter Island in 1827, confirming its exceptionally steep slopes, but not recording any eruptive activity.[29] Captain Benjamin Morell saw Manam activity in 1830, and made the doubtful claim that he had seen seven active volcanoes in all, five of which were in eruption and the other two 'smoking'.[30] His book is an example of an entertaining style of nineteenth century travel reporting in which inaccuracies and exaggerations, such as this, cast doubt on the veracity of other observations that may in fact be accurate.

An early and remarkably bold attempt at European settlement was made in 1848 on the northern coast of Umboi Island in Dampier Strait by French missionaries of the Roman Catholic Marist order. Ritter was seen producing '… thick smoke which emerges from three places almost at the highest point of the cone'.[31] The mission was withdrawn in 1849, re-established in 1852–1855, but apparently

---

27 Duyker & Duyker (2001), pp. 259–60. Jean-Baptiste-Philibert Willaumez was *enseigne de vaisseau* on the *Recherche*.
28 Stanley (1923).
29 Dumont D'Urville (1832).
30 Morrell (1832).
31 Cooke (1981), pp. 117–18. See also Laracy (1976).

without any further significant records being made of Ritter or its eruptive activity. This is not surprising as the would-be missionaries had other concerns on their mind: they were plagued by malaria and other physical hardships, and were singularly unsuccessful in convincing the local Melanesians of the biblical messages of salvation they wished to impart. Significant European settlement by missionaries and traders would not take place until the 1870s, more than 300 years after the Spanish first entered Near Oceania. Melanesians themselves during these 300 years and indeed for the previous tens of millennia, had seen volcanoes in a quite different light.

European and Melanesian Viewpoints

What can be drawn from the preceding review of European volcano observations from 1528 up to the start of European settlement beginning after about 1870? First, eruptions were reported for five volcanoes — Manam, Karkar, Ritter, Rabaul and Bagana — although the Karkar observation is questionable because of the probable misinterpretation of Tasman's reports, and Parker Wilson may not have realised that the lava on Bagana was flowing, although it probably was. Ritter is perhaps the best known volcano of the period on account of its conspicuous position in the middle of the navigable Dampier Strait, and Dampier's record of its 1700 activity and cascading pyroclastic flows is the most insightful from a volcanological perspective. Plumes or clouds of vapour and gas were being emitted from another three volcanoes during 1528–1870 — Savo, Ulawun, and possibly Kadovar — but whether these contained ash and whether, therefore, the volcanoes were in actual eruption, is unclear.

The observations at all of these named volcanoes were brief and were made in transit from points at sea. The records, too, are fleetingly short and some are inaccurate and confusing, which is hardly surprising bearing in mind the concerns of earlier voyagers in particular — ship safety, navigation in unchartered waters, the deteriorating condition of hulls and rigging, onboard discipline, inadequate diet, polluted food and water, odorous bilge waters, disease, and fear and apprehension of indigenous people. Neither is there any indication in the reports of the dangers posed by such volcanoes to either indigenous Melanesians or to future potential settlers from Europe. Nevertheless, these limitations notwithstanding, the European voyagers demonstrated in their records that Near Oceania was clearly a volcanically active area, and that the chain of islands off the New Guinea north coast contained many of the active volcanoes of the region. They were also, in the main, able to relate the times of their observations to a day on the calendar.

Melanesian viewpoints throughout the period 1528–1870 were rather different from European ones. Information about past volcanic activity can be transferred by Melanesians from generation to generation through the verbal telling of memorable events, but the absence of a traditional Melanesian calendar means that 'dates' cannot be assigned to specific happenings, except possibly in a general way by counting back the number of generations. Generations potentially can 'drop out' or be added to such sequences, although this is not so much of a problem for recent eruptions that are fresh in community memory. The accuracy of information transferred through long, even multiple, genealogical chains is also an issue in which content and emphasis change. The volcanological historian seeking so-called 'hard' facts in these stories must remember, too, that their genesis lay in a purpose different to the scientific recording of events.[32] Volcanological oral history, nevertheless, has significant value, particularly where stories can be checked for veracity against the geological record.

Two examples may be given here. Both refer to eruptions in the mid-nineteenth century that were not witnessed by Europeans, yet which are consistent with the geological record. The first example is a metaphoric story of battling spirit beings from the Rabaul area. A Roman Catholic priest in 1937 informed the *Rabaul Times* about meeting elderly villagers who remembered the volcanic activity that produced Sulphur Creek volcano to the north-west of Tavurvur and west of the Mother. His main source was a prominent village leader, To Mulue, who said in the retelling by the priest:

> All that land rose during a heavy earthquake. I was a young man when it happened, and now I am, but for an old woman at Davuan, the only living witness … the earth broke in eruption. The crater is called Kururung maqe — it is close to the hot water creek in Rabaul. A big crab had a quarrel with a snake and caused the eruption. Stones were thrown inland … [and] new land rose on the mainland …[33]

This young volcano, Sulphur Creek, is clearly identifiable today at the head of the straight, canal-like inlet that runs eastwards from the harbour. The priest estimated that the eruption must have taken place sometime in 1845–1850. Other recorders of the same eruption have suggested 1840.

---

32 See, for example, the discussion of Tolai stories of Rabaul eruptions by Sack (1987). Neumann (1992), in particular, uses oral narratives to reconstruct Tolai history and reviews some of the development of oral history as a discipline of serious inquiry.

33 Boegershauser (1937), p. 15. The same eruption is also mentioned in an earlier account by Boegershauser (1906), in which he said that the Sulphur Creek eruption took place at the same time as activity from Rabalanakaia volcano. The two reported eruptions, however, may refer to the same one, as Sulphur Creek and Rabalanakaia are close together, and there are no other known eyewitness records of an eruption at Rabalanakaia at this time.

Oral traditions about an eruption on Savo Island sometime in 1830–1840 have been collected by at least three people. The most recent collection was by Thomas Toba, a Solomon Islands seismologist, who was born on the island and who documented *kastom* stories about Savo eruptions through a series of villager interviews.[34] All of the accounts of the eruption are broadly consistent with one another. There were periods of total darkness over the island caused by ash clouds that spread as far the north-western coast of Guadalcanal Island and which continued for a long time. Savo villagers evacuated to Guadalcanal. Explosions and 'great fiery rocks' were observed, and flows down valleys on Savo produced the smooth surfaces that, from the sea, looked like roads leading to the middle of the island. Pyroclastic flows burned people and other villagers sank into mudflows. Information about another, even more catastrophic event several generations previously, has also been handed down, and many people were killed as a result of it. The eruption is referred to as *toghavitu* meaning 7,000 or 1,007 — depending on translation — and referring to the number of fatalities. This oral history from Savo is a particularly rich one — much richer than the Sulphur Creek example — and is consistent with what is known about the volcanic geology of Savo.

An interweaving of traditional knowledge and scientific observations characterises the reconstructed histories of other volcanic eruptions in Near Oceania. Documentation of oral traditions started to emerge as European settlement and, eventually, colonisation by the Germans and British began to take hold in the region. They would be incorporated also with the results of later, twentieth century geological studies of the deposits of eruptions.

References

Beautemps-Beaupré, C.F., 1807. *Atlas du Voyage de Bruni-Dentrecasteaux, Contre-Amiral de France, Commandant les Frigates la Researche et l'Esperance, fair par Ordre du Gouvernement en 1791, 1792 et 1793 … Dépôt Général des Cartes de la Marine et des Colonies, Paris.*

Boegershauser, G., 1906. 'Das Erdbeben auf der Insel Matupi und die Zerstoerung der Nonduper Kirche'. *Monatshefte zu Ehren Unserer Lieben Frau vom Heiligsten Herzen Jesu*, 23, pp. 111–15.

——, 1937. 'Eruption of a Volcano at Rabaul', *Rabaul Times*, 6 August (no. 35), p. 15.

---

34   Toba (1993). See also Petterson et al. (2003).

Bultitude, R.J., 1981. 'Literature Search for pre-1945 Sightings of Volcanoes and their Activity on Bouganiville Island', in R.W. Johnson (ed.), *Cooke-Ravian Volume of Volcanological Papers*. Geological Survey of Papua New Guinea Memoir, 10, pp. 227–42.

Cooke, R.J.S., 1981. 'Eruptive History of the Volcano at Ritter Island', in R.W. Johnson (ed.), *Cooke-Ravian Volume of Volcanological Papers*. Geological Survey of Papua New Guinea Memoir, 10, pp. 115–23.

Dumont D'Urville J.S.C., 1832. *Voyage de la Corvette L'Astrolabe execute par Ordre du Roi, pendant les Annees 1826–1827–1828–1829, sous le Commandement de M.J. Dumont Durville, Capitaine de Vaisseau. Histoire du Voyage*. Tome Quatrieme. Tastu, Paris.

——, 1843. *Voyage au Pole Sud et dans l'Oceanie sur les Corvettes L'Astrolabe at La Zelee, execute par Ordre du Roi pendant les Annees 1837–1838–1839–1840, sous le Commandement de M.J. Dumont D'urville, Capitaine de Vaisseau ... Histoire du Voyage*. Tome Cinquieme. Gide, Paris.

Duyker, E. & M. Duyker (eds & trans), 2001. *Bruny d'Entrecasteaux: Voyage to Australia and the Pacific, 1791–1793*. Melbourne University Press, Carlton South.

Ehrenberg, R.E., 2006. *Mapping the World: An Illustrated History of Cartography*. National Geographic Society, Washington D.C.

Guppy, H.B., 1887. *The Solomon Islands: Their Geology, General Features and Suitability for Colonization*. Swan Sonnenschein, Lowrey and Company, London.

Hawkesworth, J., 1773. *An Account of the Voyages Undertaken by the Order of His Present Majesty, for Making Discoveries in the Southern Hemisphere, and Successfully Performed by Commodore Byron, Captain Wallis, Captain Carteret, and Captain Cook in the Dolphin, the Swallow, and the Endeavour; Drawn up from the Journals which were Kept by the Several Commanders and from the Papers of Joseph Banks, Esq*. 1. W. Strahan and T. Cadell, London.

Hogan, Captain, 1801. 'Instructions for Entering Dampier's Straits from the Eastward', in *The Oriental Navigator; or, New Directions for Sailing to and from the East Indies, China, New Holland ...* . 2nd edn. pp. 601–07. Laurie and Whittle, London.

Hunter, J., 1968 (1793). *An Historical Journal of Events at Sydney and at Sea 1787–1792*. Angus and Robertson, Sydney.

Jack-Hinton, C., 1969. *The Search for the Islands of Solomon 1567–1838*. Clarendon Press, Oxford.

Jacobs, T.J., 1844. *Scenes, Incidents, and Adventures in the Pacific Ocean, or the Islands of the Australasian Seas, during the Cruise of the Clipper 'Margaret Oakley' under Capt. Benjamin Morrell*. Harper & Brothers, New York.

Labillardiere, J.J.H. de, 1800. *Voyage in Search of La Perouse Performed by Order of the Constituent Assembly, During the Years 1791, 1792, 1793, and 1794, and Drawn up by M. Labillardiere, Correspondent of the Academy of Sciences at Paris, Member of the Society of Natural History, and one of the Naturalists Attached to the Expedition*. John Stockdale, London.

Laracy, H., 1976. *Marists and Melanesians: A History of Catholic Missions in the Solomon Islands*. The Australian National University Press, Canberra.

Morrell, B., Junior, 1832. *A Narrative of Four Voyages to the South Sea, North and South Pacific Ocean, Chinese Sea, Ethiopic and Southern Atlantic Ocean, Indian, and Antarctica Ocean from the Year 1822 to 1831*. Harper, New York.

Neumann, K., 1992. *Not the Way It Really Was: Constructing the Tolai Past*. Pacific Islands Monograph Series 10. University of Hawaii Press, Honolulu.

Petterson, M.G., S.J. Cronin, P.W. Taylor, D. Tolia, A. Papabatu, T. Toba & C. Qopoto, 2003. 'The Eruptive History and Volcanic Hazards of Savo, Solomon Islands', *Bulletin of Volcanology*, 65, pp. 165–81.

Rossel, M., 1808. *Voyage de Dentrecasteaux, Envoye a la Reserche de La Perouse*. L'Imprimerie Imperiale, Paris.

Sack, P., 1987. 'The Emergence and Settlement of Matupit Island: Vulcanological Evidence, Oral Tradition and "Objective" History in Papua New Guinea', *Bikmaus*, 7, pp. 1–14.

Sharp, A., 1960. *The Discovery of the Pacific Islands*. Clarendon Press, Oxford.

——, 1968. *The Voyages of Abel Janszoon Tasman*. Clarendon Press, Oxford.

Spriggs, M., 1997. *The Island Melanesians*. Blackwell, Oxford.

Stanley, E.R., 1923. 'Report on the Salient Geological Features and Natural Resources of the New Guinea Territory including Notes of Dialectics and Ethnology', *Report on the Territory of New Guinea*, Commonwealth of Australia Parliamentary Paper 18, Appendix B.

Tasman, A.J., 1898. *Abel Janszoon Tasman's Journal of his Discovery of Van Diemens Land and New Zealand in 1642 ... with an English Translation and Facsimiles of Original Maps to which are added Life and Labours of Abel Janszoon Tasman by J.E. Heeres*. Frederik Muller & Co., Amsterdam.

Thomas, N., 2010. *Islanders: The Pacific in the Age of Empire*. Yale University Press, New Haven.

Toba, T., 1993. 'Analysis of Savo Custom Stories on the Eruption of Savo Volcano', Seismological Unit, Water and Mineral Resources Division, Ministry of Energy, Water and Mineral Resources, Honiara, Solomon Islands, Technical Report TR4/93.

Tyssen-Amherst, Lord W.A. & B. Thomson (eds), 1901. *The Discovery of the Solomon Islands by Alvaro de Mendana in 1568*, 1. Hakluyt Society, London.

Villiers, J.A.J. de (trans.), 1906. *The East and West Indian Mirror, being an Account of Joris van Speilbergen's Voyage around the World (1614–1617), and the Australian Navigations of Jacob le Maire*. Hakluyt Society, London.

Wallis, H. (ed.), 1965. *Carteret's Voyage Round the World 1766–1769*. 1. Hakluyt Society, Cambridge.

Whittacker, J.L., N.G. Nash, J.F. Hookey & R.J. Lacey, 1975. *Documents and Readings in New Guinea History: Prehistory to 1889*. Jacaranda Press, Milton.

Wilson, D.P., 1839–1843. 'Log and Private Journal of D. Parker Wilson, Ship's Surgeon of the South Sea Whaler *Gypsy* on a Voyage from 23 October 1839 to 19 March, 1843'. Microfilmcopy No. M198 of the unpublished manuscript, Pacific Manuscripts Bureau, The Australian National University, Canberra. Original diary in Royal Geographical Society Library and Archives, London.

Wright, I.S., 1968 (1945). 'Early Spanish Voyages from America to the Far East, 1527–1565', in *Greater America: Essays in Honor of Herbert Eugene Bolton*. Books for Libraries Press, Freeport, New York, pp. 59–78.

3. European Intruders and the 1878 Rabaul Eruption: 1870–1883

In the evening the sight became more than grand — it was awful; every few moments there would come a huge convulsion, and then the very bowels of the earth seemed to be vomited from the crater into the air; enormous stones, red hot, the size of an ordinary house would be thrown up almost out of sight, when they would burst like a rocket, and fall hissing into the sea.

Wilfred Powell (1883)

Blanche Bay and the Tolai

The landscape surrounding the vast bay at Rabaul had long been occupied by Melanesian people when the British naval surveyors Captain C.H. Simpson and Lieutenant W.F.A. Greet entered the bay on 17 July 1872, naming it after their vessel, HMS *Blanche*. The Melanesians of the Rabaul area are known today as the Tolai, a populous group made up of different matrilineal descent lines or *vunatarai*.[1] They had migrated there at different but undated times from New Ireland, some through the Duke of York Islands, had settled successfully by taking agricultural advantage of the rich soils of Rabaul's volcanoes, and are thought to have displaced the Baining people, who moved into the mountains to the west. The Tolai language is Tinata Tuna, although it is most commonly referred to as Kuanua, a word from the Duke of York Islands meaning 'over there'.

Numerous features of Blanche Bay — volcanic peaks and ridges, gullies and rocks, prominences and islands, reefs and inlets — have been named individually by the Tolai. Even small, seemingly insignificant locations are named after particular plants, trees, or other natural features. These special places are of overall importance because of their intimate association with past events or ancestors, and a spirit world that is a part of the Tolai natural environment and belief system. Naming places and incorporating them in genealogies are important aspects of Tolai descent-line claims of land ownership. Land still holds special meaning for the Tolai, as it does for many other Melanesian groups, reflecting a fluid system of collective land tenure. Early European intruders in search of 'purchasable' land, and who attempted to individuate, codify and regulate land tenure, were generally slow to understand the complexities — religious, historical, social and political — lying behind this indigenous view of land ownership.

---

1   See, for example, A.L. Epstein (1969), T.S. Epstein (1968), Salisbury (1970) and Neumann (1992).

Fire Mountains of the Islands

Volcanic activity was acknowledged by the Tolai both as an agent of landscape creation and change, and as a manifestation of the workings of the spirit world. The Kuanua word *kaia* refers to the different spirits that appear, most commonly, as giant snakes called *valvalir* or *kaliku*, and the most prominent *kaia* live inside the craters of volcanoes and cause volcanic eruptions. *Rakaia* — 'the Spirit' — was the name given by the Tolai to the volcano that emerged from Blanche Bay in 1878, even though more than one *kaia* were thought to live there.[2] Some *kaia* can be befriended and some may be of assistance. One story is of two Tolai sorcerers instructing two *kaia*, who were disguised as snakes, and sending them to Tavurvur to start the 1878 eruption. The sorcerers also made two magic sticks, which were thrown into a fire and any crackling was to be the signal for the *kaia* to use their powers again and stop the eruption. Many earthquakes of volcanic origin are felt in the Rabaul area and the Kuanua word for earthquake, *guria*, is now part of the vocabulary of Tok Pisin, one of the national languages of modern Papua New Guinea.

Beings in the physical guises of *tubuan* and *dukduk* are created by the Tolai in the form of painted, conical, volcano-like heads and bodies of leaves that cover all of the enclosed person's body, except for feet and calves. Tubuan heads rise from the leafed bodies as if they are volcanic peaks rising from a forested landscape, and the bodies rustle and vibrate during tremor-like shaking movements, as allegories of volcanoes in eruption and for ground-shaking by earthquakes. Tubuan are 'raised' in commitment to the rituals of male secret societies which are also called *tubuan*. The Tolai have complex rites at initiation, mortuary, funeral, and other ceremonies, which were — and to a diminishing extent still are — part of the intricate tapestry of traditional cultural activity in the Rabaul area.

Simpson and Greet produced the first map of Blanche Bay, naming waters in the north of the bay after themselves — Simpson Harbour and Greet Harbour. Their map is sketchy and distorted, but it does give a first impression of the imposing scale of Blanche Bay and the prominence of the peaks of North Daughter, Mother, and South Daughter, the Tolai names for which are Tovanumbatir, Kombiu or Kabiu, and Turagunan. The beauty of the bay impressed them, but Simpson wrote, more pragmatically, that '… one cannot look at it a moment without being struck at the natural strength of the position in a military point of view … [there is] water in it for the navies of the world to anchor in, perfectly sheltered from all winds'.[3] The fleet of the Japanese South Seas Force was one such navy that, in 1942, would take military advantage of Blanche Bay and its harbours, even under volcanic threat, during an advance southwards towards Australia during the Second World War.

---

2 Neumann (1992), quoting the German missionaries Josef Meier and August Kleinitschen. The story of the two sorcerers is from Meier (1908), an English translation of which is given by Neumann (1996).
3 All quotations in this and the following paragraph are from pp. 4 & 5 of Simpson's report (1873).

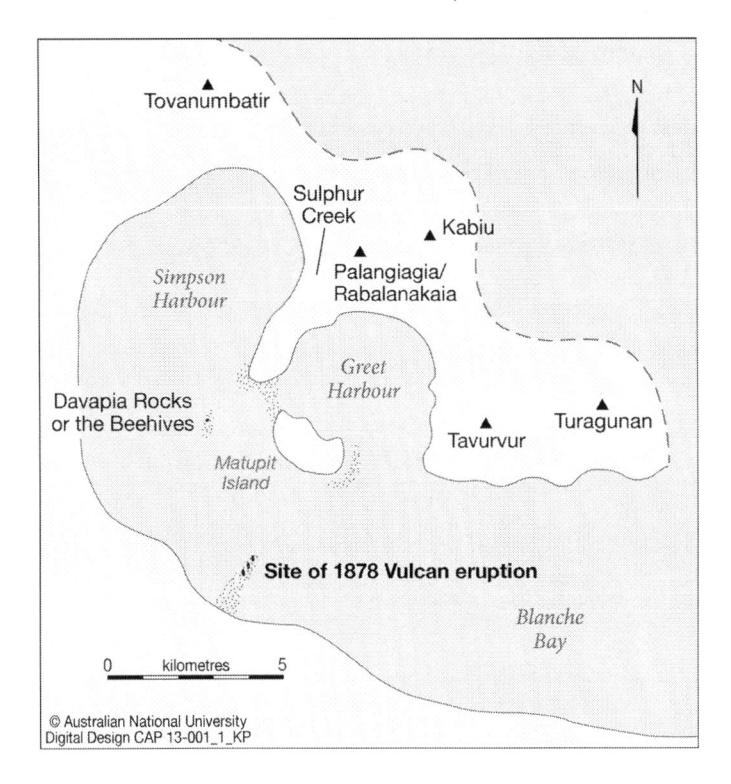

3. European Intruders and the 1878 Rabaul Eruption: 1870–1883

Figure 13. This map has been traced and adapted from the sketch map made by Simpson and Greet in 1872. The triangles represent mapped peaks. Names are those used today, including Sulphur Creek, which was not identified by Simpson and Greet.

Source: Adapted from part of Chart 797 (B1), 1873 (A3191, Pacific Folio 5). Hydrographer of the Navy, Taunton, Somerset, United Kingdom. The map was referred to by Simpson (1873, p. 4) as 'annexed plan 4'.

Simpson and Greet were aware of the volcanic origins of the three main peaks at Rabaul and they identified two further volcanoes — one just west of the Mother, the other west of South Daughter. The Tolai call the former Palangiagia and the latter is Tavurvur which, wrote Simpson, '… has been much more lately active … [and] with the smell of sulphur'. Simpson and Greet were impressed also by two precipitous rocks rising from Simpson Harbour west of Matupit Island. They named them the Beehives, although today they are known also as Davapia Rocks. Simpson and Greet were surprised to find on the larger Beehive, 'a village containing perhaps 200 inhabitants … many of their houses are built in the water on piles, they had numerous canoes moored around them'. Perhaps their most significant observation, however, bearing in mind the volcanic eruption that would follow in 1878, was the presence of 'a reef of rocks ending in three or four detached islands' extending from the south-western shore of the bay. This reef was not obviously a volcano, but its submarine foundation had likely been formed by volcanic activity. A new island would form there within six years.

The 1872 survey by Simpson and Greet is not the only example of volcanologically significant coastal mapping by the British Navy in Near Oceania in the mid- to late nineteenth century. The towns and cities of eastern Australia were growing and trade needed to be conducted with Asia, safe passages through the Melanesian islands needed charting, and there was now interest in potential places for European settlement. A survey by Captain John Moresby in the British HMS *Basilisk* is another example of shoreline-mapping and harbour-seeking. Moresby traversed the south-eastern coast of New Guinea in 1874, finding the harbour on whose shores a settlement would root, take his family name, and later become a capital city — Port Moresby. He also sailed the passage between the D'Entrecasteaux Islands and the New Guinea mainland, recognising the volcanic nature of Fergusson Island, including its hot springs. Moresby named Dawson Strait between Fergusson and Normanby Islands, where there are young volcanic cones — Lamonai, Oiau and Dobu — that were, however, not recognised by him as eruptive centres. These volcanoes, together with felt earthquakes in the early 1950s, would so concern volcanologists that a permanent volcanological observatory would be built at Esa'ala on the Normanby shoreline at Dawson Strait.

Moresby also mapped further west '... a lofty promontory ... [where a] double peaked mountain rises 4000 feet high ... the features were so striking that I resolved to honour them with great names. The Cape is therefore Cape Nelson, the two summits of the mountain are Mounts Victory and Trafalgar, and the great bay thus formed, is now Collingwood Bay'.[4] Moresby did not recognise that Victory and Trafalgar were two separate volcanoes. He did, however, succeed in scattering names from the enduring British victory over the Franco–Spanish fleet at the Battle of Trafalgar in 1805, in a remote part of New Guinea overlooking the island group named previously by the Frenchman D'Entrecasteaux. The battle indeed had introduced a century of *Pax Britannica*, a period of relative peace and international stability and trade that was underpinned by the prowling British Royal Navy, providing benefits to far-flung colonies such as Australia, and including more numerous and safer passages through Melanesian waters. Russian warships, however, had been testing their own approach to imperialism in the Pacific.

---

4   Moresby (1876), p. 269.

3. European Intruders and the 1878 Rabaul Eruption: 1870–1883

SHOOTING A WALLIBY, NEAR THE BOILING SPRINGS, ON THE D'ENTRECASTEAUX ISLANDS.

Figure 14. A landing party from the *Basilisk* visited the geothermal area on the north-western side of Fergusson Island in 1874. They are here depicted shooting a wallaby near thermal emissions of water vapour.

Source: Moresby (1876; plate following p. 254). National Library of Australia.

Miklouho-Maclay

The armed Russian corvette *Vityaz* came into New Guinea waters in September 1871 carrying a Russian naturalist, Nikolai Miklouho-Maclay.[5] The young scientist asked Captain Nazimov to leave him near Melanesian villages on the north coast of New Guinea, south-east of present-day Madang, where he stayed for 15 months during this first of three visits to New Guinea. Miklouho-Maclay was a charismatic figure who became mythologised by the Russians as a humanitarian hero dedicated to the welfare of humankind. His main scientific interests were in zoology, anthropology, ethnology, and comparative anatomy — including the differences in brain physiology between human races — but he was also a talented artist and maintained a broad interest in all aspects of the natural world. He made some observations of earthquake effects and volcanic eruptions.

5   Webster (1984).

43

Fire Mountains of the Islands

Figure 15. Miklouho-Maclay, in explorer mode, is here posing for a staged photograph, probably in the early 1870s.

Source: Queensland Digital Library, State Library of Queensland.

Miklouho-Maclay, on his second visit to the north coast of New Guinea in June 1876, was told by villagers of several earthquakes that had taken place since his first sojourn there. Falling trees had caused deaths to villagers and damage to huts. Miklouho-Maclay saw for himself how landslides had denuded trees on hill summits and how coastal stretches of forest had been destroyed by tsunamis. Even the depth of the sea, which had previously been sounded by the *Vityaz*, had changed, as well as the direction of small streams which had been diverted by sedimentation caused by the tsunamis. Miklouho-Maclay left the

New Guinea coast by commercial schooner in 1877 after his second visit and, on the evening of 11 November, he observed from the vessel eruptive activity from Manam volcano:

> I saw a red fire in the north-west … The fire occurred in intervals of several minutes and lasted each time for only about half to two minutes. At a considerable distance of more than 60 sea miles it was similar to the periodical light of a lighthouse. During the whole night one could watch the flickering of the red light … [On the following night as] … it grew darker I saw for several times mighty fork lightnings flash through the darkening clouds … [and] white smoke masses, periodically thrown out changed into columns of fire which flared up in the same way as they had done the day before. It was an impressive view.[6]

There was no time to investigate the eruption because the schooner's skipper had to adhere to a schedule, but Miklouho-Maclay was able to produce a watercolour rendition of it at dawn the next morning. Furthermore, his description is sufficiently detailed that it can be regarded as the first known report from Near Oceania of the **strombolian** type of volcanic eruption. Strombolian eruptions characterise much of Manam's known eruptive activity, as well as that of other volcanoes to the east along the length of the Bismarck Volcanic Arc. Strombolian eruptions involve periodic explosions of lava high in the conduits beneath volcanoes. They are caused when volcanic gases froth out of solution, and break up the lava, ejecting it as incandescent pieces which, individually, can be many metres in diameter but commonly much less. The pieces are thrown out on smooth parabolic trajectories that, in night-time, time lapse photographs, look like the stalks of a brilliant bunch of orange-red flowers. Some pieces spin through the air and create spindle-life forms by the time they cool and land on the volcano's flanks. Others 'splat' onto the ground forming cow-pat bombs, or break up on impact spraying glowing fragments around as secondary explosions. And still others reach the ground at such a great rate and in such large amounts without appreciable cooling, that they reconstitute and form lava flows that move off down the flanks of the volcano.

Miklouho-Maclay in his reports on Manam in 1877 also briefly mentioned eruptions from Bam volcano to the west of Manam. Furthermore, during his third visit to the region, he observed, from an anchorage on the north coast of Manus Island in the Admiralty Islands, a volcanic eruption on 28 March 1883. This may have been from the general submarine area of Tuluman Volcano, although Miklouho-Maclay himself thought it

---

6   Miklouho-Maclay (1878), pp. 409, 410.

might, very likely, have been the volcano on the small island called by the natives Loo [Lou], and from which they obtain the obsidian for their weapons and implements ... I could see a large halo as from an immense fire, and two or three times heavy thunderlike rolling noises were heard, followed by distinct flashes like columns of fire on the horizon.[7]

Figure 16. Miklouho-Maclay sketched Manam Island in his field notebook from the west at daybreak on 12 November 1877, and showed clearly both the main and southern craters in eruptive activity. The night-time incandescence, so characteristic of strombolian activity, would not have been clear in the daylight, but there is a suggestion of strombolian jetting from the southern crater on the right.

Source: Miklukho-Maklai (1952; figure between pp. 286 & 287). Image provided by E. Govor.

Traders, Missionaries and a Gentleman Explorer

Natural resources in Near Oceania had been obtained by foreign traders and whalers well before the first significant European settlement of the Duke of York Islands and Rabaul area in the 1870s. Traders bartered with Melanesian villagers during passage through the region, offering mainly metal goods in exchange, and both sides learnt skills in obtaining mutually acceptable deals. The German

---

7   Miklouho-Maclay (1885), p. 965.

company J.C. Godeffroy & Sohn, however, pioneered direct European trading in the south-west Pacific, opening for business in Samoa in 1857. The company expanded northwards and by 1873 two Godeffroy traders plus four non-European companions had established trading posts at Nonga on the north coast north of Tovanumbatir, and on Matupit Island in Blanche Bay.[8] The Nonga group was driven out by the Tolai and retreated to Matupit from where, coming again under attack, they fled to the Duke of York Islands. Godeffroy's agent in Samoa informed the authorities in Berlin of the incident and a German naval vessel, the SMS *Gazelle*, on a surveying expedition and under the captaincy of G.E.G. von Schleinitz, visited Blanche Bay on 12–17 August 1875. Schleinitz published a chart of Blanche Bay as a result of his 1875 visit that, like the one prepared by Simpson and Greet, has the prominent reef extending out into the bay from the south-western shore, from where eruptive activity would take place in 1878.[9] A Godeffroy trader, William Hicks, would later witness the 1878 eruption.

1875 was also the year that Methodist missionary the Reverend George Brown and the German merchant Eduard Hernsheim first came, separately, to the Rabaul area. Both would witness the effects of the 1878 eruption. Brown and a team of Fijian and Samoan teacher–missionaries arrived on the *John Wesley* at Port Hunter in the Duke of York Islands in August 1875, while Schleinitz was exploring Blanche Bay.[10] This event — the arrival of the *lotu*, or church — is significant in Tolai history as it marks the introduction and eventual acceptance of Christianity in the New Britain region and, from a more secular viewpoint, establishment of the first quasi-permanent European settlement there. Brown himself was an energetic if not dogged leader, a diarist, and a keen observer of the natural world — indeed, a polymath whose interests extended well beyond proselytism alone.

Eduard Hernsheim came into the area in 1875 from Micronesia to the north in order to seek and exploit new commercial opportunities in Melanesia, but found there instead — as he had in Micronesia — the competing presence of Godeffroy & Sohn. Hernsheim was, nevertheless, to become one of the most important merchants in the region and to play a role in political events that led to German acquisition of colonies in the New Guinea region in 1884. Hernsheim also would eventually establish a substantial commercial base on Matupit Island, including a copra-processing factory. He left in 1892, however, broken in health and having experienced the year before 'nerve-wracking, continual earth tremors and mysterious subterranean rumblings' that not only resulted in many of his staff leaving but causing in himself 'an inexplicable fear ... in the daily expectation of a volcanic eruption [from Tavurvur] and the disappearance of my island'.[11]

---

8 See, for example, Neumann (1992).
9 See, for example, Schleinitz (1889).
10 Brown (1908).
11 Hernsheim (1983), p. 114.

Fire Mountains of the Islands

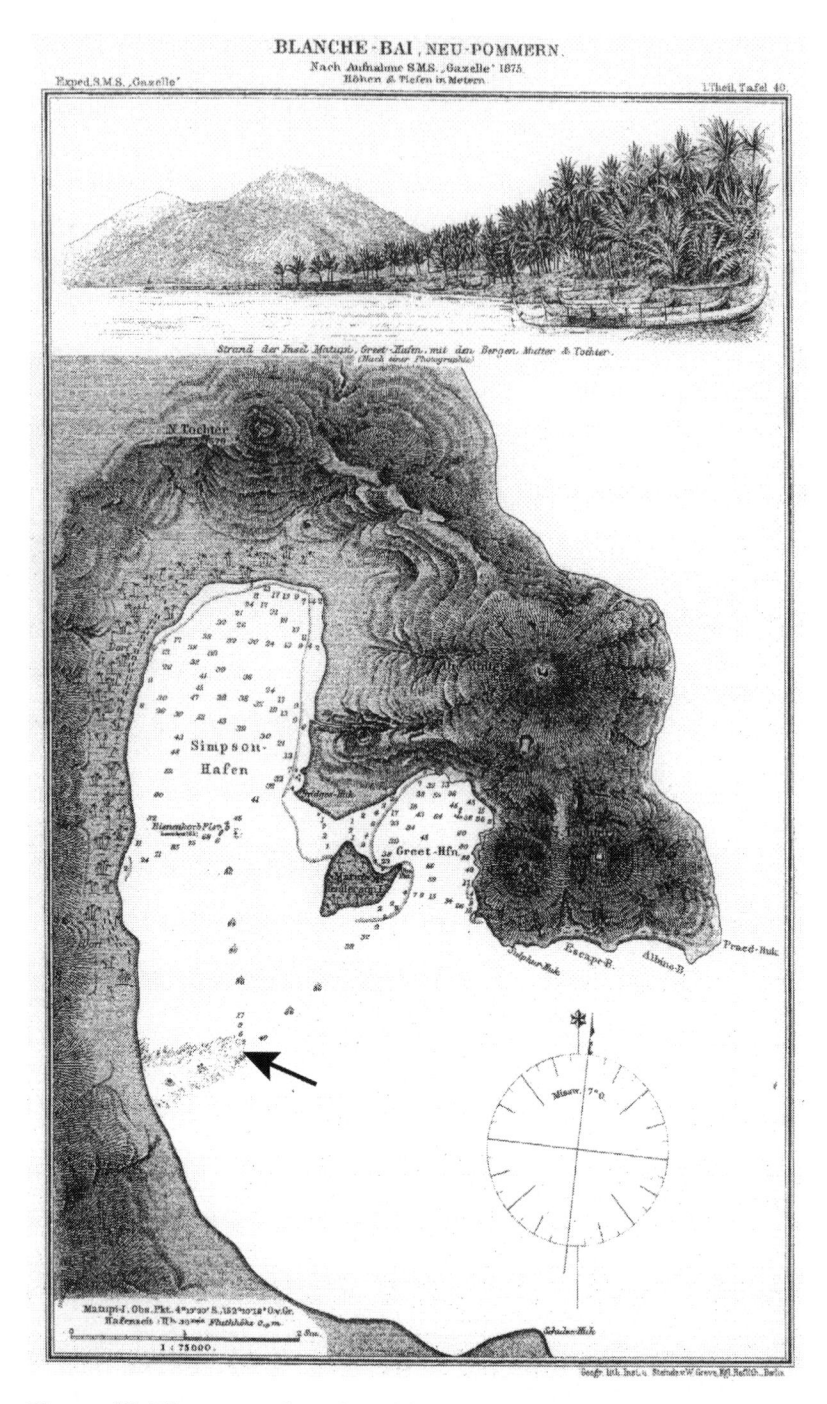

Figure 17. The area of reef and islets running out from the western shore of Blanche Bay is the site of the Vulcan eruption in 1878.

Source: Schleinitz (1889, tafel 40, opposite p. 240).

3. European Intruders and the 1878 Rabaul Eruption: 1870–1883

One other European observer who was on hand to witness directly the 1878 eruption at Rabaul was an adventurous Englishman, Wilfred Powell, one of the 'landed gentry' in Britain, and who is perhaps best described as a 'gentleman-explorer'.[12] He was a cousin to Lord Baden Powell, founder of the Boy Scout movement. Powell spent a good deal of 1877–1880 exploring New Britain in the ketch *Star of the East*.

Figure 18. Wilfred Powell (1853–1942).

Source: Royal Commonwealth Society, London.

---

12 Royal Commonwealth Society (1962).

1878 Eruption at Rabaul

Three Europeans — Brown, Powell and Hernsheim — reported on the 1878 Rabaul eruption, or at least on its immediate effects, and information from the Matupit-based trader Hicks, was collected not long afterwards. These form the basis for a summary of the events at Rabaul in 1878.[13] There is, however, an additional and fascinating, but rarely told story of the 1878 eruption by a Tolai eyewitness. This account was reported in English in 1951 after first being translated from an oral version told in Tok Pisin to an Australian more than 30 years previously — that is, about 40 years after the 1878 eruption itself. There are some differences between all these accounts, but the main features of the eruption are clear.

The Tolai eyewitness was To Maran, or Tomaran, of Matupit Island. To Maran recalled that the weeks and months prior to the eruption were 'a time of famine and hunger' for the Matupits because the ground on their island had been 'getting steadily hotter — it was this that had killed our gardens and started the famine'.[14] Severe tremors then began to be felt 'so that at times our people were thrown down and none could remain standing'. The wealth of the Matupits lay in shell money and one day a large group of them decided to cross the bay to Keravia to buy much needed food, but on the way 'with a noise like a great cannon, there burst forth out of the sea [at Vulcan] not far from us a great explosion which threw the sea-water into the air'. They returned to Matupit, one canoe being caught up in tsunamis from the submarine disturbances. Then, next morning while at Matupit watching the Vulcan eruption across the bay:

> suddenly a new opening appeared along the beach near the sea, throwing fire high up the mountain side [at Tavurvur]. Stones were shot high into the air, also dense smoke [from which] fell much ash, so that some of the people of Talawat, nearby, were killed.

The Matupits decided to evacuate to Malaguna, 'the big place' to the north-west where, however, fighting broke out between them and the villagers at Malaguna because of the extreme hunger of the Matupits: 'Yes, our men stole from their gardens so that trouble arose and they cried out that they would kill and eat us.' Eventually the eruptions ceased and the Matupits returned to their island. However, the 'ownership of the land gave us some trouble to define, for the old boundaries could not be seen. Trees were broken, rocks covered up, and the beach was not the same shape it had been. Many owners had died …' Village discussions were held and 'the land was re-alloted [sic] and the ownership then decided has held good till this day.'

---

13   Johnson et al. (1981).
14   All quotations in this and the following paragraph are from p. 67 of Tomaran (1951).

3. European Intruders and the 1878 Rabaul Eruption: 1870–1883

This dramatic account can be set against a summary of the European records of the time.[15] Brown recorded that frequent and locally strong earthquakes preceded the 1878 volcanic outbreak, particularly overnight on Sunday 3 February. Two tsunamis eroded the shorelines of Blanche Bay on the next morning and, soon afterwards, 'clouds of steam were observed rising from the Bay in a direct line' between Tavurvur and the south-west shore of the bay. This phenomenon was described even as a 'line of fire' by Hicks. A submarine volcano then developed about 1.5 kilometres from the shoreline at the south-western end of the line, Tavurvur across the bay 'burst out with terrific power' a few hours later, and inhabitants of the bay shores and Matupit Island fled to higher ground, reported Brown. Hicks put these events a day later, on the 5 February, and also said that there had been six tsunamis, initially four metres high. Brown stated clearly that the Vulcan eruption preceded the outburst from Tavurvur by a few hours — as did Tomaran — although Brown did not witness the two eruptions himself. Brown, however, had earlier noted in his diary on 30 January that he had received a report that the 'Volcano at Matupit' — presumably meaning Tavurvur — was in eruption. The significance and accuracy of the reporting on this early date of 30 January remains obscure, but in any case, eruptions at both volcanoes were well established by 5 February.

Brown noted that by 13 February 'The whole channel in front of our house [in the Duke of York Islands] and seemingly for miles to the southward is full of immense fields of pumice stone from the volcano.' Hernsheim, too, who had arrived at the Duke of York Islands on 9 February, noted that '… thick layers of pumice covered the ocean as far as the eye could see', and that the passage between the islands and New Britain mainland was closed by a blanket of pumice up to 1.8 metres thick. Powell recorded that 'huge blocks' of floating pumice surrounded his ketch anchored at Makada in the Duke of Yorks and that 'it really appeared as though one could walk to New Britain on it.' These large volumes of pumice would eventually drift eastwards, washing up on the shores of the Solomon Islands and even as far east as the Ellice Islands in the Pacific. Fishing was disrupted, and the hull of at least one ship was scraped clean of paint by the abrasive pumice.

Powell was imprecise about the times of these events at Rabaul in 1878, but he produced a dramatic description of Tavurvur's activity, which he saw from the summit of the Mother. He saw its incandescent explosions and the ejection of huge rocks, and recorded that

> At the same time angry flames would dart up, almost to the altitude on which we stood, and of the most dazzling brightness. Then all would die

---

15   The primary references and quotation sources used in this and the following paragraphs are given by Johnson et al. (1981).

down to a low sulphureous breathing, spreading a blue flame all over the mouth of the crater, whilst over us and all the country near hung a panoply of thick black smoke, broken only by the falling of red-hot stones in showers, which destroyed all the vegetation to leeward to a distance of about two miles.[16]

Volcanic activity at Vulcan is thought to have lasted about three or four days, whereas Tavurvur continued in eruption for almost a month. A new island, Vulcan, a few kilometres in diameter, was formed where the south-western reef had been, but parts of which still survived. The island was visited at different times by Brown, Powell and Hicks. It had been built up by the accumulation of pumice, yet Brown seems to have been more struck by the island having been created by upheaval and being 'thrown up'. This was in contrast to the Beehives that he '… saw were gradually sinking as the Houses which were some feet above the high water mark on my previous visit are now quite flooded at high water'. New Vulcan Island had a crater containing boiling water and the ground was so hot that visitors had to keep moving to avoid their feet being burnt, wrote Powell. Brown, however, noted that this had not prevented the opportunistic trader Hicks from taking 'possession of it by planting Cocoa Nuts even before it was quite cool … [and that] at the head of Simpsons Harbour Mr Hicks assured us that the water was all at scalding heat for several days. The fish were all killed and the Turtles were so much cooked that when the Natives got them the Shell (Tortoise Shell) had dropped off'.

Several conclusions can to be drawn from these accounts of the 1878 Rabaul eruption. First, the eruption is the earliest in Near Oceania to be described in any sort of detail in eyewitness accounts. Most descriptions of previous eruptions are by individual eyewitnesses of short-duration eruptive periods, in contrast to the several observations made over a month at Rabaul. Secondly, and more significantly, the described 'eruption' is in reality a *double* eruption — that is, more or less synchronous eruptive activity from two separate volcanoes. This is an unusual phenomenon, even globally, yet is one that would be repeated at Rabaul in both 1937 and 1994, on both occasions with damaging results. These repeated 'double' eruptions are separated by periods of 59 and 57 years respectively. The double eruption, naturally, has led to speculation about whether the two volcanoes are linked underground by some sort of geological fault, a notion fuelled perhaps by the claim of a 'line of fire' running between them in 1878. Geophysical data that would help to define the nature of Rabaul's underground 'plumbing' system would not be obtained for more than another century.

---

16   Powell (1883), p. 113.

Figure 19. Tolai people had re-established themselves on the larger of the Beehives by the time this photograph was taken in 1883, but in lesser numbers than those before the 1878 eruption. The Beehives consist mainly of layers of pumice, such as that produced by Vulcan.

Source: Methodist Church of Australia, Department of Overseas Mission Papers, held by the State Library of New South Wales. Published courtesy of the Secretary for the World Mission.

Both of the 1937 and 1994 eruptions at Rabaul took place at a time of year when the south-east trade winds were blowing, but the north-west monsoon had set in by the time of the February 1878 eruption, meaning that the floating pumice was immediately blown out of Blanche Bay into St Georges Channel to the south-east, rather than piling up in Simpson Harbour to the north. Deaths resulting from the 1878 eruption are thought to have been very few, unlike in 1937. But one can imagine that the tsunamis at least must have been threatening to lives in 1878 — for example, those of the villagers living on the Beehives to the north of Vulcan, where bush-material huts were destroyed. Neither are there post-1878 records of so many people ever again living on these vulnerable rocks after the 1878 eruption, and what little remains of the Beehives today does not support any sort of settlement.

The chemical compositions of the rocks produced by Vulcan and Tavurvur are known to be similar, yet the styles of eruption at each volcano are remarkably different. **Pumice** is commonly produced by Vulcan, but not normally by Tavurvur. Pumice represents frothed volcanic rock that contains so many now empty gas cavities that its density is less than that of water, and so is the only common type of rock that will float. The gas has come out of solution in the magma as bubbles so rapidly — like the froth discharging from an opened bottle of beer — that the magma breaks into inflated pieces which are scattered by water, air, or in pyroclastic flows. Pumice is the product of many large and catastrophic eruptions that accompany the formation of large calderas. It is an important product of many other significant eruptions in Near Oceania.

Tavurvur eruptions produce volcanic ash and large lava blocks — some up to the 'size of an ordinary house', as noted by Powell. These are flung out in repeated explosions that may last from minutes to hours, in a general eruption style known as **vulcanian**. The explosions generate ash clouds that may rise several kilometres, but the ash is not distributed as widely as in some other styles of explosive eruption. Vulcanian explosions are thought to take place from the deeper parts of the volcano's conduit, compared with Strombolian eruptions, and in some cases may represent the disruption of a frozen cap of lava, deep in the vent, caused by the build-up of underlying pressure. A distinctive feature of some vulcanian eruptions are 'bombs' of fresh lava, the interior of which has frothed somewhat, causing cracking of the chilled outer skin and the formation of a 'breadcrust' structure. Crater glow is commonly seen at night. Some vulcanian eruptions appear to produce no fresh magma, but only rocks from the walls of the conduit, as if the volcano is 'clearing is throat'.

The distinction between vulcanian and strombolian eruptions is not always clear and, indeed, there may be in practice a continuum between the two during the course of an eruptive period at a single volcano. The description of 1878 Tavurvur activity provided by Powell may refer to such a transitional example, but one which is more towards the vulcanian end of the 'spectrum'.

Powell's Voyage and a Possible Eruption 'Pulse'

Powell's and Hernsheim's volcanological stories do not finish here. Powell in particular — later in 1878 and possibly into 1879 — made important volcano observations along the north coast of New Britain. He documented especially a fundamental geological feature — one that would be explained a century later by the theory of plate tectonics — that the active volcanoes of New Britain all seemed to be located on the northern side of the island and that there were none on the southern side. He was the first European, also, to identify the considerable extent of the volcanic terrain between Open Bay and Du Faur volcano immediately west of the present-day town of Kimbe, including the 'islands' to the north of Du Faur. And then, further west, he made the keen observation that there was a volcanic 'gap' between Du Faur and the western end of New Britain. All of these observations represent the first documented recognition of a major part of the 1,000-kilometre-long, discontinuous chain of volcanoes stretching from Rabaul in the east to the Schouten Islands in the west and known today as the Bismarck Volcanic Arc.

Figure 20. Powell sketched Lolobau Island from the east in 1878. A volcanic plume rises from the crater of the volcano on the left. The high peak on the right is an extinct volcanic peak.

Source: Powell (1883, plate following p. 220).

Powell noted the volcanic nature of Lolobau Island, which had been named Du Portail by D'Entrecasteaux in 1793. He also described and illustrated the towering stratovolcanoes of Ulawun and Bamus, which had not been identified as such by earlier voyagers, together with the lower volcano of Likuruanga to Ulawun's north-west. Powell named these three the Father, South Son and North Son — not, evidently, as his own complement to Carteret's naming of the Mother and Daughters triplet at Rabaul, but probably because he heard versions of myths told by the Nakanai people of the area. Following is a composite of extracts from versions of these myths that were reported some years later by a Catholic missionary:

> A stranger called Ulawun came down from the mountain one day smoking tobacco and met Simolo, a Nakanai woman, who was so pleased with the effects of the tobacco, that she married Ulawun. The two of them went back up the mountain 'and sat there forever smoking and stomping about causing the ground to tremble and the mountain to breathe fire'. They later gave seeds to the villagers for planting, which is how tobacco came to be grown in Nakanai gardens. Ulawun was a snake but Simolo was a true woman, and she bore a son who, when he grew up, was told by his parents to find another place to live. He settled on a nearby mountain which he named after himself, Bamus. 'He

stayed up there constantly smoking and breathing fire and throwing up stones to such an extent that the villagers in the valley were devastated and people were killed.' Bamus and Ulawun one day invited all the volcanoes to a singsing and gave out gifts of betel nut and shell money to their guests, except for the unpopular Vuna Kiku who then became angry. Vuna Kiku, who alone of the volcanoes had a tail, drove it into Likuruanga, or North Son, and kept on driving it in until the earth trembled and shook for a long time. The mountain finally collapsed into the sea and stopped smoking. This was Likuruanga's last eruption.[17]

Figure 21. The volcanoes Ulawun, on the left, and Bamus, on the right, were sketched by Powell from Lolobau Island in 1878. Both are shown emitting vapour plumes, which, in the case of Ulawun at least, appears to contain some volcanic ash. The summit areas of both volcanoes were reported to be free of vegetation, an indication of recent eruptive activity, whereas Bamus today is covered by vegetation. The prominent peak on the extreme left is Likuruanga volcano.

Source: Powell (1883, figure on p. 219).

Powell unfortunately misidentified Du Faur as an island, and repeated the likely error made by D'Entrecasteaux that Willaumez, Raoul and Gicquel were also islands, probably because he was too distant; although, Powell did add specifically to his map that 'All the passages between these islands appear to be dangerous'. The islands are part of Willaumez Peninsula and Du Faur is an extinct volcano near the southern end of it and is part of New Britain. Powell also made the clearly exaggerated statement that at the western end of New Britain there were:

> innumerable volcanoes, small and large, all in violent eruption … there must have been some hundred or more all belching fire and smoke, indeed the land seemed all on fire … When night came on the sight was wonderful. Flames seemed to cover the mountainous point of land, and it would have been easy to read a book by the light; the air was full of fine ashes covering and making everything a light grey colour; indeed it

---

17 Dormann & Meier (1909), pp. 214–17.

was difficult to breathe comfortably. Tupinier [sic; Ritter] Island was in eruption also, and the noise made by them all was like low continuous thunders.[18]

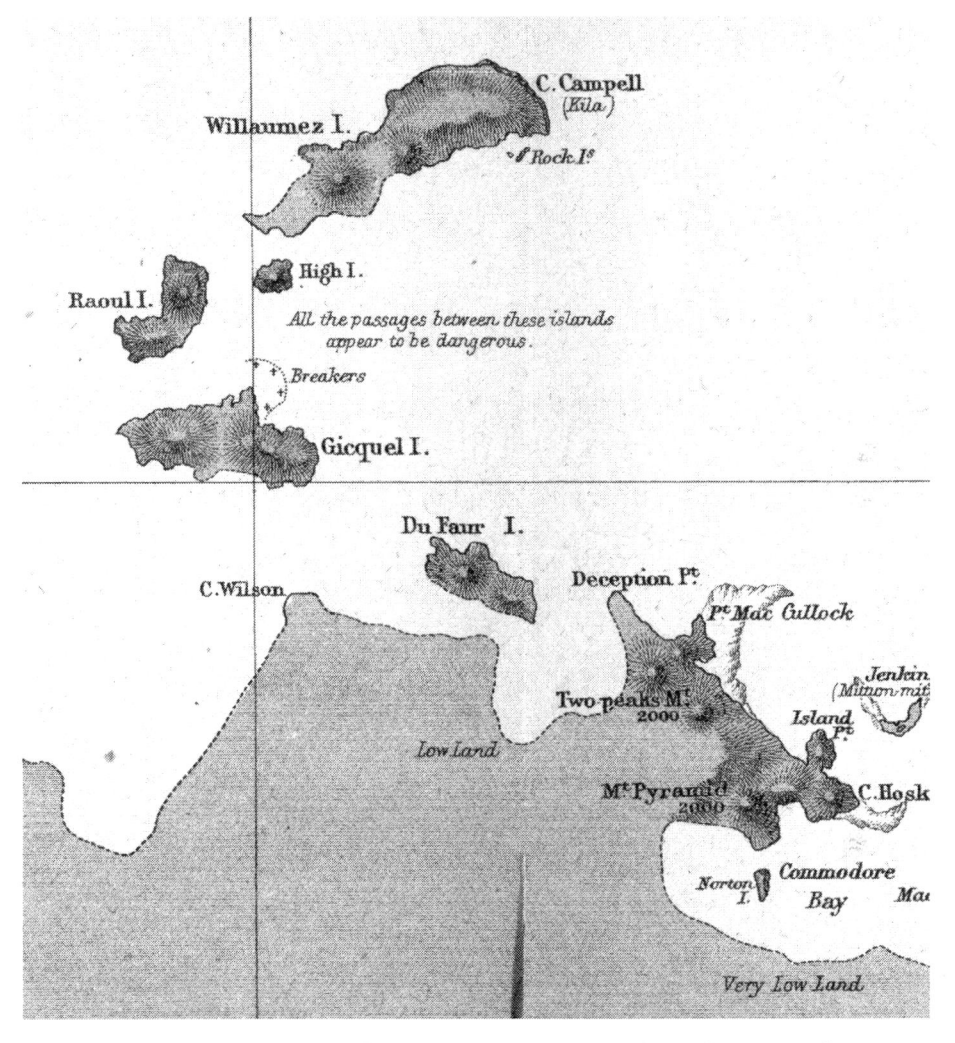

Figure 22. The peaks in Wilfred Powell's map are the volcanoes of Willaumez Peninsula and the Cape Hoskins area of the central-north coast of New Britain. Powell was mistaken in believing that most were islands separate from the mainland. He has used the French names of islands that were first recognised, probably falsely, by D'Entrecasteaux in 1793.

Source: Powell (1883); detail from Sketch Survey of the North-East Portion of New Britain 1878-9. National Library of Australia.

There is, however, only one active volcano at the western end of New Britain — Langila. Powell may have mistaken an active field of lava flows from Langila,

18   Powell (1883), p. 230.

which as they flowed broke open in many places to reveal, at night, incandescent interiors. The lava flows also may locally have set fire to nearby vegetation, as commonly happens in such circumstances, even in the tropics. Powell also probably mistook Tupinier, or Sakar, which is an inactive volcano, for Ritter, the only active island volcano in the area.

The combined observations of Powell and Miklouho-Maclay in the second half of the 1870s provide some evidence for a possible 'pulse' or 'time cluster' of volcanic eruptions in Near Oceania at this time — that is, a period of volcanic activity when more volcanoes were in eruption than 'normal'. Six volcanoes were definitely active between 1875 and 1878, and possibly three others, if the vapour plumes witnessed by Powell at Lolobau and Bamus are taken as evidence of recent eruptive activity, and if sluggish lava flows were still being emitted at Bagana when viewed in 1875.

Table 1. Volcanoes in Eruption in Near Oceania from 1875 to 1878

Bam 1877. Reported by Miklouho-Maclay to be in 'full eruption' and synchronous with activity at nearby Manam
Manam 1877. The eruptive activity was described by Miklouho-Maclay.
Ritter 1878. Powell noted noisy eruptions but misidentified the volcano as Sakar.
Langila 1878. Powell gave a vivid but inaccurate description of eruptive activity.
Ulawun 1878. Powell observed an eruption and illustrated a dense, presumably ash-laden plume in a sketch.
Rabaul 1878. Near simultaneous eruptions at Tavurvur and Vulcan volcanoes.
Bagana 1875. The volcano was identified as 'active' by Schleinitz. Observed emissions were apparently of vapour only, although lava flows may well have been active, despite being emitted sluggishly.
Bamus 1878. Powell showed a plume rising from Bamus in the same sketch as the one showing Ulawun in eruption, but the plume does not appear to have been significantly ash-laden, if at all.
Lolobau 1878. Powell provided a sketch of Lolobau Island and a vapour plume rising from one of its volcanic centres.

The evidence for a late-1870s volcanic pulse, however, is not conclusive, bearing in mind the brevity and inaccuracy of some of the observations. Nevertheless, it is intriguing because two, much more obvious, volcanic pulses would be documented in the twentieth century, and possibly one in the 1890s. Furthermore, a more explicit connection between tectonic earthquakes in the early 1870s and eruptions later in the same decade would be proposed in the

1950s by volcanologist G.A.M. 'Tony' Taylor in attempting to generate an integrated theory for the origin of eruptions in the region, as well as support for his explanation for a volcanic 'pulse' in 1951–1957.

Volcanological Events Elsewhere

Hernsheim was well settled at Matupit Island by 1883. He enjoyed the evening view within Blanche Bay 'when the green gradually changed to red and then to deep violet and every single tree stood out against the background in the clear air, [turning] this spot into a scene of greatest beauty, which I can never forget. These colour effects were particularly magnificent in the year 1883'.[19] Hernsheim was not alone in appreciating the unusual sunsets at this time, as they were being witnessed in many parts of the world. A major and catastrophic explosive eruption had taken place in August of that year at Krakatau Island in the Dutch East Indies, which had lofted ash and gas into the stratosphere, perhaps to heights of 40–50 kilometres. The main cause of the vivid sunsets was the presence of aerosols of sulphuric acid formed by the hydration of sulphur-dioxide gas from the volcano.

The eruption at Krakatau in 1883 is probably the most significant event in the history of volcanology in terms of the extent of the scientific reporting in the decades afterwards, and of its impact on understanding volcanic processes in general.[20] More than 36,000 people were killed by the great tsunamis — some at least 30 metres high — that resulted from the eruption, and most of the original island disappeared, forming a caldera. The explosions were heard on the other side of the Indian Ocean and in central Australia, and worldwide barographs recorded the passage of the airwaves, some several times as the waves bounced back and forth around the world. News of the eruption spread throughout the world at an unprecedented speed by means of the growing telegraph system.

The 183 years between William Dampier's 1700 voyage in the *Roebuck* and the 1883 Krakatau eruption mark extraordinary changes in European thought and society, including an understanding of volcanoes. The foundations of modern science had been laid down, the powers of Church and State were being challenged by thinkers of the Enlightenment in a new Age of Reason, and new technologies were beginning to transform lives of drudgery as a result of the Industrial Revolution. Beliefs that the Earth's rocks had formed in an imagined world of mysterious catastrophes, supernatural cataclysms and biblical floods were disappearing.

---

19   Hernsheim (1983), p. 81.
20   See, especially, Verbeek (1885) and Simkin & Fiske (1983).

The foundations of the new science of geology were created largely in the late eighteenth century, led by the great Scottish scientist James Hutton, although many others — such as Hutton's contemporary, the French geologist Nicholas Desmarest and the pioneering Danish anatomist Nicholas Steno — contributed fundamental geological insights.[21] Charles Lyell in 1830 then produced his seminal *Principles of Geology*, which in turn set the stage for Charles Darwin's revolutionary *On the Origin of Species*. Hutton was a principal proponent of the Plutonist school, which successfully undermined the beliefs of the Neptunists that rocks precipitated from oceans. The Plutonists, in contrast, concluded that the processes forming the rocks of previous times were the same as those seen operating today. Volcanic rocks were previously magmas formed when materials deep in the Earth were melted, and then erupted and cooled, and not by precipitation from ancient oceans or the combustion of flammable materials. Furthermore, the Earth had to be a great deal older than the age of 4,004 BC that had been calculated by Bishop James Ussher on the basis of scripture.

Europeans in the nineteenth century had also become more aware of the disastrous effects of the eruption at Vesuvius, which in 79 AD had destroyed the Roman towns of Pompeii and Herculaneum on the Bay of Naples, Italy. This increased awareness was largely because of the treasure-seeking that could be undertaken there by visitors, as well as the volcanological writings of Sir William Hamilton — who arrived in Naples in 1764 in time to observe eruptions at Vesuvius the following year — and the archaeological excavations and discoveries that were later undertaken at the buried towns. The 79 AD eruption was already well known to historians through letters written to the Roman historian Cornelius Tacitus by the Younger Pliny whose uncle, the Roman naval commander, the Elder Pliny, had perished in the eruption. One of the letters contained a description of the Vesuvius eruption cloud that volcanologists now recognise as ***plinian***.

The Younger Pliny wrote that the cloud was '… like an umbrella pine, for it rose to a great height on a sort of trunk and then split off into branches …'.[22] The high-rising 'trunk' is a narrow column of hot ash, pumice and gas that is propelled upwards from deep within the volcano by forceful jetting, as if from a powerful, vertically directed hosepipe. The column then ascends much further through ingestion of air and by thermal convective expansion, until winds — at different heights and with different velocities —capture it and form horizontal layers from which the solid material falls forming plinian deposits on the ground. Large plinian eruptions can reach many tens of kilometres into the stratosphere, such as at Krakatau in 1883. Plinian eruptions would also take place at Rabaul in 1937 and 1994.

---

21 See, for example, Geike (1905).
22 Radice (1969), p. 166.

References

Brown, G., 1908. *George Brown, D.D. Pioneer-Missionary and Explorer. An Autobiography. A Narrative of Forty-Eight Years Residence and Travel in Samoa, New Britain, New Ireland, New Guinea, and the Solomon Islands.* Hodder and Stoughton, London.

Dorman & J. Meier (comp.), 1909. 'Aus der deutschen Suedsee. Mitteilungen der Missionare vom heiligsten Herzen Jesu'. Band 1: P. Matthaeus Rascher, M.S.C. und Baining (Neu-Pommern) Land und Leute. Aschendorff, Munster.

Epstein, A.L., 1969. *Matupit: Land, Politics, and Change among the Tolai of New Britain*. The Australian National University Press, Canberra.

Epstein, T.S., 1968. *Capitalism, Primitive and Modern: Some Aspects of Tolai Economic Growth*. The Australian National University Press, Canberra.

Geike, A., 1962 (1905). *The Founders of Geology*. 2nd ed. Dover, New York.

Hernsheim, E., 1983. *Eduard Hernshiem: South Sea Merchant*, ed. & trans P. Sack & D. Clark. Institute of Papua New Guinea Stdies, Boroko.

Johnson, R.W., I.B. Everingham & R.J.S. Cooke, 1981. 'Submarine Volcanic Eruptions in Papua New Guinea: 1878 Activity of Vulcan (Rabaul) and Other Examples', in R.W. Johnson (ed.), *Cooke-Ravian Volume of Volcanological Papers*. Geological Survey of Papua New Guinea Memoir, 10, pp. 167–79.

Meier, J., 1908. 'A *kaja* oder Der Schlangenaberglaube bei den Eingebornen der Blanchebucht (Newpommeren)', *Anthropos*, 3, pp. 1005–29.

Miklouho-Maclay, N., 1878. 'Ueber vulkanische Erscheinungen an der nordoestlichen Kueste Neu-Guinea's', *Petermanns Mitteilungenaus Justus Perthes geographischer Anstalt*, 24, pp. 408–10.

——, 1885. 'On Volcanic Activity on the Islands near the North-East Coast of New Guinea and Evidence of Rising of the Maclay-Coast in New Guinea', *Proceedings of Linnean Society of New South Wales*, 9, pp. 963–67.

——, 1952. *Sobranie sochinenii v 5-ti tomakh* [*Collected Works in 5 volumes*], 3, no. 2, *Stat'i po zoologii, geografii i meteorologii* [*Articles on Zoology, Geography and Meteorology*], Izd-vo Akademii nauk SSSR, Moscow-Leningrad.

Moresby, J., 1876. *Discoveries and Surveys in New Guinea and the D'Entrecasteaux Islands. A Cruise in Polynesia and Visits to the Pearl-shelling Stations in Torres Straits of H.M.S. Basilisk*. John Murray, London.

Neumann, K., 1992. *Not the Way It Really Was: Constructing the Tolai Past*. University of Hawaii Press, Honolulu.

——, 1996. *Rabaul: Yu Swit Moa Yet: Surviving the 1994 Volcanic Eruption*. Oxford University Press.

Powell, W., 1883. *Wanderings in a Wild Country; or, Three Years amongst the Cannibals of New Britain*. Sampson Low, Marston, Searle, & Rivington, London.

Radice, B. 1969. *The Letters of the Younger Pliny*. Penguin Books, London.

Royal Commonwealth Society, 1962. 'The Wilfred Powell Collection — I. Library Notes with List of Accessions', New Series 64, London.

Salisbury, R., 1970. *Vunamami: Economic Transformation in a Traditional Society*. Melbourne University Press.

Schleinitz, G.E.G. von, 1889. 'Die Forschungsreise S.M.S. "Gazelle" in den Jahren 1874 bis 1876', I. Theil, *Der Reisebericht*, pp. 1–307. Ernst Siegfried Mittler und Sohn, Berlin.

Simkin, T. and R.S. Fiske, 1983. *Krakatau 1883: The Volcanic Eruption and its Effects*. Smithsonian Institution Press, Washington D.C.

Simpson, C.H., 1873. 'Hydrographical Extract from a Six Months' Cruise among the South Sea Islands', letter dated 1 November 1872, HMS *Blanche*, Sydney, New South Wales to the [British] Hydrographer. Hydrographic Notice No. 1 of 1873 (Pacific Notice No. 23), 1–8.

Tomaran, 1951. 'A tale of '78', *Pacific Islands Monthly*, 22, no. 4, p. 67.

Verbeek, R.D.M., 1885. *Krakatau*. Landsdrukkerij, Batavia.

Webster, E.M., 1984. *The Moon Man: A Biography of Nikolai Miklouho-Maclay*.

4. Volcanic Events of the German Era: 1884–1914

There is 'the question whether it was advisable to establish the new capital of the territory at Rabaul in this endangered area of Simpson Harbour. It is of course possible that the volcanic force will lie dormant for decades, even centuries, but it is also possible that it will soon become active again; nothing could be more unpredictable'.

Karl Sapper (1910c)

Colonial Partitioning

A diverse mix of new Europeans became established in the St Georges Channel area in the years following the 1878 eruption at Rabaul. Roman Catholic missionaries, for example, arrived in 1882, gaining a mission foothold in competition with the Methodists, and Ludwig Couppé came later as bishop, strengthening the Catholic base at Vunapope at present-day Kokopo. Trader Thomas Farrell and his partner, Emma Coe, arrived there too from Samoa, acquiring large tracts of land from the Tolai, and eventually creating a successful plantation economy and great personal wealth for the legendary 'Queen Emma'. German traders in general — particularly Eduard Hernsheim and those requiring New Guinea labour in Samoa — had lobbied the Reich in Germany for establishment of government protection for their commercial activities and interests, but Chancellor Otto von Bismarck was reluctant to raise the German flag and claim colonies — at least until 1884.[1]

The governments of the Australian colonies had also lobbied the British Government in London, urging a claim on at least part of New Guinea in order to counter the potential threat of foreign powers — in particular Germany — becoming established on Australia's doorstep, but there was resistance in London too. These different pressures came to a head in 1884 when agreement was reached between the German and British governments to partition, and establish protectorates over, the areas east of 141 °E. The Dutch, much earlier, had claimed New Guinea west of this meridian. Germany took north-eastern New Guinea together with the Bismarck Archipelago and Bougainville Island, and the British established a protectorate in south-east New Guinea — British

---

1 See, for example, Whittacker et al. (1975, Documents D22–36) and Gash & Whittacker (1975) for background information on colonial partitioning in Near Oceania.

New Guinea. Thus began a remarkable 30 years of German colonial rule,[2] during which there would be expeditions, much scientific activity and significant volcanological happenings.[3]

The British by 1884 had already been obtaining volcanological information from the Solomon Islands, which extended from the volcanically active Bougainville Island in the north-west to the older San Christoval Island in the south-east. This was largely the result of Royal Navy vessels making periodic calls to the islands following Carteret's visit there in 1767, but European missionaries, traders, and travellers, as well as 'blackbirders' exploiting Melanesian labour, had also visited the islands during the course of the mid- to late nineteenth century. Britain finally exerted control over part of the Solomon Islands in 1893 when it established a protectorate over the south-eastern islands. Ownership of islands in the north-western Solomons, which had been claimed originally by Germany, was transferred to Britain in 1900 by means of a treaty, leading to establishment of the separately administered British Solomon Islands Protectorate, but excluding Bougainville Island which remained under German control. The Melanesians themselves had no negotiating involvement in any of these colonial machinations and decisions.

Dr H.B. Guppy was appointed surgeon to the British naval vessel HMS *Lark* in 1881, in part because of his interests in natural history, and in this capacity he undertook visits to the Solomon Islands over the next three years. Guppy noted that Bagana on Bougainville Island was the only active volcano in the Solomon Islands chain. He saw Bagana from west of the Shortland Islands on several occasions during 1884 and gave the only known report of deaths by eruptions from this volcano:

> A white column of vapor appeared to be constantly issuing from its summit … it has the appearance of an isolated conical mountain somewhat truncated … At the end of April 1884, I learned from Gorai, the Shortland chief, that about four months before there had been a great explosion in this volcano by which a number of natives were killed. From what I can gather from various sources, it would seem that this vent has been in continual eruption for at least fifteen or twenty years.[4]

Guppy also described Savo volcano in the strait north of Guadalcanal, based on information provided by others, and noted both its potentially active state and the local stories of its past eruptions. He also hinted at the parsimony of British interest in the region, remarking — after 'stifling' his British patriotism — that

---
2   Sack (1973), Hempenstall (1978), Firth (1983), Mackenzie (1987) and Hiery (1995).
3   A systematic search in the vast holdings of German colonial literature for references to volcanic activity in German New Guinea has never been undertaken, so new information is almost certainly yet to be discovered.
4   Guppy (1887b), pp. 21, 22.

> I cannot but think that the presence of Germany in these regions will be fraught with great advantage to the world of science … and conducted with that thoroughness which can only be obtained when, as in the case of Germany, geographical enterprises become the business of the State.[5]

The German Reich initially, in 1885, provided an Imperial Charter for the running of the new protectorate to the *Neu Guinea Compagnie*, which concentrated its surveying, development, and settlement efforts along the north-eastern coast of New Guinea Island — called *Kaiser-Wilhelmsland* by the Germans. The company's aim was to establish a plantation colony that would be run commercially in the interests of a European minority, and the European and Tolai communities on the shores of distant St Georges Channel were left much to themselves, at least for the time being. Ornithologist and anthropologist Otto Finsch had come to the area in 1884 and, from a base in the Duke of York Islands, led ostensible natural-science expeditions along the Kaiser Wilhelmsland coastline. His primary aim, however, was to identify coastal lands that were both suitable for plantations and settlements, and close to harbours from which produce could be exported.

Finsch had several opportunities to observe and describe the volcanoes off the New Guinea north coast, including eruptive activity at Manam, thus adding to the series of European observations of this volcano that had begun in the sixteenth century. He described for the first time a feature that would later intrigue volcanologists studying Manam — its sound effects:

> We went along the west coast close to shore and saw the island in all its imposing beauty before us. Once the engine was stopped, one could hear the subterranean forces at work, a powerful booming becoming constantly louder, that passed into a weaker moaning and groaning, until it became completely silent for a while, and then soon started up anew. The wondrous, uncanny noise was reminiscent of a giant pair of bellows, and up above, from the flue, vast masses of white smoke came rolling out; truly one of Nature's subterranean workshops, which fills the observer with mute admiration.[6]

Ritter Island Disaster

Administrative headquarters for the New Guinea Company were established at Finschhafen on the mainland at the southern entrance to Dampier's Strait, and G.E.G. Schleinitz — who had sailed into Blanche Bay at Rabaul in 1875 —

---

5 Guppy (1887a), p. xii.
6 Finsch (1888), p. 367.

arrived at his new headquarters in June 1886 with his wife and children as the first *Landeshauptmann* or administrator. Schleinitz was active in mapping the coastlines of northern New Britain and Kaiser Wilhelmsland and, in so doing, observing volcanoes and their activity. His mapping of north-coast New Britain established clearly that the 'islands' mapped by D'Entrecasteaux in 1793 and Powell in 1878 were in fact part of New Britain Island, and that Willaumez was indeed a volcanic peninsula.[7]

Figure 23. Otto Finsch reproduced this picture of Manam volcano in his 1888 memoir. The two hills to the left of one of Manam's four radial valleys are small satellite volcanoes near the south coast of the island. Vapour is emerging from the southern crater of the volcano.

Source: Finsch (1888, between pp. 366 & 367.)

The challenges facing the New Guinea Company were substantial and eventually became insurmountable, causing failure in both the commercial and colonial sense. There was confusion, for example, over who was responsible for law enforcement — the company, which was a commercial enterprise, or the German navy, which could bombard recalcitrant villages from the sea, but had little capability to penetrate on foot into the rainforest. Tropical diseases took a devastating toll, and nearly 50 of the small number of Europeans in Kaiser Wilhelmsland in 1887 died, including Schleinitz's wife.[8] The experienced Hernsheim, who seems to have had little time for Finsch anyway, blamed the naturalist for the recommendations on settlement that had been made to Berlin.[9]

---

7   Schleinitz (1896, 1897).
8   Sack (1973).
9   Hernsheim (1983).

4. Volcanic Events of the German Era: 1884–1914

Schleinitz himself had had enough of dealings with Berlin and left on 19 March 1888, only days after a catastrophic event at Ritter volcano[10] in nearby Dampier Strait added to the list of concerns for the Germans at Finschhafen.

The steep, frequently active 'volcano island' that William Dampier had seen in impressive eruptive behaviour in 1700 was a well known feature to ships passing through Dampier Strait, but what happened shortly before daybreak on 13 March 1888 will never be known fully. Geologists and others for years afterwards deduced that the volcano must have behaved just as Krakatau had done in 1883 in the Dutch East Indies, and certainly, like Krakatau, most of Ritter Island did indeed disappear suddenly beneath the sea, producing a devastating tsunami. The sparse and distributed contents of German records of the event, however, did not receive careful attention by volcanologists until R.J.S. Cooke compiled information in the late 1970s.[11]

The steepness of Ritter volcano was perhaps its most conspicuous feature before 1888. Several mariners had commented on this. Some of them even produced illustrations of it, portraying slopes well in excess of 50° which is probably too steep in some cases, given the general penchant of illustrators of those times to exaggerate the slopes of volcanoes. An angle of 50° is much greater than the slopes of volcanoes of this type elsewhere in the Bismarck Volcanic Arc. Nevertheless, the general steepness can be accepted as real in view of what happened on 13 March 1888. Observations of eruptions at Ritter had been made previously and, by 1888, it was perhaps the best known volcano in the region except, possibly, for Manam. Ash had fallen on Finschhafen in February 1887, accompanied by earth tremors that, according to Schleinitz, caused 'wall clocks to stop at once, and their pendula and clock weights, and hanging lamps, to swing violently'.[12] The source of the ash could have been Ritter, but there are in any case no other reports of Ritter eruptions after this and before 13 March 1888. The following are translated extracts of observations made at Finschhafen on the morning of 13 March:

> a noise like thunder was heard shortly after 6.30 in the morning, and at the same time the sea and the water in the harbour started to move with surging rapidity in such a way that it flowed up and down and the ships in the harbour were in danger. The water fell so sharply that the reef south of the wooded island Madang was dry in 2 minutes and stood 5–6 feet out of the water. Then the water came back with the same force. The time between the lowest and highest level was 3 to 4 minutes, the speed of the current was reckoned to be 8 to 10 knots … After the arrival of the tidal wave [tsunami], some observers noticed a fine, barely perceptible rain of ash.[13]

---

10 The Germans named the volcano after the pre-eminent German geographer, Carl Ritter (1779–1859).
11 Cooke (1981).
12 Anonymous (1887), p. 211.
13 Anonymous (1888), p. 76.

Little damage was done to European infrastructure at Finschhafen, but Melanesian people there lost canoes and shoreline houses. Some of them believed Governor Schleinitz had supernatural powers, knew that he was about to leave New Guinea within a few days, and suspected that he had caused the tsunami by casting a spell. A small New Guinea Company expedition from Finschhafen consisting of two Germans, four Malaysians and 12 Melanesian labourers from the Duke of York Islands, had arrived on the west coast of New Britain on 6 March, planning to survey sites for a coffee plantation, but the German captain of the ship, attempting a rendezvous with the party on the evening of 15 March, found the coastline completely changed by the destructive impact of the tsunami. He returned to Finschhafen with flags at half mast on 16 March, when Schleinitz organised a relief expedition to be sent back immediately to New Britain. The aim was to look for any survivors, but evidently was motivated primarily by the suspected loss of the two Germans. Only five young labourers from the party had survived. They had been carried into trees by the wave and had then clung to branches as the waters receded. Trees that remained standing had signs of the waves reaching heights of about 10 metres.[14]

Figure 24. This simple profile of Ritter Island, produced by Schleinitz in about 1887, is perhaps the most realistic of all the sketches of Ritter provided by pre-1888 observers. Note that the slope is a good deal less than is shown in Dampier's sketch of 1700.

Source: Schleinitz (1896, plate 8). See also Cooke (1981, figure 2D).

There is no clear indication from the account of the relief party of major losses of life having resulted from the Ritter tsunami along the shorelines of Kaiser-Wilhelmsland. Indeed, the relief effort apparently represented more a commercial company interested in the immediate fate of its employees than representatives of a government being concerned about the condition of its subjects. One of the members of the relief party identified 17 species of orchid during the expedition. This is suggestive of an element of distraction from what might have been expected to be the main task at hand, and even some disregard by the white settlers towards the black Melanesian villagers who perished as a result of the tsunami. Yet the total number of deaths must have been substantial as many villages were scattered along the coastlines of west New Britain, Umboi

---

14   Steinhauser (1891–1892).

4. Volcanic Events of the German Era: 1884–1914

and Sakar islands, and on nearby mainland New Guinea. Population figures are not known for the area, but missionaries had been attracted to the villages of Dampier Strait for religious work amongst the coastal Melanesians since the 1840s.[15] Fatalities in 1888 are likely to have numbered in the hundreds, and possibly more than 1,000.

The captain of a German vessel, returning later to the area, wrote:

> On Tupinier [Sakar] and Rook [Umboi] Island in particular the devastation caused by the tidal wave must have been terrible. Even today, more than two years later, a sharply defined strip, approximately 40 to 50 feet [12–15 metres] above sea level and running parallel with the coastline, clearly defines the path of the masses of water. Collapsed forests and mountain sides which have slid down will bear witness to the disaster for some time to come. Many people must have perished on populous Rook Island. At Marienhafen [on the southeastern end of Umboi] where I had seen masses of canoes on a previous visit, no trace of natives could be found.[16]

Another report contains the statement that 'The formerly populated Lutherhafen [at the north-western tip of Umboi] was completely abandoned. According to the old headman hundreds of people have perished. The survivors have fled to the mountains and do not dare to come down to the beach.'[17]

Notable features of the event in the German accounts are, first, the almost complete disappearance of Ritter Island, leaving today only a scalloped remnant and islets and, second, the apparent absence of evidence for major explosive eruptions accompanying the island's disappearance. This is unlike Krakatau in 1883 when huge volumes of pumice were produced. There was at Ritter, in contrast, no noise, no earthquakes, no visible eruption column, no incandescence, no ashfall — apart from the 'fine, barely perceptible rain of ash' at Finschhafen — and no floating pumice. This apparent discrepancy was not addressed until after volcanological lessons began to be learnt from the Mount St Helens eruption in the western United States in 1980.

The northern flank of Mount St Helens had been observed bulging for many weeks prior to the disastrous eruption of 18 May 1980 when the flank gave way, creating a giant rock slide. This huge sector of the volcano quickly disintegrated into what geologists call a ***debris avalanche*** of rock, dust and air that flowed northwards into river valleys, where the avalanche changed into mudflows,

---
15   Wiltgen (1979) provided sketch maps showing many pre-1888 villages bordering Dampier Strait and on Umboi Island.
16   Anonymous (1891), p. 61.
17   Anonymous (1890), p. 84.

disgorging downstream. An eruption was triggered and a giant amphitheatre was formed on the volcano itself, made up of a steep back wall and two spurs that extend outwards and gradually decline in height in a roughly horseshoe-like form to the north. *Avalanche amphitheatres* of this type are different from true calderas, where the confining cliffs completely encircle the collapse depression. Geologists had described such avalanche amphitheatres well before 1980, but the Mount St Helens event triggered considerable subsequent research on their origin.[18]

Figure 25. The small double escarpment at Ritter Island today is the highest part of the avalanche amphitheatre formed in 1888 and which extends out onto the sea floor to the left. The vegetated sloping area on the far right represents part of the outer slopes of the original Ritter Island.

Source: E.E. Ball, 8 November 1974.

Could Ritter have produced an avalanche amphitheatre in 1888? I was able to survey the area around Ritter Island in 1985 using the commercial single-beam echo sounder of a schooner borrowed from an archaeological research project. The collapse was found to be of avalanche-amphitheatre type, rather than a caldera, and the rough features of the upper end of a debris avalanche were detected.[19] Steep-sided Ritter volcano had been unstable in 1888 and had indeed collapsed, but without an accompanying explosive eruption like the one at Mount St Helens. Then, in 2004, a major marine survey used multi-beam side-scan sonar to map the extent of the submarine avalanche and its related deposits. Material from the 1888 avalanche was shown to have flowed westwards out onto the sea floor for as much as 75 kilometres from the volcano. The Ritter event was described as 'the largest lateral collapse of an island volcano in historical time'.[20]

---

18 Siebert (1984).
19 Johnson (1987).
20 Ward & Day (2003), p. 891, and Silver et al. (2005, 2009).

Hahl and Sapper

The German Reich by 1902 had taken over Imperial administration of the protectorate, following the failure of the New Guinea Company in Kaiser Wilhelmsland. Headquarters were now at Herbertshöhe — present-day Kokopo — immediately south-east of the volcanically active Simpsonhafen at Rabaul, near the Catholic mission at Vunapope, and close to Emma Coe's plantation at Ralum facing majestically across St Georges Channel. Dr Albert Hahl, also in 1902, began 11 years as governor of German New Guinea based at Herbertshöhe.[21] Hahl was trained in law and had come to the protectorate first as Imperial judge in 1896, when the New Guinea Company was still running the colony, and well after Finschhafen had been abandoned as the main administrative centre. Hahl's career in the colony would be characterised by a paternalistic concern for the future of the Melanesians, but also by firm and at times quite brutal control of them. This reflected, at least in part, the pragmatism that Hahl was obliged to practice in response to the interests of the commercially driven planters, who required plantation labour and who had influence in the German Colonial Office in Berlin.

Hahl, in his memoirs, acknowledges the valuable geographical work that had been undertaken previously in the protectorate by both Finsch and Schleinitz, noting too the contributions to exploration made by botanist and geographer Dr Carl Lauterbach. Hahl would oversee during his own governorship an escalation in the numbers of visitors and expeditions to the region, made up of mainly German naturalists, scientists, travellers, observers and the generally curious. In particular, the extensive and well-funded scientific activity of the Hamburger Wissenschaftliche Südsee-Expedition — the Scientific South Sea Expedition of Hamburg[22] — took place in 1908 and, in 1912–1913, the Kaiserin Augusta Fluss or Sepik River Expedition, whose members included the pioneering geomorphologist W. Behrmann.

The German navy also paid visits for oceanographic and hydrographic surveys, including the well-equipped SMS *Planet*. Its crew plumbed exceptionally deep water west of Bougainville Island. The 'Planet Deep' is part of the great submarine trench that runs eastwards off the south coast of New Britain and turns sharply down the south-western side of Bougainville Island. This submarine trench would be of key importance in understanding the origin of the volcanoes of both New Britain and Bougainville when the theory of plate tectonics emerged in the late 1960s. German technology was also advancing vigorously on other fronts of relevance to volcanology. Seismographs were being developed and improved, particularly by Dr Wiechert at the University of Gottingen. The Wiechert instrument would later be used for a time in Near Oceania.

---

21 Firth (1978), Sack (1980) and Hahl (1980).
22 See, for example, Reche (1954).

Fire Mountains of the Islands

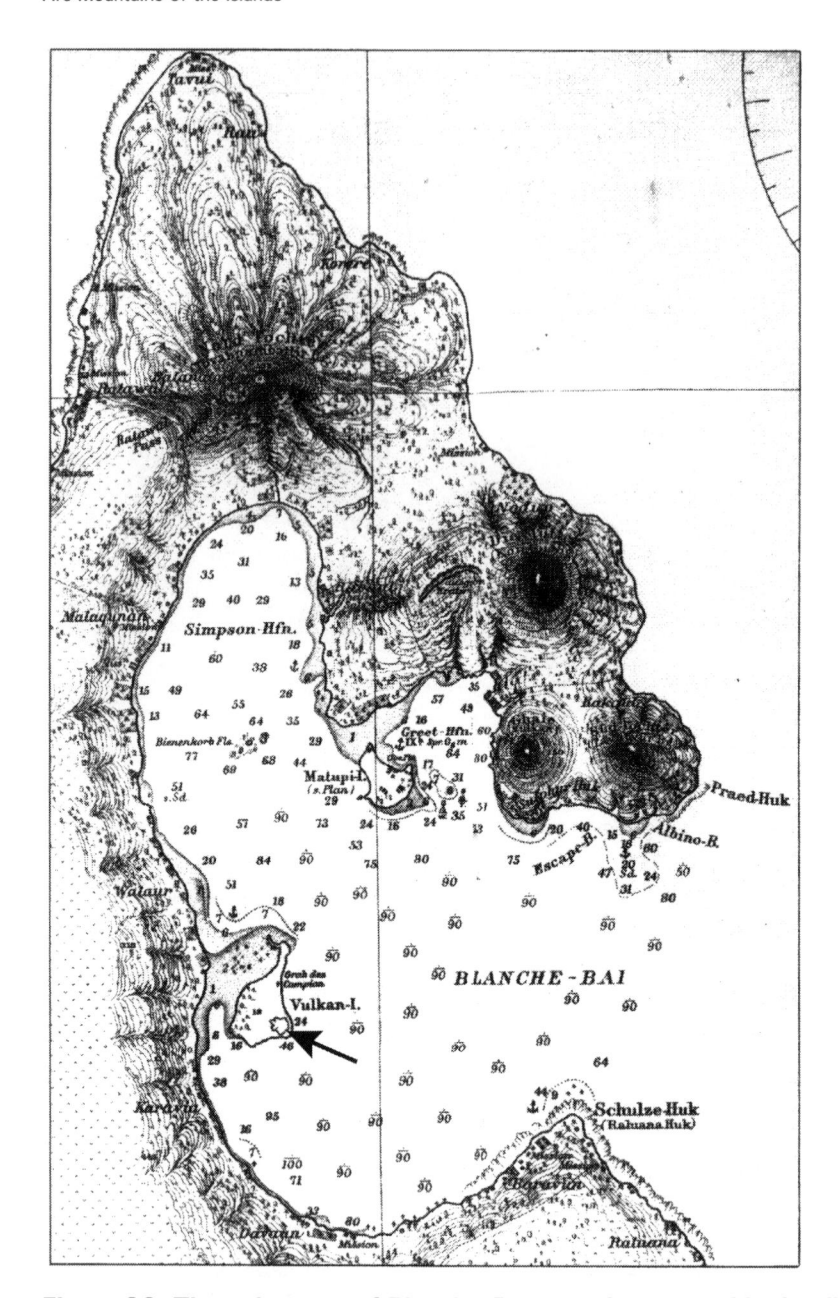

Figure 26. The volcanoes of Blanche Bay are shown on this detail from a German chart drawn from surveys after the 1878 eruption and before Rabaul town was built at the northeastern shoreline of Simpson Harbour. Water depths are in metres. The arrow indicates the site of the 1878 crater of Vulcan Island.

Source: Gazelle Halbinsel und Neu-Lauenberg published in Nachrichten über Kaiser-Wilhelmsland und den Bismarck-Archipel, 1888, 1:100,000 scale chart.

4. Volcanic Events of the German Era: 1884–1914

German geoscientific work of particular significance here is that of the Sapper-Friederici Expedition of 1908–1909, which surveyed throughout the Bismarck Archipelago and in the Solomon Islands. It was led by geologist and volcanologist Professor Karl Theodor Sapper and it would provide a broad context for a specific focus on the dangers of the Rabaul volcanoes. Retired captain and anthropologist G. Friederici accompanied Sapper during much of his fieldwork. Hahl himself did not make many notable observations of volcanic activity, but he is a key figure in the series of developments that eventually led the German administration to transfer its headquarters from Herbertshöhe to the shore of Simpson Harbour and so create the volcanically vulnerable town of Rabaul.

Figure 27. Karl Sapper (1866–1945) (left), from 1917, when he was on the faculty of the University of Strasburg, Germany. Sapper's reputation continued to grow after his 1908 visit to German New Guinea and he became an established figure in international volcanology, particularly after publication of his landmark *Vulkankunde* in 1927. Albert Hahl (right), from about 1896.

Source: Sapper, supplied by V. Lorenz. Hahl, frontispiece in Hahl (1980); supplied to P. Sack and reproduced courtesy of the Hahl family.

Sapper came to German New Guinea in 1908 with volcanological expertise on the nature of explosive volcanic eruptions and their hazardous impacts.[23] He was 42 years old in 1908 and had spent his early years in volcanically active Central America and southern Mexico, first with his brother helping establish coffee

---

23 Termer (1966) and McBirney & Lorenz (2003).

estates there, but later developing experience through extensive fieldwork in ethnology, geography and geology, including volcanology and geomorphology. He also made a visit in 1902–1903 to Mont Pelée volcano on the small island of Martinique in the Caribbean to investigate the disastrous eruption that had recently taken place there. The town of St Pierre was overwhelmed by a searing cloud of ash, rocks and dust which had raced down the flanks of the volcano as a block-and-ash pyroclastic flow on 8 May 1902, killing about 29,000 people.

French volcanologist A. Lacroix gave the name **nuée ardente**, or 'glowing cloud', to the type of destructive cloud that destroyed St Pierre in 1902. The name is of imprecise etymology but it is now embedded in general literature, although less so today in volcanological reports. This decline in usage is because of ongoing controversies about the complex origin of pyroclastic rocks in general and uncertainty about which explosive process produces what kind of pyroclastic deposit. Careers in volcanology are built or broken on such matters. A nuée ardente is a form of pyroclastic flow — such as seen by Dampier in 1700 at Ritter — that produces block-and-ash deposits and, indeed, to a large extent the term 'block-and-ash flow' is now preferred. The name **peléean** was introduced also as a result of the 1902 eruption and applied to eruptions that produce nuée ardente types of explosive activity. But use of the term peléean has also declined, largely because pyroclastic flows can be produced by processes other that the specific one deduced for Mont Pelée in 1902. Semantic and confusing discourses of this type, however, cannot deflect in any way from recognition of the importance of the 1902 Pelée eruption in the development of volcanology during succeeding decades.

Sapper, then, had travelled widely by 1908. He had been 'ordered' by the Reich to go to German New Guinea in 1908, a directive that likely reflects more on his growing stature as a natural scientist who could assist in expanding the knowledge base of the colony, than on any reluctance on his part to go there. Sapper and Friederici used Herbertshöhe as a base, but much of their time was spent surveying the geology and geography of New Hanover and New Ireland, as well as the offshore volcanic islands of the Tabar-Feni group.[24] Hahl and Sapper together crossed Bougainville Island from east to west in mid-July 1908, supported by 50 Melanesians — 20 soldiers and 30 carriers — and accompanied by American ethnologist G.A. Dorsey and government officer A. Doellinger.[25] Balbi and Bagana volcanoes were seen during the crossing. Both of these volcanoes had been observed at times in the 66 years since Parker Wilson's observations from the *Gypsy* in 1842, leaving no doubt about the ongoing volcanically active nature of Bagana. Indeed, local people informed the Hahl–Sapper group that the volcano had been in eruption just two days before the

---

24 Sapper (1910a) assisted by C. Lauterbach.
25 Sapper (1910b) and Hahl (1980).

4. Volcanic Events of the German Era: 1884–1914

arrival of the European party.[26] Activity at Balbi volcano, on the other hand, was restricted to water vapour emissions from the summit. Nevertheless, in 1908, the experienced Sapper was impressed by Balbi, the highest point on Bougainville Island, referring to it as a magnificent volcano and calling it a 'complex' made up of several cones in the summit area.

Herbertshöhe had no suitable harbour, yet a superb natural one existed at the head of Simpsonhafen within Blanche Bay to the north. Hahl, as early as 1902, had an ambition to develop a better harbour for large vessels, and so encourage expansion of commerce in, and increase exports from, the colony. He negotiated with the Bremen-based Norddeutscher Lloyd shipping line for what, for them, was basically a monopoly that cut out its Australian competitor, the Burns Philp Company. Land was acquired at Rabaul, swampy ground was drained — *rabaul* means 'the mangroves' — and, by 1905, a pier, store and houses had been established. Thus began an inexorable shift of the colony's capital away from Herbertshöhe to Rabaul that was well underway by 1908. There is no known record of the conversations between the governor and the professor about Rabaul, volcanoes and volcanic hazards, but there is no doubt that Sapper had serious concerns about the volcanic risks involved in developing the new capital at Rabaul. He wrote that the Rabaul eruption of 1878 was 'a warning to the inhabitants of Blanche Bay', and drew attention to both the clear volcanic risk that existed at Rabaul and to the 'unpredictable' nature of the volcanic eruptions there.[27]

Engineer Ludwig Kohl in 1909 also warned of future activity and referred to the need for geodetic measurements: 'In all likelihood the whole area — the Mother–Matupi–Vulcan Island area — is involved in a relatively marked movement of elevation, so that frequent surveys will be necessary. Movements of elevation frequently and suddenly occur in association with earthquakes.'[28] Neither Sapper nor Kohl was in a position in 1908–1909 to argue that the shift to Rabaul be reversed, but Sapper did offer the following practical advice for volcanic-risk reduction:

> provision should be made to alleviate the effects of further volcanic eruptions or devastating earthquakes by specially constructed dwellings, which should be kept low and built of timber to offer maximum resistance to earthquakes; on the other hand, they would have to have steep roofs so as to render harmless a rain of ash or pumice [which] cannot settle on a steep roof and will slide off.[29]

---

26 Bultitude (1981).
27 Sapper (1910c), p. 193.
28 Cilento (1937), p. 39.
29 Sapper (1910c), p. 193.

Hahl wrote that Herbertshöhe by 1909 'had become a sleepy hollow' and that he had been 'forced to transfer some Government offices [from Herbertshöhe] and to increase the staff' in order to service the increased trade and communications at Rabaul. Hahl remained that year 'in lonely splendour in Herbertshöhe', but he allocated funding — without prior approval from a later disapproving colonial office in Berlin — to complete the move to Rabaul, which in 1910 became the new capital of German New Guinea.[30] A neat network of tree-lined streets and roads was laid out beneath and between Tovanumbatir and Rabalanakaia volcanoes, forming the basis for a tropical town that was the home of departments of German administration for agriculture and health in particular, but which included no provision for a volcanologist or volcanological observatory. This is the town that would be totally evacuated in 1937 as a result of volcanic activity, completely destroyed by Allied bombing in the Second World War while occupied by the Japanese and, after being rebuilt in the same place, largely destroyed again by further volcanic activity in 1994.

There seems to have been little concern, either, from the German authorities and from the Australian military administration that followed them in 1914, about other active volcanoes in the colony and the effects that eruptions were having on local populations. Eruptions took place at Karkar volcano in 1885[31] and again in 1895 when, reported a German missionary on the island, there were:

> thick clouds of smoke, often of terrible appearance … and at night the entire upper cone of the mountain is sometimes bathed in fire. A constant thundering and rolling noise reminds us of the danger we are in. So far the Lord has shown mercy. One day, when the air was heavy with ash, we expected the worst.[32]

Pago volcano along the central New Britain coast also broke out into explosive activity in 1911, lasting until 1918. Bishop Couppé reported for 1911–1912:

> an incessant, dreadful din … and columns of smoke would be carried along by air currents and would cause terrible damage … Forests are singed and native gardens destroyed over a vast area. To escape famine the inhabitants … were forced to move to neighbouring villages or even further inland.[33]

An elderly village leader named Boas witnessed some of the explosive activity at Pago as a child and recalled it in the 1960s, confirming the damage to gardens and lack of food, and adding that there were many explosions, but neither strong earth tremors nor fatalities.[34]

---

30 Hahl (1980), pp. 132–33.
31 Zöller (1891).
32 Kunze (1901), p. 62.
33 Couppé (1912), p. 247.
34 Blake & Bleeker (1970).

4. Volcanic Events of the German Era: 1884–1914

Time Cluster of Eruptions

German natural scientists, including Karl Sapper, were voracious collectors of information in New Guinea and prolific in their production of lists and compilations, books and papers, encyclopedias and lexicons. Many of Sapper's publications on German New Guinea are, or include, historical compilations of eruption observations.[35] More eruptions, and more volcanoes in eruption, were reported in 1884–1914 than in any previous period in the history of Near Oceania. This, of course, is not surprising at first sight, given the presence of an increasingly large number of literate observers over the 30 years of German colonial rule.

The 1888 collapse of Ritter was the most disastrous volcanic event of the German period. Manam had by far the largest number of recorded eruptions, and Bam, Karkar, Langila, Pago, Ulawun, and Bagana volcanoes were all in explosive eruption. More significantly, however, eruptions took place at another four volcanoes — Bamus, Lolobau,[36] Makalia, and Victory — none of which has been active in the more than 90 years since German times. These four volcanoes are today shrouded in vegetation and appear 'extinct' to the casual observer. There is a dearth of written observations by eyewitnesses of the eruptive activity at the four volcanoes, and knowledge of the eruptions comes from observations made of the appearance of the volcanoes shortly after the activity or from oral history supplied by local people.

The number of volcanoes active during the narrower period of 1884–1899 — the first 16 years of German control — is especially striking. There were possibly as many as nine volcanoes in eruption, including three of the four that have not been in eruption since that time. Why should so many volcanoes have been active in the late nineteenth century? Should these eruptions be dismissed simply as an artifact of reporting intensity? Or do they represent another time cluster or 'pulse' of eruptive activity, like the one suspected for the mid- to late 1870s and, more particularly, like those that would be well documented in the mid-1950s and early to mid-1970s?

Hahl left Rabaul early in 1914, never to return. The First World War broke out later that year and, in September, the Australians, with instructions from the British, sent from the south a large expeditionary force and invaded German New Guinea. The Reich had long recognised that their tiny colony — half a world away from Berlin — would be strategically and militarily unimportant in any major war centred on Europe, and so made virtually no preparations for

---

35   See also, for example, Hammer (1907).
36   Several German reports refer to previous eruptions from Lolobau Island in 1904–1908 and possibly 1912.

its defence. Germany lost a small jewel in its imperial crown — a useful status symbol on the stage of international politics but ultimately a limited colony of no great economic or strategic value to them.

Table 2. Volcanoes in Eruption in Near Oceania from 1884 to 1899

Ulawun 1898. An eruption was seen from 210 kilometres away, and eruption devastation — probably caused by pyroclastic flows — on the volcano, was observed the following year.[a]
Bamus*. Descriptions of the volcano in 1894 are consistent with an eruption having taken place sometime during the previous several years.[b]
Makalia*. Information collected in 1963 from villagers at the northern end of Willaumez Peninsula refers to an eruption from nearby Makalia volcano, within Dakataua Caldera, that must have taken place in about 1880–1890.[c]
Langila 1884–1890. Eruptive activity is known from a few short reports.[d]
Ritter 1888. Evidence for eruption of new magma at the time of the cone collapse has not been found, but intrusion of magma into the volcano is one of the possible triggering mechanisms for the collapse.
Karkar 1885, 1895. Explosive eruptions were reported briefly by missionaries.[e]
Manam 1885–1899. Several observers recorded eruptions.[f]
Victory*. A series of observations of the mountain, in British New Guinea, in the late nineteenth century is consistent with eruptive activity having taken place sometime in the late 1880s.[g]
Bagana 1884–1899. The volcano was identified in 1884 as 'active', but unequivocal explosive eruptions were not recorded until 1894–1899.[h]
Note that Tuluman volcano can be added to this list if the range is expanded to 1883–99 and if its reported eruption of 1883 is a correct identification.

a. Parkinson (1999; originally published in German in 1907).
b. Couppé (1896).
c. Branch (1967).
d. Palfreyman et al. (1981).
e. McKee et al. (1976).
f. Palfreyman & Cooke (1976).
g. See, for example, Macgregor (1890–1891).
h. Bultitude (1981).

* The asterisks refer to volcanoes that have not been in eruption since the late ninteenth century.

The Germans' 30 years in New Guinea were, without question, scientifically successful. German scientists did not, however, venture far into the interior of the New Guinea mainland, except on voyages up the Sepik and Ramu rivers, so they had no knowledge of either the volcanoes or the tens of thousands of Melanesians in the highlands region of New Guinea south of the Sepik and Ramu. Also, and more particularly, German authorities seem to have held no special concern about volcanic eruptions from coastal volcanoes that might impact seriously on their investment and lives, and on the lives of Melanesians. This was notwithstanding both the Ritter disaster of 1888 and Sapper's poignant warnings about Rabaul's volcanic risk.

Figure 28. A Roman Catholic brother made this sketch of Bamus during the voyage made by Bishop Couppé in 1894. Couppé wrote that 'During a recent particularly severe eruption the fiery flow of lava [probably a block-and-ash flow] went beyond the foot of the mountain and ran into the valley [between Bamus and Ulawun] causing great damage ... the formerly majestic trees on its banks are left upright but dead and dried up' (1896, p.119).

Source: Couppé (1896; the drawing is on p. 150 of the issue containing Couppé's paper).

References

Anonymous, 1887. 'Aschenfall in Neu-Guinea', *Annalen der Hydrographie und Maritimen Meteorologie*, 15, pp. 210–11.

Anonymous, 1888. Untitled note, *Nachrichten über Kaiser Wilhelmsland und den Bismarck-Archipel für 1888*, pp. 76–79.

Anonymous, 1890. 'Arbeiter-Anwerbung im Schutzgebiet', *Nachrichten uber Kaiser Wilhelmsland und den Bismarck-Archipel für 1890*, pp. 81–85.

Anonymous, 1891. 'Die Ritterinsel und die Fluthkatastrophe in der Dampierstrasse am 13. März 1888', *Mitteilungen aus den Deutschen Schutzgebieten*, 4, pp. 59–61.

Blake, D.H., & P. Bleeker, 1970. 'Volcanoes of the Cape Hoskins Area, New Britain, Territory of Papua and New Guinea', *Bulletin Volcanologique*, 34, pp. 385–405.

Branch, C.D., 1967. 'Volcanic Activity at Lake Dakataua Caldera, New Britain'. Bureau of Mineral Resources, Canberra, Report 107, pp. 21–25.

Bultitude, R.J.S., 1981. 'Literature Search for pre-1945 Sightings of Volcanoes and their Activity on Bougainville Island', in R.W. Johnson (ed.), *Cooke-Ravian Volume of Volcanological Papers*. Geological Survey of Papua New Guinea Memoir, 10, pp. 227–42.

Cilento, R., 1937. 'The Volcanic Eruption in Blanche Bay, Territory of New Guinea, May, 1937', *Journal of the Historical Society of Queeensland*, 2, pp. 37–49.

Cooke, R.J.S., 1981. 'Eruptive History of the Volcano at Ritter Island', in R.W. Johnson (ed.), *Cooke-Ravian Volume of Volcanological Papers*. Geological Survey of Papua New Guinea Memoir, 10, pp. 115–23.

Couppé, L., 1896. 'Brief des Hochwürdigsten Bischof Couppé an den Hochwürdigen P. Generalobern', *Kalender zu Ehren Unserer Lieben Frau vom Heiligsten Herzen Jesu*, 7, pp. 99–128.

——, 1912. 'Bericht des Missionsbischofs Ludwig Couppé an den hochw. P. Provinzial in Hiltrup', *Hiltruper Monatshefte*, 29, pp. 245–48.

Finsch, O., 1888. *Samoafahrten. Reisen in Kaiser Wilhelms-Land und Englisch-Neu-Guinea in den Jahren 1884 u. 1885 an Bord des deutschen Dampfers Samoa*. Ferdinand Hirt & Sohn, Leipzig.

Firth, S., 1978. 'Albert Hahl: Governor of German New Guinea', in J. Griffin (ed.), *Papua New Guinea Portraits*. The Australian National University Press, Canberra, pp. 28–47.

——, 1983. *New Guinea under the Germans*. Melbourne University Press.

Gash, N. & J. Whittaker, 1975. *A Pictorial History of New Guinea*. Robert Brown & Associates, Brisbane.

Guppy, H.B., 1887a. *The Solomon Islands and Their Natives*. Swan Sonnenschein, Lowrey and Company, London.

——, 1887b. *The Solomon Islands: Their Geology, General Features, and Suitability for Colonization*. Swan Sonnenschein, Lowrey and Company, London.

Hahl, A., 1980. *Governor in New Guinea*, eds & trans P.G. Sack & D. Clark. The Australian National University Press, Canberra.

Hammer, K.L., 1907. *Die geographische Verbreitung der vulkanischen Gebilde und Erscheinungen im Bismarckarchipel und auf den Salomonen.* Münchow'sche Hof- und Universitäts-Druckerei, Giessen.

Hempenstall, P.J., 1978. *Pacific Islanders under German Rule: A Study in the Meaning of Colonial Resistance.* The Australian National University Press, Canberra.

Hernsheim, E., 1983. *Eduard Hernsheim: South Sea Merchant*, eds & trans P. Sack & D. Clark. Institute of Papua New Guinea Studies, Boroko.

Hiery, H.J., 1995. *The Neglected War: The German South Pacific and the Influence of World War I.* University of Hawai'i Press, Honolulu.

Johnson, R.W., 1987, 'Large-scale Volcanic Cone Collapse: The 1888 Slope Failure of Ritter Volcano', *Bulletin of Volcanology*, 49, pp. 669–79.

Kunze, G., 1901. *Im Dienst des Kreuzes auf ungebahnten Pfaden.* 2nd edn. Heft 4: Kleine Zuge aus dem Missionsleben auf Neu-Guinea.

Macgregor, W., 1890–1891. 'Despatch Reporting Visit of Inspection to Northeast Coast of the Possession'. *British New Guinea Annual Report for 1890–91*, Appendix D, Despatch No. 100, Brisbane, 16 September 1890, pp. 10–18.

Mackenzie, S.S., 1987 (1927). *The Australians at Rabaul: The Capture and Administration of the German Possessions in the Southern Pacific.* University of Queensland Press, St Lucia, and the Australian War Memorial, Canberra.

McBirney, A.R., & V. Lorenz, 2003. 'Karl Sapper: Geologist, Ethnologist, and Naturalist', *Earth Sciences History*, 22, pp. 79–89.

McKee, C.O., R.J.S. Cooke & D.A. Wallace, 1976. '1974–75 Eruptions of Karkar Volcano, Papua New Guinea', in R.W. Johnson (ed.), *Volcanism in Australasia.* Elsevier, Amsterdam, pp. 173–90.

Palfreyman, W.D. & R.J.S. Cooke, 1976. 'Eruptive History of Manam Volcano, Papua New Guinea', in R.W. Johnson (ed.), *Volcanism in Australasia.* Elsevier, Amsterdam, pp. 117–31.

Palfreyman, W.D., D.A. Wallace & R.J.S. Cooke, 1981. 'Langila Volcano: Summary of Reported Eruptive History, and Eruption Periodicity from 1961 to 1972', in R.W. Johnson (ed.), *Cooke-Ravian Volume of Volcanological Papers.* Geological Survey of Papua New Guinea Memoir, 10, pp. 125–33.

Parkinson, R., 1999. *Thirty Years in the South Seas: Land and People, Customs and Traditions in the Bismarck Archipelago and on the German Solomon Islands.* Crawford House Publishing, Bathurst.

Reche, O., 1954. *Ergebnisse der Südsee-Expedition 1908–1910. II. Ethnographie: A. Melanesien, Band 4. Nova Britannia 1.* Teilband Ludwig Appel, Hamburg.

Sack, P.G., 1973. *Land Between Two Laws: Early European Land Acquisitions in New Guinea.* The Australian National University Press, Canberra.

——, 1980. 'Editor's Introduction', in A. Hahl, *Governor in New Guinea*, eds & trans P.G. Sack & D. Clark. The Australian National University Press, Canberra, pp. ix–xix.

Sapper, K., 1910a. *Wissenschaftliche Ergebnisse einer amtlichen Forschungsreise nach dem Bismarck-Archipel im Jahre 1908. 1. Beiträge zur Landeskunde von Neu-Mecklenburg und seinen Nachbarinseln.* Mitteilungen aus den Deutschen Schutzgebieten. Ergänzungsheft No. 3.

——, 1910b. 'Eine Durchquerung von Bougainville', *Mitteilungen aus den Deutschen Schutzgebieten*, 23, pp. 206–17.

——, 1910c. 'Beiträge zur Kenntnis Neupommerns und des Kaiser-Wilhelms-Landes', *Petermanns Mitteilungen aus Justus Perthes' geographischer Anstalt*, 56, pp. 189–93, 255–56.

——, 1927. *Vulkankunde.* J. Engelhorns (Nachfolger), Stuttgart.

Schleinitz, G.E.G., 1896. 'Begleitworte zur Karte der Nordkuste des westlichen Teils der Insel Neu-Pommern', *Zeitschrift der Gesellschaft fur Erdkunde*, 31, pp. 137–54.

——, 1897. 'Begleitworte zur Karte des östlichen Teils der Insel Neu-Pommern', *Zeitschrift der Gesellschaft fur Erdkunde*, 32, pp. 349–59.

Siebert, L., 1984. 'Large Volcanic Debris Avalanches: Characteristics of Source Areas, Deposits, and Associated Eruptions', *Journal of Volcanology and Geothermal Research*, 22, pp. 163–97.

Silver, E. et al., 2005. 'Island Arc Debris Avalanches and Tsunami Generation', *Transactions of the American Geophysical Union*, 86, pp. 485, 489.

Silver, E., S. Day, S. Ward, G. Hoffmann, P. Llanes, N. Driscoll, B. Applegate & S. Saunders, 2009. 'Volcanic Collapse and Tsunami Generation in the Bismarck Volcanic Arc, Papua New Guinea', *Journal of Volcanology and Geothermal Research*, 186, pp. 210–22.

Steinhauser, R., 1891–1892. 'Flutwelle und die Hilfsexpedition von Finschhafen nach der Südwestküste von Neu-Pommern', *Westermanns Illustrierte deutsche Monatshefte*, 71, pp. 265–75.

Termer, F., 1966. 'Karl Theodor Sapper 1866–1945: Leben und Wirken eines deutschen Geographen und Geologen', *Deutsche Akademie der Naturforscher Leopoldina, Lebensdarstellungen deutscher Naturforscher*, 12, pp. 8–89.

Ward, S.N. & S. Day, 2003. 'Ritter Island Volcano — Lateral Collapse and the Tsunami of 1888', *Geophysical Journal International*, 154, pp. 891–902.

Whittacker, J.L., N.G. Nash, J.F. Hookey & R.J. Lacey, 1975. *Documents and Readings in New Guinea History: Prehistory to 1889*. Jacaranda Press, Milton.

Wiltgen, R.M., 1979. *The Founding of the Roman Catholic Church in Oceania 1825 to 1850*. The Australian National University Press, Canberra.

Zöller, H., 1891. *Deutsch-Neuguinea und meine Ersteigung des Finisterre-Gebirges*. Union Deutsche Verlagsgesellschaft, Stuttgart.

5. Australian Colonists and the Volcanoes of Mainland New Guinea: 1849–1938

The elder men in Wanigera will tell you of a time when the 'burning mountain' burst asunder, and sent flaming streams of lava flowing down to the sea, and they remember how the people dwelling on the higher ground made haste to build new and safer homes more near to the shore, and how from that time onwards travellers and huntsmen have been careful to keep away from the slopes of Keroro.

A.K. Chignell (1911)

First Impressions

Melanesians have occupied the interior of New Guinea Island probably for more than 30,000 years, but nineteenth century Europeans regarded it as *terra incognita*. Many visiting Europeans would have doubted why anyone would want to live in such rugged, intimidating, and inhospitable mountainous terrain. Ridge after ridge ascends steeply into the impenetrable clouds of the central, forest-covered mountain ranges, so abruptly that possibly habitable valleys are invisible from the coast. Some access to the base of the high ranges is afforded by sailing up the island's large rivers — the Sepik and Ramu in the north and the Fly/Strickland in the south — but movement overland between the rivers is difficult and dangerous. New Guinea, therefore, remained a 'mystery' island until well into the twentieth century, an unknown prospect for European imaginings in which volcanoes were part of the anticipation of what actually may be found there. Many of these imaginings found their way into print and there was an enthusiastic market for stories, based on fact or entirely fanciful, of mysterious and awe-inspiring New Guinea.

Captain J.A. Lawson reported on an epic return journey made in 1872–1873 across the widest part of New Guinea, starting on the south coast. He encountered two active volcanoes during his expedition, as well as other volcanic landscapes, and climbed to the crater of one volcano — the 'column of smoke that ascended from the centre was enormous, and hung in a dense cloud above, that quite excluded the heavens from our sight'.[1] Lawson, some days later, encountered a range of snow-capped mountains, one of which was the 16,743-foot Mount Vulcan — a 'Papuan Etna'. This active volcano, however, was overshadowed by

---

1 Lawson (1875), pp. 135, 137.

the nearby, 32,783-foot Mount Hercules. Lawson's descriptions are believable in their volcanological precision — perhaps he had visited Etna in Italy — but his book is a fabrication. He also described exotic wildlife including giant monkeys, spiders and scorpions, and made the improbable claim that snow-covered Mount Hercules was the highest known mountain, higher even than Everest.

The Reverend Henry Crocker of St Ann's Parsonage, Weremai, New Zealand, published in 1888 his edited narrative of the adventures of Louis Trégance, a French sailor, who had spent nine years amongst the Orangwok tribe of New Guinea. Trégance, having been shipwrecked with others in south-eastern Papua, went on to make an extraordinary journey westwards to the city of Kootar in central New Guinea, and over the Tannavorkoo, or Tannavakoo, Mountains where he saw an active volcano — 'Tannavorkoo's kitchen' — in eruption, 'one of the many in active operation in the interior of the country'. Trégance eventually reached the goldmining city of Watara, in western New Guinea, where he introduced mining techniques that he had learnt on the Victorian goldfields in Australia. He witnessed many dramatic events in his travels, including an eruption from a volcano that '... began to discharge its molten flood, which ran for miles over the country, doing a vast amount of injury, and causing the loss of many lives. A great panic fell upon the population of Watara as the lava threatened to reach the town itself'.[2] This narrative too is entirely fictitious, including the existence of Kootar, Watara and the Orangwoks, which leaves the reader to wonder about the motives of the mischievous Crocker and the true identity of the adventurous Trégance. Both narratives are examples of the genre of imaginary voyages that exists within English travel literature that had been popular since the beginning of the Enlightenment, particularly in Great Britain.

Volcanologist Karl Sapper took a more serious and scientific approach to identification of active volcanoes in New Guinea. Yet, in his 1917 compilation of eruptive activity in Melanesia,[3] he was able to identify, and then only tentatively, just two volcanoes — both in the Bird's Head or Vogelkop region of western New Guinea, and both now known to be non existent. The first identification was based on the false report by Dampier of volcanic activity in 1700 at the western end of the Bird's Head. The second one derived from a report of 'fire' and creation of a large, smoking, vegetation-free area following an earthquake on the night of 21–22 May 1864, further east in the Arfak Mountains. Lengthy discussions on the origin of the fire, which turned out to be not volcanic in origin, were published in 1921–1922 in Sapper's *The Volcanic Mountains of New Guinea*.[4] Sapper also took the opportunity to include descriptions of volcanoes in south-eastern New Guinea, at the 'tail' of the Bird, where the British and Australians had long known of the existence of youthful volcanoes.

---

2 Crocker (1888), pp. 108–09, 132.
3 See, Sapper (1917).
4 See Sapper (1921–1922). The possible existence of a volcano, later called Umsini, in the Arfak Mountains of the Vogelkop continued to be reported in the volcanological literature for at least another 50 years, until M.

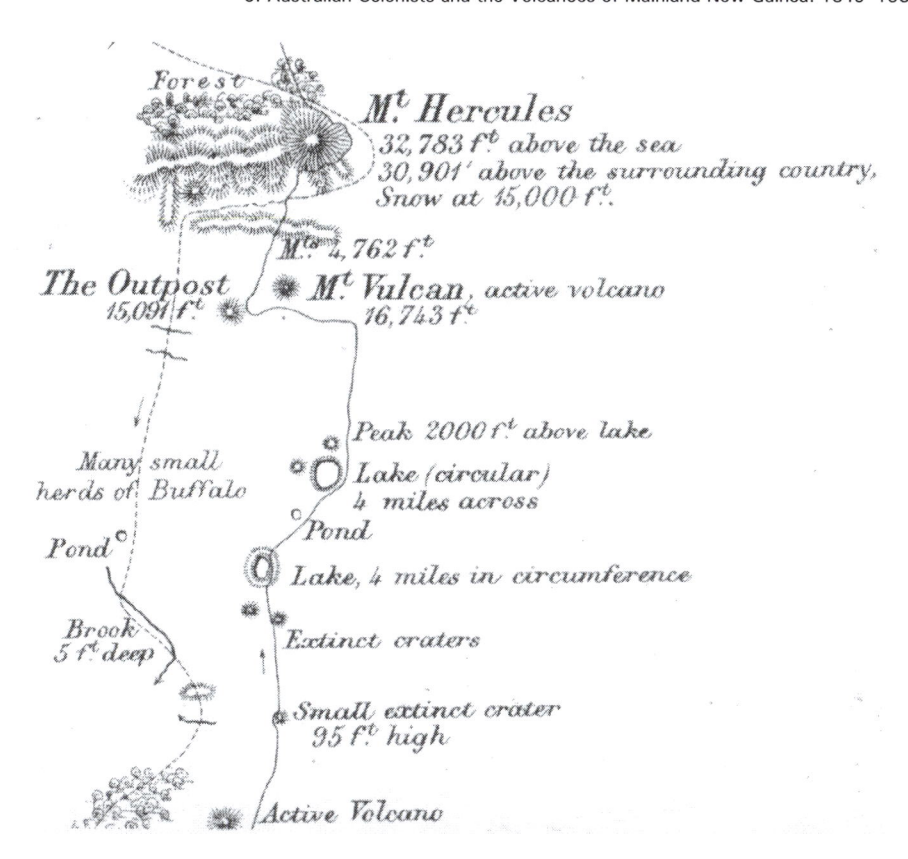

5. Australian Colonists and the Volcanoes of Mainland New Guinea: 1849–1938

Figure 29. Numerous volcanic features were noted by Captain Lawson in his imaginary journey into the highlands of New Guinea.

Source: Lawson (1875; fold-out sketch map).

British New Guinea and Victory Volcano

South-eastern New Guinea had become a British Protectorate in 1884, the year that Germany claimed the north-eastern portion of New Guinea calling it Kaiser-Wilhelmsland. There had been earlier raisings of the Union Jack in order to claim all of eastern New Guinea — for example, in 1873 when Captain John Moresby did so, and again in 1883 when magistrate Henry Chester, under orders from the Queensland Government, raised the flag and read a proclamation at Port Moresby. The British Government ignored these ceremonies. There had also been previous Australian interest in this part of New Guinea from explorers, naturalists, gold diggers, labour recruiters and missionaries, and even some

---

Neumann van Padang (1976) — author of the 1951 Indonesian part of the then definitive *IAVCEI Catalogue of the Active Volcanoes of the World* — finally discredited its identification.

early settlement — most notably the missionary W.G. Lawes at Port Moresby in 1874. Now, however, the 1884 British proclamation of the protectorate meant that there was official government involvement in the running of the new territory. It did not, however, long remain a protectorate, because of disabling jurisdictional restrictions and, by 1888, Britain had claimed full sovereignty, had annexed the territory and had named it the colony of British New Guinea.[5]

William McGregor in 1888 became British New Guinea's first Administrator, a position he occupied for the next ten years. McGregor was the ambitious son of a poor Scottish crofter and had become a doctor of medicine, in Scotland, before embarking on his successful career as a British colonial administrator.[6] He had interests in science and the classics, was an energetic and scientifically curious traveller throughout the colony, and appears to have been the first European to infer that Mount Victory in eastern Papua was a recently active volcano. He observed in late July 1890 that the mountain had 'great masses of bare rock' near its summit and that

> Its sides were scored and marked by brown lines from near the summit to its base; these at first looked as if caused by lava running down the mountain, but the closest inspection could detect no presence of lava, so that it was concluded that these lines had been caused by recent great earthslips ... a few days later we had the opportunity in the early morning of seeing numerous columns of steam rising, some from the very tops of the two crests of Mount Victory ... [where] vegetation is very scant ... Mount Trafalgar, on the contrary, is covered over the summit with dense forest. Flame was not at any time seen by us on Mount Victory, nor could we obtain from the natives any information regarding it.[7]

In addition, Anglican missionary A.K. Chignell in December 1909 observed that Victory volcano had 'white steam, or spirals of darker smoke ascending from a dozen fissures in its rugged crown', and went on to record local stories of past eruptions and the transfer of vulnerable homes on the mountain to safer places near the coast.[8]

And Resident Magistrate W.M. Strong in 1916 reported:

> Reliable native accounts [telling of] ... an extensive eruption — one or more villages were overwhelmed — and the Awanabairia people, who then lived on its slopes, fled to their present home at Lakwa.[9]

---

5   Whittacker et al. (1975, Documents D1–57).
6   Joyce (1971).
7   MacGregor (1890–1891), p. 14.
8   Chignell (1911) pp. 1–2.
9   Strong (1916), p. 409.

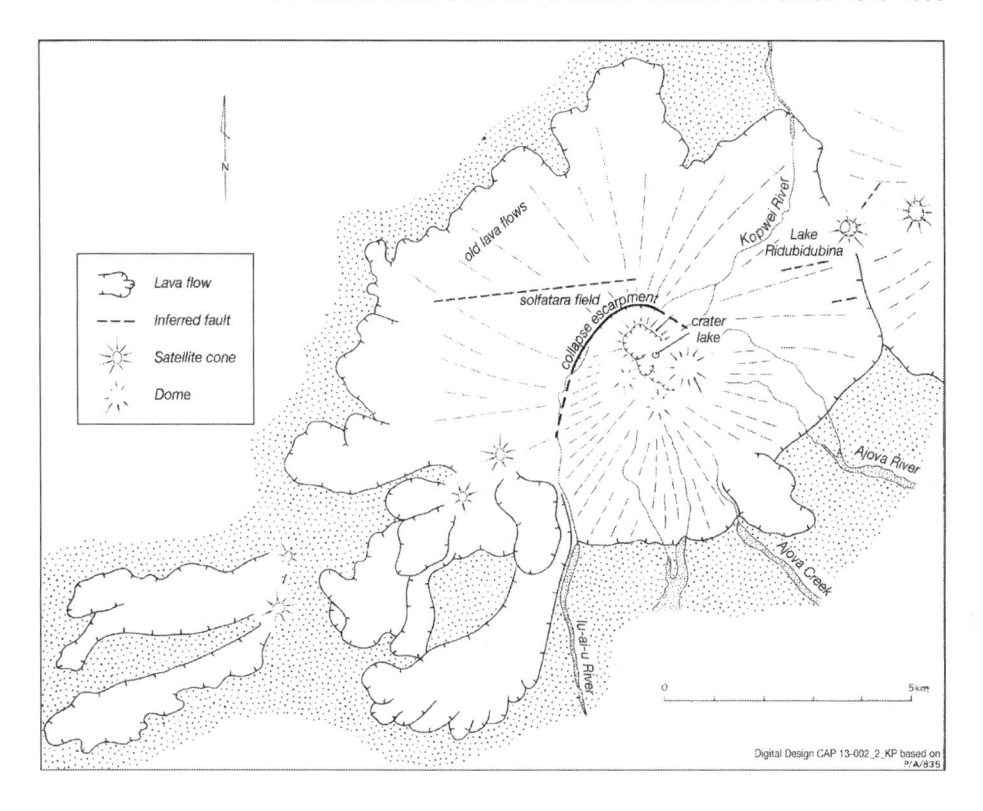

Figure 30. The main volcanic features of Mount Victory are shown in this sketch map, including what appears to be a debris-avalanche escarpment produced by a previous collapse of the volcano towards the south-east.

Source: Adapted from Smith (1981, Figure 2).

There are other reports, several of them derivative, and oral history too, about the Victory eruption or eruptions that produced these effects. Some refer to night-time glow from the volcano and others to how ships used the glow for navigational purposes, and there are reported observations of water vapour being emitted from the volcano as late as the 1930s. No documents by eyewitnesses of the Victory eruptions have been found, however, and therefore separating fact from fiction is difficult. Nevertheless, the available evidence is indicative of explosive eruptions having taken place at Victory, probably in the late 1880s, impacting on some villages, possibly disastrously, and almost certainly producing pyroclastic flows rather than the 'landslips' inferred by McGregor.[10]

---

10 A great deal of information on the nineteenthth century eruption at Victory was compiled in the 1970s by J.R. Horne, a former Territory of Papua New Guinea agricultural officer in the Northern Province of Papua. I am grateful to Mr Horne for access to his unpublished work and to the valuable references contained in it. See also Smith (1981).

The amount of geological knowledge that had emerged from British New Guinea by 1890 was limited and, indeed, had been obtained largely incidentally by explorers and general naturalists primarily interested in other matters. The Queensland Government geologist A. Gibb Maitland visited some coastal areas in 1891, using the British New Guinea Government yacht *Merrie England* as his base, but his fieldwork was restricted by the movements and requirements of the administration staff on board.[11] Nevertheless, Gibb Maitland summarised in his valuable report the known geological findings up to that time, and he included his own descriptions of the youthful volcanic nature of the D'Entrecasteaux Islands, where there are hot springs and obsidian, as well as volcanic cones such as make up Dobu Island.[12] None of the D'Entrecasteaux volcanic cones, however, are as large as the towering Victory volcano on the mainland that MacGregor had observed the year before. Gibb Maitland thought the nearby, lofty, Mount Dayman was also the remains of a volcanic cone, but he was mistaken.

Gibb Maitland pronounced the geological pioneer of British New Guinea to be John MacGillivray, the talented naturalist who had accompanied Captain Owen Stanley during the British scientific voyage of the HMS *Rattlesnake* to the Coral Sea coast in 1849. This claim is despite the restrictions on landings for fieldwork along the New Guinea coast that were imposed on the frustrated scientists by the wary Stanley, who considered that '… the natives are warlike, very numerous, well armed and very treacherous … [and] will certainly consider the first party who attempt to advance towards the interior, whether by land or water [to be invaders]'.[13] MacGillivray was one of a large team of surveyors and scientists on board the *Rattlesnake*, which also included the young T.H. Huxley, later to become a major proponent of Charles Darwin's theory of evolution. No observations of volcanological significance, however, were made from the *Rattlesnake*. This is hardly surprising as no volcanoes are visible from the New Guinea coastline that was visited by the vessel, although young ash layers are now known to mantle much of this part of the New Guinea cordillera that, somewhat ironically, was named after the reluctant captain — the Owen Stanley Range.

Evan R. Stanley in Papua

A young Australian graduate in geology, Evan R. Stanley — he was no relation of Owen Stanley — arrived in Papua from the University of Adelaide in 1911 as the territory's first government geologist. That year is additionally significant because Europeans in 1911 became aware of oil and gas seepages in the lower

---

11 Gibb Maitland (1891–1892).
12 Smith (1981).
13 Goodman (2005), p. 245.

Vailala River area, west of present-day Kerema on the Gulf of Papua.[14] This discovery marks the beginning of oil and gas exploration in south-eastern New Guinea and the triggering of periods of intensive geological mapping and exploration that led, secondarily, to the discovery of many volcanoes on the mainland. New South Wales Government geologist, J.E. Carne, investigated the oil find in 1912, as well as earlier reports of coal in the Purari River.

The colonies of Australia federated in 1901 to form the new nation of the Commonwealth of Australia, and consequently Britain gave up its New Guinea possession to the new nation, which, nevertheless, remained very much a part of the British Empire. South-east New Guinea in 1906 was proclaimed a colony — the Territory of Papua — following passage of the *Papua Act* by the Australian Parliament in late 1905. Hubert Murray became the new Territory's Lieutenant Governor, occupying this influential position until 1940, and sending his young officers — assistant resident magistrates and patrol officers also known as *kiaps* — into the unexplored Western and Delta districts of the Territory. The Australian Government was aware of the success of oil exploration in Persia. It was aware, too, of the exploration support being provided by the British Government, whose navy in 1914 was switching from coal to oil. Australia therefore secured the services of geologists from the Anglo-Persian Oil Company in assessing the Territory's oil-producing potential.

Figure 31. Evan R. Stanley with his family in Port Moresby in 1919.

Source: The Stanley family and H.L. Davies.

---

14   Rickwood (1992).

Fire Mountains of the Islands

Evan Stanley became involved with the oil explorers, but his main purpose was geological mapping and assessment of mineral potential in general.[15] Australian gold prospectors had been attempting the retrieval of alluvial gold in south-eastern New Guinea since the 1870s, and miners had been active at officially declared goldfields during British New Guinea times, including at Sudest, and on Misima and Woodlark islands, as well as on the mainland at Milne Bay, Yodda and Lakekamu.[16] None of the gold miners came away as wealthy men. Many were simply 'gully rakers' and 'tucker men' — those prepared simply to scratch out enough gold to maintain the barest of livings yet retaining dreams of wealth.

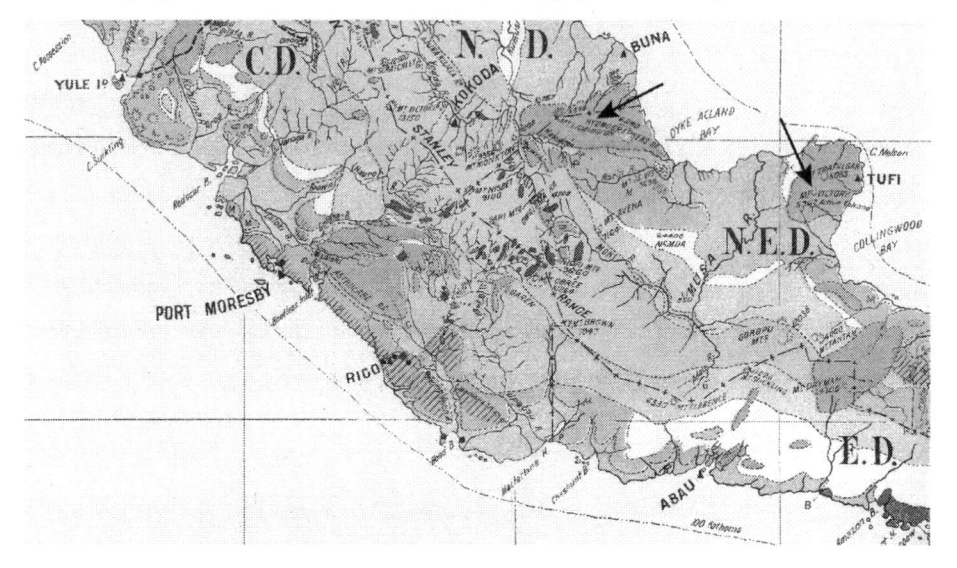

Figure 32. Areas in eastern Papua mapped by Evan R. Stanley and consisting of geologically youthful volcanic rocks, include Mount Victory, which is labelled on his map as an active volcano — see the arrow on the right. Mount Lamington and the Hydrographers Range (left) are grouped together and are not identified as volcanically active. C.D., N.D., and so on refer to the names of administrative districts.

Source: Stanley (1924, including the geological map).

Stanley was an energetic and conscientious field geologist, and a prolific reporter of his geological findings. He published in 1924 a landmark report of his and others' work, accompanied by a coloured geological map, and including a review of what was known about the volcanoes of the Territory of Papua. The volcanic nature of the D'Entrecasteaux Islands, as described by Gibb Maitland, was confirmed, but Gibb Maitland's mistake — that Mount Dayman was a volcano — was repeated. Victory and Trafalgar were shown as being made up

---

15   Davies (1987).
16   Nelson (1976).

of young volcanic rocks, and Victory was identified as 'apparently the only active volcano in the Territory'.[17] A significant point, particularly in view of what would happen in 1951, is that Stanley mapped the Quaternary volcanic rocks of the Hydrographers Range and Mount Lamington — both to the west of Victory — as a single geologically youthful unit, although did not point out that Lamington is younger than the rocks of the Hydrographers Range. Nor did he mention the presence of a crater at the summit of Mount Lamington, or the youthful, solidified, lava masses near the crater. Presumably, however, he must have realised that an eruptive centre, or centres, lay somewhere within the mapped area of Mount Lamington. Stanley would have been able to see all of these young volcanic features from the air, if aircraft or aerial photographs had been available to him.

There is, finally, another significant aspect of Stanley's volcano-related work. He identified the near-coastal Aird Hills, at the head of the Gulf of Papua in the far west of the territory, as volcanic and mapped a young volcanic area to the north of it that includes Mount Favenc. These are indications of discovery of the huge Fly-Highlands volcanic 'province' that extends northwards into the high cordillera of New Guinea — so high, indeed, that there were glaciers on the summits of its impressive volcanoes during the Pleistocene. Stanley's map, therefore, hints at the existence of two separate volcano groups. First is the then better known one in eastern Papua, which includes Victory and Lamington volcanoes. Second is a western group that includes the southerly volcanoes of the Aird Hills and Favenc.

Australians in the Territory of New Guinea after 1920

The Australian military gave up its control of what had been German New Guinea in 1921, as a consequence of the signing of the Treaty of Versailles following the end of the First World War. It handed over the newly named Mandated Territory of New Guinea to an Australian civil administration, which would be accountable to the Australian parliament and the League of Nations. Thus were formed separate Australian administrations of two contiguous colonial territories, which, for the next 20 years, would be characterised by a strong degree of mutual independence, a lack of close collaboration, competition at times, and even resentment when the old goldfields in Papua became worked out and gold diggers in the Mandated Territory discovered the rich auriferous gravels of the Bululo River system. These riches were in addition to the militarily acquired plantations, road networks, administrative system and ready-made

---

17   Stanley (1924), p. 34.

towns, such as Rabaul, that had been taken from the Germans. Some Australian administration officials became either 'Papuans' or 'New Guineans' in their loyalties. From 1921, however, both administrations, would start to benefit from a new event — the arrival of aviation in Papua and New Guinea — and so would the study of volcanoes.

Captain Frank Hurley, traveller and photographer, led a privately funded, public-interest expedition to New Guinea in 1922. He did not achieve his aim of flying across the island, but one of his aircraft — a Curtiss Seagull biplane flying boat — took off from Port Moresby's harbour on 5 September 1922 in the first known flight of an aircraft in Papua and New Guinea.[18] Use of aircraft for aerial reconnaissance would later greatly assist the Europeans of both territories in their explorations, and the oil geologists would benefit and learn the skills required from aerial photography and airborne observations to determine the geological structures of the country, including faults, folds and dipping strata. Volcanologists, furthermore, would see the way in which volcanoes were laid out and identify features, such as craters and calderas, that were not visible from the forest floor. They would be able to observe eruptions during aerial inspections, and pilots would become front-line observers of eruptive activity. Some would have to learn techniques of avoiding high-rising, aircraft-damaging, ash plumes.

The legendary 'Sharkeye' Park, an Englishman, together with Jack Nettleton, who had been part of the Australian expeditionary force that had invaded Rabaul in 1914, found gold in a young volcano on Koranga Creek on the Bulolo in 1922, and worked it for a year before outsiders heard of the discovery.[19] But a goldrush to the area did not start until 1926 when Bill Royal climbed high into the Upper Edie Creek south of Koranga and found New Guinea's richest alluvial field. One of the world's last gold rushes was triggered, and hundreds came to the Morobe Goldfield in search of wealth. Amongst them was Michael J. 'Mick' Leahy, who had been cutting wood for railway sleepers in Queensland and who had left his truck by the side of the road in Townsville when he heard about the Upper Edie gold strike. Leahy did not achieve wealth from gold but he prospected for mining companies and this allowed him, together with his brothers Pat, Jim and Dan, his friend Mick Dwyer, and Patrol Officer Jim Taylor, to achieve fame through exploration of valleys in the highlands, making remarkable 'first contacts' with Melanesian communities, cutting rough airstrips for hair-raising landings by supply aircraft, and also identifying volcanoes.[20]

---

18 Sinclair (1983).
19 Nelson (1976).
20 Leahy (1936, 1994).

Leahy and Dwyer in 1930 looked out into wide grass-covered valleys in the eastern highlands and realised that they

> had discovered a totally new country. This land was populated by tens of thousands of Stone Age natives, whose village fires at night extended in the distance as far as the eye could see across the grass valleys and ranges. [Then, looking down on the eastern end of a highland valley river system, and from] the bloated bodies of natives floating aimlessly in the current ... [and from] the innumerable skulls and bones littering the sands, we concluded that there must be an even larger valley or valleys ... coming in from the west and that the native population must be huge.[21]

The Melanesians of these populated valleys met Leahy and his companions in the following years in a dramatic series of 'first contacts' during which the gold-seeking Europeans would find out more about the great Wahgi Valley, its peoples and three extinct volcanoes that dominate the central highlands — Hagen, Keluwere and Ialibu. Leahy and his party climbed to the summit of Keluwere — which is today better known as Giluwe — on 15 June 1934, measuring there a height of 4,100 metres above sea level. Leahy wrote in his diary for that day: 'There is no doubt that it [Giluwe] was one of the many gigantic volcanoes which in the dim past formed this portion of New Guinea'. He also wrote in his later memoirs that:

> It must have been a tremendous mountain in its active days. The whole Hagen district must have been a frightening, awe-inspiring area when Hagen, Keluwere, and Ialibu, three points of a volcanic triangle, were active. The local legend of dust from above, which accumulated on the roofs of houses and made them collapse, was amply supported by physical evidence in the present ash beds above lava flow level. Still, the country appeared to have been free from such activity for so long that the legend seemed unbelievable.[22]

Leahy is here referring to a legend that is widespread throughout the highlands, yet whose volcanological significance would not be properly addressed until the 1970s. Young ash layers are common throughout the region, but which, if any, of the highlands volcanoes produced them? What volcanological information is contained in the legends? When did the ash falls take place? Are all the highlands volcanoes as extinct as Hagen, Giluwe, and Ialibu seem to be?

---

21 Leahy (1994), pp. 9–11. The bodies and bones were presumably the result of tribal warfare.
22 Leahy (1994), p. 201. There is some uncertainty as to whether Leahy heard about the dust-fall legend in the early 1930s, as it is not mentioned in the edited version of his journeys published in 1936 or in his diary for 1934 which is held by the National Library of Australia in Canberra (NLA MS384).

Figure 33. The strikingly craggy, alpine-type scenery of the summit of Giluwe volcano was formed by glaciation about 15,000 years ago. Glacial moraines are seen dominating the foreground and middle slopes.

Source: D.H. Blake. Geoscience Australia (GIL69-4-10).

The name 'Hagen' was used by Leahy for this imposing volcano because he believed it to be the mountain that German geomorphologist W. Behrmann had seen from ranges south of the Sepik River in 1912. Leahy later, however, began to have doubts about this usage because he could not see the Sepik Basin when he climbed the volcano, and Lutheran missionaries later told him that the mountain was too far inland to be seen from the low country. Mount Hagen — Hagen Gebirge — was in fact named in 1896, but not recognised as a volcano, by the German botanist Carl Lauterbach, who had been on an expedition to the Ramu River valley. The mountain was named after Kurt von Hagen, a Prussian military officer who was briefly *Landeshauptmann* in German New Guinea, until he was shot dead by a Buka policeman in 1897. Behrmann himself was part of the Sepik River Expedition of 1912–1913, and took part in land expeditions southwards from the Sepik. He came within viewing distance of the grass-covered lower slopes of Mount Hagen, which, like Lauterbach, he did not recognise as a volcano. Easy access to the heavily populated highlands would have been possible from that point, but Behrmann turned back because his supplies were running out.[23]

---

23 Löffler (1977) and Firth (1983). I am grateful to Dr Löffler for additional information about the naming of Hagen volcano used in this paragraph.

5. Australian Colonists and the Volcanoes of Mainland New Guinea: 1849–1938

Australians in the Territory of Papua before 1938

Europeans were making further volcano discoveries on the Territory of Papua side of the border with New Guinea, although the mountains here were being identified mostly as prominent peaks rising impressively from the flat Fly River plateau. The Mackay-Little Expedition of 1908, for example, was in the Upper Purari looking for coal deposits that had been reported earlier by McGregor, and the party struck out westwards reaching and naming Mount Murray. This inactive volcano would also be used as a geographical marker for later European explorations. Two government patrols of particular significance were undertaken in the mid-1930s — the first by J.G. 'Jack' Hides and L.J. 'Jim' O'Malley in 1935, on the Strickland-Purari Patrol; and the second by Ivan F. Champion and C.T.J. 'Bill' Adamson in the following year, on the Bamu-Purari Patrol.[24]

Hides and O'Malley travelled eastwards from the headwaters of the Strickland River, between Sisa and Bosavi mountains — both of which are now known to be Pleistocene volcanoes — and then north into the populous Tari River valley where the Huli people encountered them in a series of 'first contacts'. Anthropologists would later collect stories from the Huli of *mbingi*, or 'time of darkness' and, with others, would interpret what these tales — the same ones mentioned by Leahy — mean in terms not only of modern volcanic eruptions and ash falls, but also of the Huli's unique, and apocalyptic, world view. Hides and O'Malley climbed eastwards out of the Tari valley between two peaks, the southern one of which they named Mount Champion, but which today is known as Kerewa, a Pleistocene volcano. The northern mountain is Doma Peaks, a volcano that much later would be discovered to have thermal activity and therefore possibly a potential for future volcanic eruptions. Hides and O'Malley completed their famous journey by patrolling south-eastwards through the headwater tributaries of the Purari to Mount Murray and then to the navigable Kikori River. An appreciative Australian public was made aware of their arduous journey when Hides published his successful book *Papuan Wonderland* in 1936. Hides did not refer to volcanoes there, but he did write that the Leonard Murray Mountains — his name for Bosavi— 'appeared like a giant Fujiyama [the famous Japanese volcano], drawing themselves together from the plateau'.[25]

Champion and Adamson undertook their well-organised government expedition in 1936. This Bamu-Purari Patrol was preceded by aerial reconnaissance, and its course was better navigated than was Hides and O'Malley's Strickland-Purari Patrol. Champion used astrofixes, for example, and a heavy radio receiver was carried for communications. Champion had accompanied Charles Karius in

---

24 Sinclair (1988).
25 Hides (1936), p. 46.

1927–1928 in an epic crossing of New Guinea from the Fly to the Sepik, well to the west of the Fly-Highlands volcanoes, and he would later become involved, as a senior officer of the Administration, in overseeing the relief effort following the 1951 Lamington disaster. Champion and Adamson, who approached Bosavi from the south-west, noted the existence, well away from the main mountain, of a crater lake and 'much loose volcanic rock'. These are part of a **volcanic field** formed by eruptions from many small volcanic centres south-west of Bosavi. The party spent many days on the northern flank of Bosavi prospecting for gold, but none was found. Champion wrote that 'This mountain massif may be likened to a giant octopus, the peaks being at the head, and the spurs the long arms which run out for miles with deep gullies between'.[26] Champion also noted the dominating presence to the north of a conical mountain, named Sisa, as the patrol moved eastwards to Lake Kutubu. Champion and Adamson travelled to the northern sides of Giluwe and Ialibu and then, on their return journey, between Suaru and Karimui mountains, down to the Purari River near Mount Favenc. All of these named mountains are Pleistocene volcanoes.

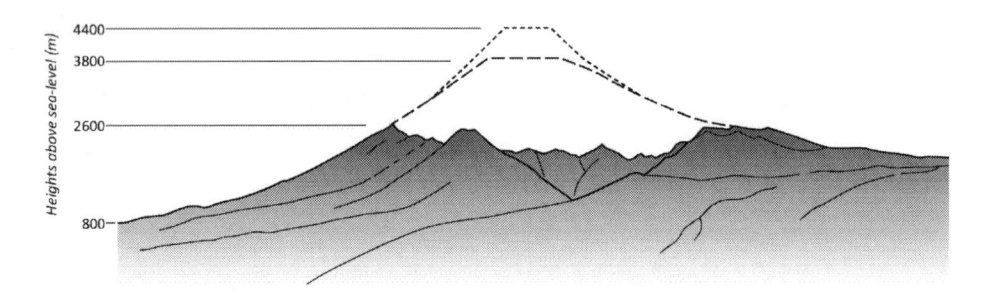

Figure 34. **The deeply eroded volcano of Mount Bosavi may originally have been as high as 3,800 or 4,400 metres above sea level, as suggested by the two dashed lines. The lower reconstructed profile may be the more likely. Bosavi today is still an impressive volcano, even in its eroded state, rising from the Fly River plateau in forested and still sparsely populated country, and occupied by a breached crater that has been enlarged by erosion. The mapped extent of Bosavi is greater than that of any other volcano in Near Oceania.**

Source: Adapted from Mackenzie & Johnson (1984, Figure 25B).

The oil company geologists were making geological discoveries, especially after Oil Search Limited in 1936 was granted Special Permit 5 enabling aircraft-supported exploration from the Gulf of Papua up to the border with the Mandated Territory. One of the company's geologists, S.W. 'Sam' Carey, published in 1938 what, for its time, was a groundbreaking and scientifically influential paper 'The

---

26 Champion (1940), p. 199.

Morphology of New Guinea'.[27] Carey, who would later achieve fame as an ardent exponent of the theories of continental drift and of an expanding Earth, took the report of the Champion and Adamson patrol as a basis for his identification of several volcanoes in what he called the 'Southern Foothills' of New Guinea, but, curiously, he did not refer to Hagen, Giluwe, and Ialibu volcanoes further north in the central cordillera. The reasons for this are unclear, but the omission gave the false impression, to readers of the peer reviewed scientific literature at least, that the impressive Fly-Highlands volcanic province was not as extensive as it is now known to be.

Geological surveys of New Guinea were interrupted by the Second World War and the full extent of the Fly-Highlands province would not be fully appreciated by the wider geoscientific community until the 1960s, when the eruptive potential of these magnificent volcanoes would also become better understood. Volcanological attention would be directed, anyway, to Rabaul volcano as a result of eruptions in 1937, and to the volcanoes in south-eastern New Guinea where the most deadly natural disaster in historical time in Near Oceania would be produced by Lamington volcano in 1951. The Fly-Highlands volcanoes were, as a result of these events, to an extent sidelined.

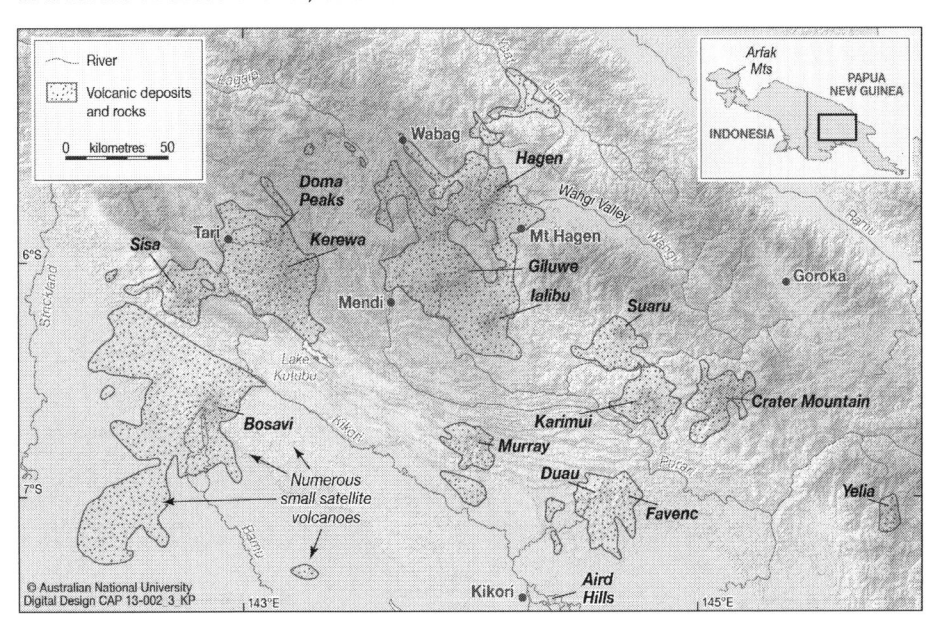

Figure 35. The considerable extents of the volcanoes of the Fly-Highlands province are shown clearly by the areas of volcanic deposits and rocks mapped by geologists during the 1960s and 1970s.

Source: Adapted from D'Addario et al. (1975). See also Mackenzie & Johnson (1984, Figure 1).

---

27   Carey (1938). See also Noakes (1939).

References

Carey, S.W., 1938. 'The Morphology of New Guinea', *The Australian Geographer*, 3, pp. 3–31.

Champion, I., 1940. 'The Bamu-Purari Patrol, 1936', *Geographical Journal*, 96, pp. 190–206, 243–57.

Chignell, A.K., 1911. *An Outpost in Papua*. Smith, Elder and Co., London.

Crocker, H. (ed.), 1888. *Adventures in New Guinea: the Narrative of Louis Trégance a French Sailor: Nine Years in Captivity among the Orangwoks a Tribe in the Interior of New Guinea*. Sampson Low, Marston & Company, London.

D'Addario, G.W., D.B. Dow & R. Swaboda, 1975. 'Geology of Papua New Guinea'. 1:2 500 000 scale map. Bureau of Mineral Resources, Canberra.

Davies, H.L., 1987. 'Evan Richard Stanley, 1885–1924: Pioneer Geologist in Papua New Guinea', *BMR Journal of Australian Geology and Geophysics*, 10, pp. 153–77.

Firth, S., 1983. *New Guinea under the Germans*. Melbourne University Press.

Gibb Maitland, A., 1891–1892. 'Geological Observations in British New Guinea in 1891', *British New Guinea Annual Report for 1891–92*, Appendix M, pp. 53–85.

Goodman, J., 2005. *The Rattlesnake: A Voyage of Discovery in the Coral Sea*. Faber and Faber, London.

Hides, J.R., 1936. *Papuan Wonderland*. Blackie & Sons, London.

Joyce, R.B., 1971. *Sir William MacGregor*. Oxford University Press, London.

Lawson, J.A., 1875. *Wanderings in the Interior of New Guinea*. Chapman & Hall, London.

Leahy, M., 1936. 'The Central Highlands of New Guinea', *Geographical Journal*, 87, pp. 229–62.

——, 1994. *Exploration into Highland New Guinea 1930–1935*, ed. D.E. Jones. Crawford House Press, Bathurst.

Löffler, E., 1977. *Geomorphology of Papua New Guinea*. The Australian National University, Canberra.

Macgregor, W., 1890–1891. 'Despatch Reporting Visit of Inspection to Northeast Coast of the Possession', *British New Guinea Annual Report for 1890–91*, Appendix D, Despatch No. 100, Brisbane, 16 September 1890, pp. 10–18.

Mackenzie, D.E. & R.W. Johnson, 1984. *Pleistocene Volcanoes of the Western Papua New Guinea Highlands: Morphology, Geology, Petrography, and Modal and Chemical Analyses*. Bureau of Mineral Resources, Canberra, Report 246.

Neumann van Padang, M., 1976. 'On Some Volcanoes Mentioned in the *Catalogue of the Active Volcanoes of Indonesia*', Berita Direktorat, *Geosurvey Newsletter VIII*, Indonesia, 8, pp. 10–11.

Nelson, H., 1976. *Black White & Gold: Goldmining in Papua New Guinea 1878–1930*. The Australian National University Press, Canberra.

Noakes, L.C., 1939. 'Geological Report on the Chimbu-Hagen Area, Territory of New Guinea'. Geological Survey of the Territory of New Guinea. Unpublished report.

Rickwood, F., 1992. *The Kutubu Discovery: Papua New Guinea, its People, the Country and the Exploration and Discovery of Oil*. Privately published.

Sapper, K., 1917. *Katalog der geschichtlichen Vulkanausbrueche: Melanesische Vulkanzone*. Karl J. Truebner, Strassburg, pp. 204–215, 300.

——, 1921–1922. 'Die Vulkanberge Neu-Guineas', *Zeitschrift fuer Vulkanologie*, 6, pp. 1–14.

Sinclair, J., 1983. *Wings of Gold: How the Aeroplane Developed New Guinea*. 2nd ed. Robert Brown & Associates, Bathurst.

——, 1988. *Last Frontiers: the Explorations of Ivan Champion of Papua: A Record of Geographical Exploration in Australia's Territory of Papua between 1926 and 1940*. Pacific Press, Queensland.

Smith, I.E.M., 1981. 'Young Volcanoes in Eastern Papua', in R.W. Johnson (ed.), *Cooke-Ravian Volume of Volcanological Papers*. Geological Survey of Papua New Guinea Memoir, 10, pp. 257–65.

Stanley, E.R., 1924. *The Geology of Papua*. Government Printer, Melbourne.

Strong, W.M., 1916. 'Notes on the North-eastern Division of Papua (British New Guinea)', *Geographical Journal*, 48, pp. 407–11.

Whittacker, J.L., N.G. Nash, J.F. Hookey, & R.J. Lacey, 1975. *Documents and Readings in New Guinea History: Prehistory to 1889*. Jacaranda Press, Milton.

6. Calderas, Ignimbrites and the 1937 Eruption at Rabaul: 1914–1940

> Great billowing clouds of smoke burst out on all sides: not only upwards, but in every direction, rolling in enormous bursts. The people rushed into the church to pray … Then came a rain of hot ash, so thick that we could not see a gleam of lamplight. I thought 'This is the end', felt my way through the darkness to the front, and asked the people, as loudly as I could, to beg God's forgiveness.

Father Nollen (1937)

Garrison Life and Volcanoes

Australian military forces occupied Rabaul and ruled the Old Protectorate of German New Guinea for almost seven years following a short skirmish near Bitapaka in 1914. Rabaul became a battle-deprived backwater of the First World War when thousands were dying at Gallipoli and in the trenches on the Western Front. Garrison life for the Australians in New Guinea, wrote one war historian, 'was a still lagoon, with the surf beating outside'.[1] The Australians slipped readily into the ruling upper layer of the racially segregated society that had been created by the Germans in Rabaul. Immigrant Chinese, Malays, Ambonese, and also Japanese who had the same legal status as Europeans, represented the middle layer. Melanesians occupied the bottom layer. The Rabaul town plan was also racially segregated. This was 'justified in the name of law, order, and disease control' and supported by the military medical officer Ray Cilento who, two decades later as Sir Raphael Cilento, would investigate the medical implications of the 1937 eruption at Rabaul.[2]

German citizens in 1914 were allowed to remain in Rabaul if they signed statements of neutrality, but wartime tensions and suspicions remained between them and the Australians. German residents, and others, predicted that there would be an eruption at Rabaul in 1917.[3] This was based on the belief that the 1878 eruption at Rabaul had taken place 40 years after a previous one in 1838 — presumably the poorly dated Sulphur Creek eruption reported by Father Boegershauser — and another eruption, therefore, was imminent. There was

---

1 Mackenzie (1987), p. 316.
2 Gould Fisher (1994), p. 17.
3 Anonymous (1917a) and Cummins (1917).

discussion about the matter in the garrison newspaper, the *Rabaul Record*. Potential evacuation routes were mentioned briefly in one article, which ended, however, on a rather fatalistic note that carried some militaristic historicity:

> Do like the gallant Roman soldier who, at the destruction of Pompeii, died in his sentry box. Remain at your post. When the critical moment approaches you may — to cheer yourself up — sing 'Let me like a soldier fall'.[4]

Figure 36. The large volcanoes of Kabiu (left) and Turagunan (right) together with Tavurvur, lower down on the left, form the background to this photograph of Australian troops and a six-inch gun mounted at Fort Raluana at the entrance to Blanche Bay in about 1918.

Source: Australian War Memorial (H01987).

Captain C.H. Massey, who was an officer in the Australian garrison and had had previous geological training, recorded the effects of an earthquake and tsunami in January 1916 which had caused damage to shipping and the wharfs at Rabaul. He noted that local earthquakes within the harbour seemed to originate from the area of Tavurvur volcano.[5] Some local residents believed that earthquakes felt at Rabaul happened during heavy rain, but Massey did not agree that there was such a direct connection. He did suggest, however, that there may be a relationship to high ocean tides and therefore to the north-west monsoon or 'wet' season, and so, in turn, to 'earth warping' caused by the changing seasonal

---

4   Anonymous (1917b), p. 3.
5   Massey (1918, 1923).

positions of the Moon and Sun relative to the Earth. He also stated his disbelief that volcanic eruptions could be predicted accurately using a precise 40-year periodicity.

Figure 37. Four Australian soldiers of the occupying force at Rabaul during the First World War, together with a local guide, are looking into the 1878 crater of Tavurvur volcano.

Source: Johnson & Threlfall (1985, figure on p. 12). Geoscience Australia (M2447-31A).

Massey concluded that the large flooded depression of Simpson Harbour was the original 'crater' — actually a caldera — of Rabaul volcano and that it was the source of the thick deposits of ash, pumice, and scoria making up the surrounding countryside to the west and south, notably at Bitapaka. These early geological observations by Massey relate to the origin of a *series* of calderas that form Blanche Bay, and to the especially large volcanic eruptions that accompanied caldera formation at Rabaul. They are matters that would occupy the attention of many subsequent volcanic geologists.

The Australian military authorities did not take any specific measures to monitor the volcanoes and earthquakes at Rabaul; although, in a letter written in September 1920, General G.R. Johnston, head of the Australian garrison, recommended the installation of a seismograph at Rabaul, evidently without a successful outcome.[6] Rabaul and the former German colony came under an

---

6 Stanley (1923), p. 44.

Australian civil rule in 1921, which lasted until 1942. The captured territory — following the Treaty of Versailles in 1919 — was a Class C Mandated Territory under the League of Nations, and this allowed the Australians to rule it almost as if it was part of Australia. There were subsequent high-level discussions about amalgamating the Mandated Territory with the Australian Territory of Papua, and about relationships with the British Solomon Islands Protectorate, but the three immediately adjacent territories of the British Empire remained administratively separate. Scientific interests of the new civil administration of the Mandated Territory were restricted largely to health, agriculture and anthropology and, as in previous German and Australian-military administrations, there was no interest in volcano monitoring — at least, until the disastrous 1937 eruption at Rabaul.

Australian Expedition along the Bismarck Volcanic Arc

Evan R. Stanley, the Government Geologist for the Territory of Papua, was a member of the Commonwealth Scientific Expedition to the new Mandated Territory in 1920–1921 — its aim, an assessment of the natural resources of the newly acquired territory. The expedition was led by Dr Campbell Brown, an authority in osmoridium, the natural metallic alloy used in the manufacture of pen nibs and surgical needles. Brown, who had connections with the Waterman Pen Company, had persuaded the Australian Prime Minister, W.M. 'Billy' Hughes, to have the Commonwealth Government fund the expedition. A 12-metre wooden ketch, the *Wattle*, was to be used as a mobile base.

The expedition was a poor replication of those that had taken place in German times, and it turned out to be an embarrassment to the Australian Government and Hughes because of its cost and non-achievement of objectives.[7] Stanley's report to the Australian Government — one of the expedition's few redeeming features — included information on the nature of the volcanic chain from Rabaul in the east, along the north coast of New Britain, to Manam and the Schouten islands, 1,000 kilometres to the west of Rabaul. His report also contained a recommendation that a 'Geophysical Observatory be established in conjunction with a Volcano Observatory' for public-safety purposes in the New Guinea region.[8] Calls for volcano monitoring at Rabaul would be made later, too, by the media, but like Stanley's recommendation, they elicited no support.[9]

---

7 Davies (1987).
8 Stanley (1923), p. 92.
9 Anonymous (1932).

Stanley visited the small caldera at Lolobau off the north coast of New Britain, but sailed past Galloseulo and Pago volcanoes further east without knowing, given his low vantage points at sea level, that these, too, occupied calderas. New Britain, in fact, has more young calderas than any other area of similar size in the whole south-east Asian and south-west Pacific region. Stanley noted that Pago, the volcano that had been in explosive activity in 1911–1918, was bare of vegetation, was emitting water vapour and had a 'breached' appearance. A series of lava extrusions had been underway at Pago since before the 1918 visit of a Roman Catholic missionary, Father Ischler, to the floor of the Witori caldera, and lava may have continued to flow until as late as 1923.[10]

Stanley was the first geologist to recognise the existence of the 13-kilometre-wide caldera of Dakataua at the northern end of Willaumez Peninsula, New Britain, where D'Entrecasteaux in 1793 had identified only Willaumez 'island'. He also noted that Wangore, or Bulu, just south of Dakataua, was an 'active' volcano because its upper parts in 1921 were completely bare and there were emanations of water vapour from near the summit.

Figure 38. Pago is here seen in explosive eruption in 1918, together with the rubbly surface of a recently erupted lava flow in the foreground. This photograph appears to be the earliest taken of any volcanic eruption in the Near Oceania region.

Source: P. Ischler, Mission of the Sacred Heart. Originally published in an article by Stamm (1930, p. 89). See also Cooke (1981, Figure 8).

---

10 Cooke (1981).

A large eruption that took place at Manam on 11 August 1919, and which was observed by local missionaries, was recorded by Stanley as a result of his visit to the Manam area in 1921. This eruption appears to have been much larger than those reported previously at Manam, but it may have been similar in size to the eruptions that took place at the volcano, for example, in 1957–1958. Stanley wrote:

> At about 2 p.m. large volumes of steam and black vapour ... soon obscured the sun, but the glow from the crater illuminated the lower clouds ... lava commenced to flow down the Eastern side and in about 5 minutes reached the sea ... causing the water to boil furiously [the 'lava' was probably a pyroclastic flow] ... When the rumbling ceased a grey-brown halo of dust encircled the mountain, which gradually spread for miles over the mainland. This dust was deposited to a depth of several inches at Potsdamhafen [on the mainland opposite] ...
>
> The [Manam] villages of Zogari and Josa were covered with dust, scoria, and hot fragments of vesicular lava. Gardens and houses were destroyed, and the natives for two or three days after had to place banana and breadfruit leaves on the tracks to prevent their feet from being burnt. At night it was possible to read a newspaper 8 miles away by the reflection of the lava on the clouds ... The natives of Manam consider these visitations are due to work of a bad ghost, and consequently they are continually making tambaran, i.e., a solemn dance accompanied by the beating of drums and the blowing of 6-ft bamboo pipes which they believe assists in appeasing the wrath of the ghost.[11]

Stanley, after such successful fieldwork, was beginning by 1923 to establish a reputation as a pioneer geologist in the New Guinea region, but on a visit to Adelaide in December 1924 accompanying his sick wife Helen, he contracted blood poisoning supervening on a facial pimple and died suddenly at the age of 39. The Territory of Papua Administration did not replace him.

The 13 years or so after Stanley's death are notable in the Near Oceania region for a deficiency of reports on volcanic activity, although not simply because of a lack of scientists, such as Stanley, to provide them. Reported eruptions seem to have been significantly less numerous than those reported in, for example, the late nineteenth century during German times, and the deficiency may indeed reflect a true situation of fewer eruptions during this particular period. But the low number of reports matches, curiously, a global trend in decreased reporting of volcanic eruptions that has been linked to the Wall Street market crash of 1929, the subsequent Great Depression of the 1930s, and to a worldwide

---

11   Stanley (1923), pp. 52–53.

preoccupation with economic and political anxieties rather than popular interest in volcanic eruptions.[12] Nevertheless, volcanology as a whole was advancing, and particularly so in relation to the origin of calderas, such as those found in New Britain, and to the discovery of rock type known as *ignimbrite*.

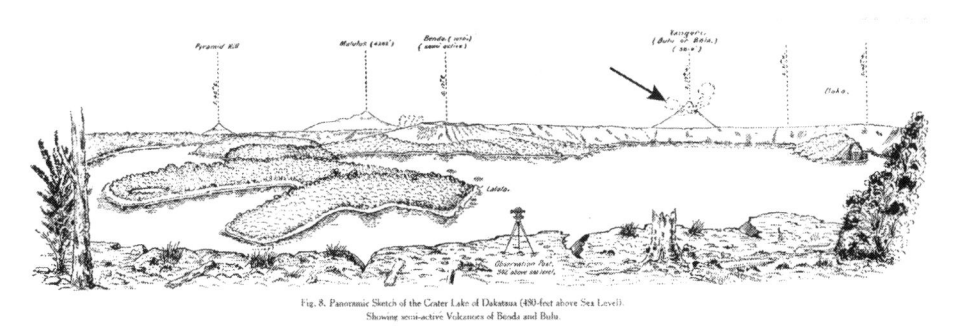

Figure 39. The caldera lake of Dakataua volcano is seen in this sketch made by Stanley in 1921 looking southwards from the north of the caldera. Makalia, or Benda, is a youthful volcano, left of centre within the caldera. The arrow on the right points to Wangore, Vangori, Bulu, or Bula volcano, which is well to the south of the caldera, and is shown emitting water vapour.

Source: Stanley (1923, Figure 8).

Calderas and Ignimbrites

A pioneering study on the origin of calderas was published in Dutch in 1885 by R.D.M. Verbeek, a mining engineer of the Netherlands East Indies, following his investigation of the catastrophic eruption at Krakatau volcano in 1883.[13] Verbeek proposed that a shallow and wide magma reservoir beneath some volcanoes is emptied so quickly during large explosive pumice eruptions that its roof collapses and disintegrates catastrophically. The formerly high 'ancestral mountain' of the volcano is engulfed and replaced by a low-rimmed but wide caldera at the surface. Other East Indies geologists and volcanologists who studied Krakatau subsequently — including the geodynamicist R.W. van Bemmelen and volcanologist Ch.E. Stehn who would later visit Rabaul — also supported this 'emptying out and breaking down theory'.[14] The theory eventually became more widely known to English-speaking geologists when an

---

12 Simkin & Siebert (1994).
13 Verbeek (1885). Simkin & Fiske (1983) provided an English translation of Verbeek's classic memoir on the occasion of the centenary of the 1883 eruption at Krakatau.
14 Bemmelen (1929; the quotation is from the English summary on p. 111) and Stehn (1929). Simkin & Fiske (1983) republished the Stehn paper in their book on Krakatau.

Fire Mountains of the Islands

American volcanologist of Welsh extraction, Howel Williams, in 1941 published the benchmark paper 'Calderas and their Origin', which included extensive discussion of many calderas of general 'Krakatau-type'.[15]

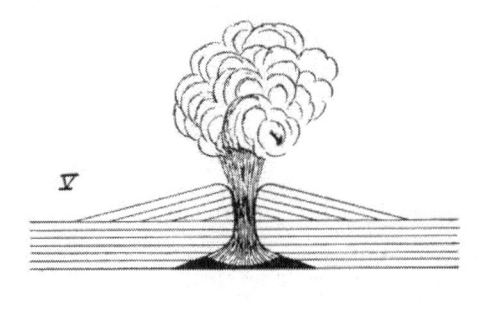

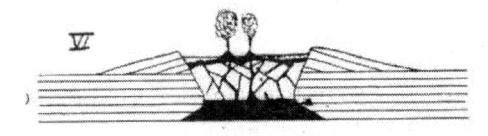

Fig. 2.
De vorming van een instortingscaldera volgens de leegblazings-instortings hypothese.

Figure 40. Magma in the shallow magma reservoir shown in the upper cross-section is erupted rapidly as a result of a powerful volcanic eruption, such as at Krakatau in 1883. This evisceration leaves the roof of the magma reservoir unsupported, and it collapses and disintegrates, as shown in the lower diagram. Subsequent geological studies of many eroded calderas have not revealed such strong disintegration of the roof rocks as shown in this cartoon, or in those shown by Howel Williams in his well-known studies of Krakatau-type calderas.

Source: Bemmelen (1929, detail from Figure 2).

A second important study on the origin of calderas was published in 1909 by three British geologists as the result of their detailed geological mapping of ancient volcanic rocks at Glen Coe in western Scotland. They proposed that a cylindrical mass of rock, several kilometres in diameter, had fallen vertically — piston-like — into an underlying reservoir of magma, which then 'welled up around the subsiding mass, like liquor in a full bottle when the stopper settles home'.[16] The upwelling magma, they suggested, produced eruptions from volcanoes arranged on the ring fault, and which filled the depression at

---

15 Williams (1941).
16 Clough et al. (1909), p. 664. Glen Coe perhaps is better known for the genocide in 1692 of members of the Maclain branch of the MacDonald clan by a Campbell regiment of government troops.

the surface — the caldera — with volcanic materials. This process was called 'cauldron subsidence'. Depressions that formed in this way later came to be called 'Glen Coe-type' calderas, and the published cartoon of the process became one of the most reproduced diagrams in the literature on volcanic geology, together with those used by Williams.

The two processes together — one based on the geology of the roots of an eroded volcano, the other on inferences from the surface on what had happened below — contain elements that are still used today in geological interpretations of the origin of calderas.

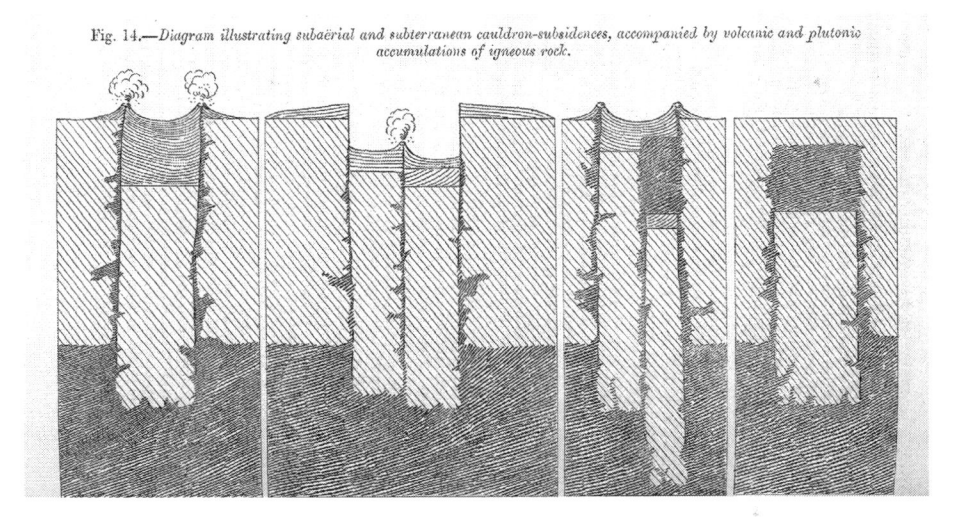

Figure 41. Volcanoes on a ring fault forming a Glen Coe-type of caldera are shown on the extreme left. Eruptions from a volcano in the centre of the caldera may later help fill it, as shown second from left. The roof above the rock cylinder on the extreme right is so strong that subterranean cauldron subsidence does not produce a caldera at the surface.

Source: Clough et al. (1909, Figure 14).

New Zealand was the site of another study that would have significant volcanological influence internationally. Pat Marshall mapped some of the geology of the Taupo Volcanic Zone in North Island and came across a distinctive type of volcanic rock that in 1935 he named 'ignimbrite'.[17] These rocks are characterised by distinctive, ubiquitous, disc-like blebs, some dark and glassy, which Marshall interpreted as collapsed pumice fragments. The fragments had been deposited from huge, pumice-bearing pyroclastic flows and subsequently squashed or flattened by the overlying weight of the deposit when the hot flow had come to rest. Ignimbrite or 'fiery igneous cloud' is similar to, and is as

---

17   Marshall (1935).

etymologically imprecise, as the French term nuée ardente or 'glowing cloud', but it has since been taken up with enthusiasm by volcanologists. 'Ignimbrite' is fundamentally a name for a pumice-bearing rock or deposit laid down by pyroclastic flows, but it is used also for the great volcanic-rock formations made up of the material, and even for the eruptive process that produced the deposit — 'ignimbrite eruptions'.

Extensive research has been undertaken on ignimbrites and their origin, and there is now general agreement that they are produced mainly by collapse of a plinian eruption column. The high-rising eruption 'stalk' of the plinian column loses its upward momentum — either because the gas pressure at the volcano's vent decreases or because the width of the vent increases by erosion caused by the inexorable expulsion of material — and the column collapses, crashing down onto the flanks of the volcano. Vast amounts of pumice, ash and gas then race outwards as gigantic and destructive pyroclastic flows, which may extend tens of kilometres from the volcanic vent. Not all of the flows are hot enough that when they come to rest they necessarily produce the flattened, disc-like, pumice fragments seen in the New Zealand examples. Ignimbrites are distinctive products of caldera-related eruptions, such as those that have taken place at Rabaul, Witori, Dakataua and Long Island.

Eruption at Rabaul in 1937

By May 1937 Rabaul had developed from its early German origins into a pleasant colonial town with shady streets, shops, sports facilities and commerce, which was dominated by the island trading companies of W.R. Carpenter and Burns Philp. It was, however, still strictly racially segregated and visibly European-controlled. The town and Territory were run by Australians for Australians, who were loyally British in their sentiments. Empire Day was celebrated on 24 May when Judge F. Beaumont 'Monte' Phillips reminded pupils at the Rabaul Public School of all the good things that the British Empire stood for. The worst of the Great Depression was over, but signs of war had again appeared following the rise of Nazism in Europe and the beginning of Japanese military expansion into east Asia.

Brigadier-General W. Ramsay McNicoll was Administrator of the Mandated Territory. He and his wife lived in Government House on Namanula Hill, where Albert Hahl had earlier enjoyed cool breezes, the westerly views of the town below and easterly ones across St Georges Channel. There was still no instrumental monitoring of the volcanoes at Rabaul, and other priorities had superceded any longer term concerns about volcanic hazards. This is perhaps

6. Calderas, Ignimbrites and the 1937 Eruption at Rabaul: 1914–1940

not surprising for Australians, as active volcanoes were unknown in Australia itself, and there had been no further volcanic activity at Rabaul since 1878 — that is, well before the Australian takeover of the former German protectorate.

A sharp earthquake shook Rabaul town at about 1.20 pm on Friday 28 May 1937. Rabaul residents later recalled that this marked the beginning of the lead-up to the 1937 volcanic eruptions at Rabaul, although tremors were said to have been felt days previously near Vulcan Island, site of the 1878 eruption.[18] Houses shook, trees swayed and crockery danced along the table on board the *Montoro*, which was discharging cargo at the Rabaul wharf. Damage and landslips were reported from out on the Kokopo Road at Karavia, near Vulcan, where the effects of the earthquake seemed greatest. Felt earthquakes became more numerous after five o'clock on the following morning, Saturday 29 May, and damage became more widespread. There were also reports of elevation and exposure of the sea floor at Vulcan Island, together with constant shaking, including at the slipway occupied by the steamer *Durour* on the nearby mainland.

A group of Europeans went out from Rabaul to investigate. Fish stranded on the exposed coral reef at Vulcan as a result of the uplift of the sea floor, were causing considerable interest, and were being collected by Tolai people from villages on the western shore of Blanche Bay, including many men who had by chance gathered at Tavana for an initiation ceremony of the *tubuan* male secret society.[19] The European party took a boat out to see the rising sea floor but hurried back to shore just in time:

> Right in our wake a blackish spiral of water was spouting 30 feet up … We had just passed over the spot … Then it started to crackle and explode and thunder, hurling up black stones and things, and was indescribable in its fury … [Minutes later the eruption was] in full force. Mountains of stones were being hurled thousands of feet high with deafening noise and sulphurous blinding smoke — black, black, black.[20]

The rapid rise of the Vulcan eruption column started at about 4.10 pm and provided little time for escape, especially from the villages close to the volcano. Pumice, gas and dust were being thrown upwards continuously from the new vent in typical plinian style. A surveyor later reported that the column reached a height of eight kilometres or more, but estimating such heights by eye from the ground nearby is notoriously difficult, and this minimum height could be an underestimate. It is, nevertheless, indicative that this plinian eruption was small compared to, say, those at Vesuvius in AD 79 or Krakatau in 1883.

---

18 Fisher (1939a) and Johnson & Threlfall (1985).
19 See, for example, Neumann (1996).
20 Chinnery (1998), pp. 210–11.

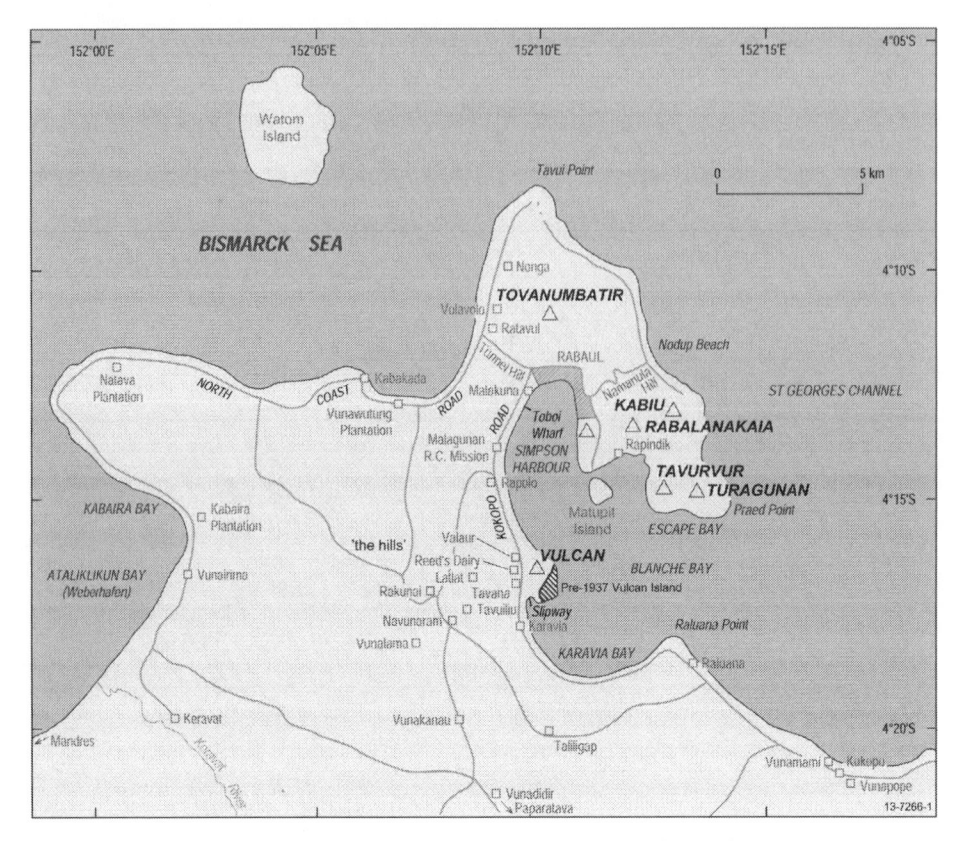

Figure 42. The main road system and settlements of the Rabaul area shortly after the 1937 eruption are shown in this sketch map. Open red triangles are volcanic centres.

Source: Adapted from Johnson & Threlfall (1985, figure on p. 5, based on pre-war sketch map).

May is at the beginning of the south-east or 'dry' season, so the winds bent the eruption cloud towards the north-west, dumping the pumice and ash onto villages and people below. Small pyroclastic flows also emerged from the base of the eruption column and almost certainly engulfed people in their path, but these flows were not identified as such until much later from photographs and by geological study of the Vulcan deposits. The eruption cloud also contained water, at least some of it drawn up from the sea, and heavy rains fell to the ground in the north-west, causing streams to flood. Those who escaped from the Vulcan eruption later gave vivid descriptions of the eruption and its effects, telling stories of terror, anxiety, and loss.[21] Father Nollen, for example, wrote of his experience of the Vulcan cloud at Malaguna on the north-west shore of Simpson Harbour. He thought about running away 'But the black wall came on so fast that it was no use to think of flight' and he feared for his life.[22]

---

21 Collections of stories were given by Arculus & Johnson (1981), Johnson & Threlfall (1985) and Neumann (1996).
22 Arculus & Johnson (1981), p. 29.

Figure 43. The image depicts the near-vertical south-eastern side of the sub-Plinian eruption column from Vulcan. The photograph seems to have been taken late in the afternoon of 29 May 1937. Fronds of coconut trees, collapsed by ash loading, and floating pumice on the harbour waters, are shown in the two smaller photographs.

Source: *Daily Telegraph*, Tuesday 8 June 1937. Australian Consolidated Press Limited.

Evacuations began when people throughout the Rabaul area realised the seriousness of their situation. Many escapees in the west moved towards shelter at settlements to the south and particularly to the Vunapope Mission at Kokopo. McNicoll was visiting the Morobe Goldfields on the mainland, Chief Justice D.S. Wanliss had been acting in charge at Rabaul, but was cut off from Rabaul by road, and so Judge Phillips took over as the officer in charge of the Administration. Phillips was 'a short sturdy man of great energy … he was in his element in a job such as this', wrote Patrol Officer J. K. McCarthy, who himself experienced the drama of the unfolding disaster.[23] People were moving spontaneously on foot and in vehicles out of the caldera using the obvious escape routes — up and over Tunnel Hill Road in the north-west, and eastwards onto Namanula Hill, site of the European hospital. Tunnel Hill evacuees who turned westwards along the north-coast road faired poorly as they had to endure ash fallout and mud rains from the Vulcan cloud.

Figure 44. Forked lighting illuminates the Vulcan eruption cloud in this photograph taken from the eastern side of Rabaul Harbour on the night of Saturday 29 May 1937.

Source: R. Davies. Johnson & Threlfall (1985, figure on p. 39). Geoscience Australia (G2008).

23   McCarthy (1971), p. 5.

The town itself did not at first receive any ash falls, as the Vulcan cloud was being blown north-westwards, but the cloud expanded as the eruption developed and local winds dumped a few centimetres of ash on Rabaul. Night fell and the many distressed evacuees in relative safety could see the ongoing eruption lit up by powerful and near-continuous lightning flashes within the cloud. Phillips was able to have messages transmitted that evening from the radio on board a copra ship, the *Golden Bear*, which was still anchored at a wharf in the harbour. He first sent news of the eruption to the Prime Minister's Department in Canberra, and later to the captain of the *Montoro*, which had left the harbour before the eruption, requesting that he return to Rabaul and standby.[24] A ship-borne evacuation was being organised.

Those who spent the night on Namanula Hill saw the extent of the ash fallout and damage to Rabaul town at dawn the next morning, Sunday 30 May. People streamed down the road from Namanula during the morning to assemble in their thousands at Nodup on St Georges Channel, where a flotilla of ships and boats was assembling to take off the evacuees. The *Golden Bear* had escaped safely from the harbour in morning daylight, dashing past Vulcan, which was still in eruption and building up a new volcanic cone. The ship reached Nodup where other vessels were gathering. *Montoro* arrived there in early afternoon, and thousands of people were ferried from the shore in small boats out to the larger, waiting vessels. The transfer was reasonably orderly although slow, but anxiety increased when the evacuees became aware of another dark eruption cloud towering up behind Kabiu peak immediately above them. An eruption had started at Tavurvur volcano at about 1.00 pm, in near-simultaneous activity with that at Vulcan — as in 1878. It was another 'double eruption'. The vessels made their way southwards towards Kokopo and the mission at Vunapope, which was becoming the main evacuation centre for the whole Rabaul area. The *Montoro*, carrying perhaps 5,000–6,000 people, stopped at night at the entrance to Rabaul Harbour, and Brett Hilder, an officer on board, recorded:

> Most of us stayed up to watch the satanic celebrations; the two volcanoes, on each side of the entrance, were throwing up a solid jet of red-hot dust and stones to a great height and the two columns appeared to meet somewhere over the town of Rabaul. The lightning was fantastic … . The noise of eruption and thunder, and the thump of falling rocks was continuous, while our mass of human cargo had plenty to look at while they stood packed on all the decks.[25]

---

24 Official government correspondence and other documents concerning the 1937 eruption are housed in files of the Territories Branch at the National Archives of Australia, Canberra.
25 Hilder (1961), pp. 54–55.

Heroic mission staff at Vunapope organised the care and shelter for the thousands of refugees now concentrated in the relative safety of Kokopo, site of the former German capital, and villages unaffected by the eruptions generously took in and fed those less fortunate. McNicoll returned from the mainland on Monday 31 May, flying in over the pumice-covered waters of Simpson Harbour to the new airstrip at Vunakanau. Australian newspapers published the Rabaul story that week as front-page news, and the editor of the *Rabaul Times* printed a special issue of his own newspaper at the Vunapope mission, detailing for residents the dramatic events of the past few days.[26]

The Administration began noticing a marked decline in activity at the two volcanoes and, within days, made a firm decision to return to Rabaul immediately, to begin cleaning up, and to encourage people to return there without delay. This prompted criticism from some of the European community, and certainly the bold decision — which turned out to be correct in the short term when eruptions ceased altogether at both volcanoes — was based neither on any scientific assessment of the potential of the volcanoes for further activity, nor on any thorough discussion on the suitability of the Rabaul site as a safe town or capital. The laborious clean-up began. Ash was removed from roofs and drains, water tanks and gutters were cleared, and health precautions were implemented. Damage to Rabaul buildings was relatively light, as the combined thickness of the pale pumiceous ash from Vulcan and the overlying dark, muddy, black ash from Tavurvur was only a few centimetres. But the ash had stripped protective vegetation from the slopes of the volcanoes and walls of the caldera, and heavy rains caused flooding in the town and the formation of deep gullies, adding to the problems of regenerating Rabaul.

The Administration's initial focus was directed mostly towards Rabaul's restoration and, evidently, much less towards determining the immediate fate of the villages close to Vulcan. Sir George Pearce, the Australian Minister for External Affairs, reported to the Australian Parliament as late as 18 June that 'The deaths of natives definitely known at present number less than twenty',[27] and the Administrator in Rabaul would not be able to inform him formally until mid-July that, in fact, hundreds had perished. The final statistic, published in 1939, was 505 villagers 'killed by being buried, crushed, asphyxiated or drowned', plus two Europeans.[28] McNicoll, in late June 1937, nominated ten Europeans for honours and decorations for their relief work during the disaster. Phillips was top of the list, and he eventually received a CBE, Commander of the Order of the British Empire. He would later become a key decision-maker in the days preceding the disastrous eruption at Lamington volcano in 1951.

---

26   Thomas (1937).
27   Pearce (1937), p. 44.
28   Fisher (1939a), p. 24.

Figure 45. Both Vulcan on the left and Tavurvur on the right are shown in reduced activity in this view from Taliligap looking northwards towards Rabaul, and photographed perhaps around 31 May 1937. Pumice covers the waters of the harbour. Vulcan has almost reached its full height of about 225 metres above sea level.

Source: Johnson & Threlfall (1985, figure on p. 85). Geoscience Australia (M2447-33A).

There were no awards for New Guineans. On the contrary, the Administration claimed that the new land that had formed at Vulcan was government-owned, even though survivors from nearby villages such as Valaur and Tavana had taken it over for traditional gardening soon after the eruption had ceased. Strong Tolai protests continued until after the Second World War in a dispute that reflected fundamental differences between European and traditional approaches to land-ownership laws. The Administration at one stage even issued a licence for a race course to be built on the new land at Vulcan. The dispute was not resolved until it was dealt with by the Supreme Court in Australia and settled in favour of the Tolai villagers.

Subsequent Investigations at Rabaul

The question of Rabaul's suitability as a capital town did not disappear and there began a series of investigations into its future. The first of these started on 10 July 1937 when Sir Raphael Cilento, tropical-medical expert and director of health and medical services for Queensland, visited Rabaul to deal with the immediate health concerns in the reoccupied town, as well as with medical aspects of any future developments involving the capital. Cilento urged a meticulous and intensified reintroduction of routine medical services, and commented on the psychological benefits to the community that had resulted from the Administration immediately reoccupying the town. He also addressed future medical needs and concluded his report with the principle that residential areas for white people should be set apart from the rest of the population, not only for health reasons but also for the physical safety of white women.[29] His proposal to settle Europeans on Namanula Hill was, however, impractical owing to the small amount of land available there.

Volcanologist Dr Charles E. Stehn, Director of the Netherlands Vulcanological Survey, and Dr W.G. Woolnough, Geological Adviser to the Commonwealth, arrived to advise the Australian Government on the suitability of Rabaul as the main administrative centre for the Mandated Territory. Woolnough was an Australian geologist, whereas the German-born Stehn had volcanological experience derived from studies undertaken, for example, at Krakatau, which could be applied to assessing the situation at Rabaul volcano.[30] These two senior men were joined by Norman H. 'Norm' Fisher, a young Australian who, since 1934, had been the Government Geologist of the Mandated Territory of New Guinea, based on the mainland at Wau on the Morobe Goldfields.[31] The three scientists undertook fieldwork together, Stehn and Fisher particularly building up an understanding of the past history of the Rabaul volcanoes. Fisher would play a major part in the aftermath of the 1937 Rabaul eruption and, indeed, would pioneer establishment of instrumental monitoring of the Rabaul volcanoes.

Stehn and Woolnough reached different conclusions as a result of their investigations, an unfortunate circumstance that did not help decision-makers in the Australian Government. Stehn, based on his experience with active volcanoes and instrumental monitoring in the Netherlands Indies, believed that a well-equipped volcanological observatory should be capable of providing warnings of impending eruptions sufficiently far in advance to enable the Rabaul population to be removed to places of safety. Woolnough on the other hand, stated bluntly that '… the advantages of retention of the capital in its present site and the provision of elaborate warning systems should not be entertained'.[32] He concluded that the required scale of such an observatory, the uncertainties of eruption prediction, the generally small industrial and commercial development in the town, and the fact that the harbour might become blocked

---

29 Cilento (1937).
30 Bemmelen (1949).
31 Wilkinson (1996).
32 Stehn & Woolnough (1937), p. 157.

by future eruptions anyway, were all reasons why an observatory should not be constructed at Rabaul and why the town should be abandoned as the capital. Woolnough's opinion, however, was ignored.

Instrumented observatories were, by 1937, a well-established feature of some of the world's best known volcanoes, the idea deriving from the much earlier concept of astronomical observatories where instruments were directed at the stars, rather than at the internal and surface behaviour of a volcanic Earth. The first volcanological observatories were built in Italy, at Vesuvius in 1841–1845, and at Etna in 1878–1881. The Americans had established an observatory at Kilauea volcano, Hawaii, in 1912, and Japanese observatories had been built on Aso by 1928 and at Asama by 1933. Furthermore, E.R. Stanley had drawn the attention of the Australian Government to the need for volcanological observatories in the Territory of New Guinea after his attendance at the Pan-Pacific Scientific Conference in Honolulu in 1920.[33] Stehn and Fisher were, therefore, building on something of a volcanological tradition.

Figure 46. Volcanologist C.E. Stehn is shown here in the foreground returning from the pumice-covered *Durour*, a vessel left stranded on a slipway south-west of the former Vulcan Island.

Source: N.H. Fisher. Johnson & Threlfall (1985, figure on p. 124). Geoscience Australia (no registered number).

33 Stanley (1923).

Other inquiries into the suitability of Rabaul as a capital followed the investigation by Stehn and Woolnough, but no definite decisions were made for four years. Furthermore, the idea of monitoring volcanic activity instrumentally at Rabaul for early warning purposes was being kept alive. Fisher began to spend more time at Rabaul after completion of the investigation and he visited Stehn in the Netherlands Indies in 1939 for volcanological training. Fisher, strongly influenced by Stehn's views, in 1939 published his conclusions on the geological history of the Rabaul area, promoting the view that a huge volcano — perhaps more than 2,700 metres high — or a group of somewhat lower cones, once covered the area now occupied by the Blanche Bay caldera. The smaller volcanoes of Kabiu, Tovanumbatir and Turagunan were simply 'parasitic cones' on the flanks of this great ancestral mountain.

Fisher wrote that 'A tremendous outburst or series of outbursts ... blew most of the mountain to fragments', scattering pieces of the central volcano around the countryside. This, he went on, was followed by subsidences around the periphery of the resulting explosion crater, forming the Blanche Bay caldera, and then a large eruption finally produced huge thicknesses of 'pumice ash'.[34] This interpretation by Fisher is not the mechanism of formation of Krakatau-type calderas, as promoted later by Howell Williams, but rather reflected the view, which was held by many geologists at that time, that calderas were the result of volcanoes that 'blew up', or exploded outwards, rather than collapsing wholesale into underlying magma reservoirs. Furthermore, Fisher and Stehn did not identify ignimbrites at Rabaul. Much later work would, however, reveal them and, indeed, a much more complex geological origin for Blanche Bay. Fisher also compiled at this time a volcano inventory for the Mandated Territory, in which the full number of young calderas in New Britain was recognised for the first time, including those at Pago and at Galloseulo where a 'crater lake' — the crater was in fact a caldera — had been recently reported by the pilot of an aircraft delivering mail.[35]

A routine of visual observations of the volcanoes and measurements of temperatures at hot springs and fumaroles around the harbour was established by Fisher at Rabaul, and funds then became available for construction of a small observatory and cellar for installation of equipment. These were built in 1940, high on the northern rim of the caldera — now called Observatory Ridge — below the south-western flank of Tovanumbatir. The observatory had spectacularly uninterrupted views overlooking the town towards Tavurvur, Vulcan and the other young volcanoes of Blanche Bay. Fisher had well-defined plans for the observatory to contain a German-made, two-component Wiechert

---

34 Fisher (1939a), p. 13.
35 Fisher (1939b). The crater lake is Lake Hargy.

seismograph, two tiltmeters of Italian manufacture for measuring changes in ground slope, an earthquake 'annunciator' that triggered an alarm when larger earthquakes took place, together with other ancillary equipment.

Figure 47. N.H. Fisher is seen here taking temperatures in an embayment on the northern side of Vulcan cone in August 1937. Temperature monitoring can, in some circumstances, be an effective way of determining whether a volcano is becoming restless or not.

Source: Johnson & Threlfall (1985, figure on p. 124). Geoscience Australia (GB-2552, detail from original photograph).

There were plans to establish observation posts in tunnels in the caldera walls near Tavurvur and Vulcan, and to equip these with tiltmeters and small pendulum devices — tromometers — to measure long-period ground motions, together with telephone lines to the central observatory.[36] Purchase of these instruments and funding to develop extra observation posts were severely curtailed by outbreak of the Second World War in Europe. Fisher finally had to design his own seismographs and annunciator for the Observatory Ridge cellar and have them built in Rabaul by the Public Works Department.[37] Nevertheless, the Rabaul Volcanological Observatory (RVO) had been established, and Fisher became the first volcanologist to be based permanently in the Near Oceania region, assisted at times during 1937–1942 by surveyor L.E. 'Les' Clout and by geologists C. L. 'Clem' Knight and L.C. 'Lyn' Noakes.

---

36 Fisher (1940).
37 N.H. Fisher (personal communications, 1982, 2006) and Johnson & Threlfall (1985).

Figure 48. The stark, utilitarian appearance of the original volcanological observatory at Rabaul, seen here sometime in 1940–1941, as well as a primitive set of monitoring instruments housed in the building's cellar, belies the historical importance of the observatory's construction and of its expectation of early warnings of volcanic eruptions at Rabaul.

Source: N.H. Fisher (likely photographer). Johnson & Threlfall (1985, figure on p. 130). Geoscience Australia (GB-1352).

Construction of the observatory at Rabaul in 1940 also marked, historically, the start of a new scientific era in using instrumental data to help understand the geophysical nature of the volcanoes in the region and to establish how eruptions might be forecast for the benefit of at-risk communities. It was, significantly and for the first time, an acknowledgement from government authorities that instrumental monitoring and early warnings of eruptions might make a difference in mitigating the effects of volcanic disasters in the Territory of New Guinea. The decades up to 1994 would be marked by a history of development of the work of RVO scientists and technicians, and of international volcanology in general. RVO would face several challenges, but particularly in 1994 when there would be a repeat of the near-simultaneous eruptions that had taken place in 1937 at Tavurvur and Vulcan.

References

Anonymous, 1917a. 'Earth Tremor', *Rabaul Record*, 2, no. 2, pp. 3–4.

Anonymous, 1917b. 'More Earth Tremors', *Rabaul Record*, 2, no. 3, p. 3.

Anonymous, 1932. 'No Scientific Records Kept of Rabaul's Quakes', *Pacific Islands Monthly*, 3, no. 5, p. 26.

Arculus, A. & R.W. Johnson, 1981. *1937 Rabaul Eruptions, Papua New Guinea: Translations of Contemporary Accounts by German Missionaries*. Bureau of Mineral Resources, Canberra, Report 229.

Bemmelen, R.W. van, 1929. 'Het caldera probleem', *De Mijningenieur, Vereeniging van Ingenieurs en Geologen*, Bandoeng, 10, no. 5, pp. 101–12.

——, 1949. 'Charles Edgar Stehn (1884–1945), Nécrologie', *Bulletin Volcanologique*, 8, pp. 133–37.

Chinnery, S.J., 1998. *Malaguna Road: The Papua and New Guinea Diaries of Sarah Chinnery*. National Library of Australia, Canberra.

Cilento, R., 1937. 'Report on the Medical Significance of the Recent Eruption in Blanche Bay, Territory of New Guinea'. National Australian Archives CRS A518, Item X836/4.

Clough, C.T., H.B. Maufe & E.B. Bailey, 1909. 'The Cauldron-subsidence of Glen Coe, and the Associated Igneous Phenomena', *Quarterly Journal of the Geological Society*, 65, pp. 611–78.

Cooke, R.J.S., 1981. 'Eruptions at Pago Volcano, 1911–1933', in R.W. Johnson (ed.), *Cooke-Ravian Volume of Volcanological Papers*. Geological Survey of Papua New Guinea Memoir, 10, pp. 135–46.

Cummins, J.J., 1917. 'Early Experiences in New Britain', *Rabaul Record*, 2, no. 7, pp. 9–10.

Davies, H.L., 1987. 'Evan Richard Stanley, 1885–1924: Pioneer Geologist in Papua New Guinea', *BMR Journal of Australian Geology and Geophysics*, 10, pp. 153–77.

Fisher, N.H., 1939a. 'Geology and Vulcanology of Blanche Bay, and the Surrounding Area, New Britain', *Territory of New Guinea Geological Bulletin*, no. 1.

——, 1939b. 'Report on the Volcanoes of the Territory of New Guinea', *Territory of New Guinea Geological Bulletin*, no. 2.

——, 1940. 'Note on the Vulcanological Observatory at Rabaul', *Bulletin Volcanologique*, 6, pp. 185–87.

Gould Fisher, F., 1994. *Raphael Cilento: A Biography*. University of Queensland Press, St Lucia.

Hilder, B., 1961. *Navigator in the South Seas*. Percival Marshall & Co., London.

Johnson, R.W. & N.A. Threlfall, 1985. *Volcano Town: The 1937–43 Rabaul Eruptions*. Robert Brown and Associates, Bathurst.

Mackenzie, S.S., 1987 (1927). *The Australians at Rabaul: The Capture and Administration of the German Possessions in the Southern Pacific*. University of Queensland Press, St Lucia, in association with the Australian War Memorial, Canberra.

Marshall, P., 1935. 'Acid Rocks of the Taupo-Rotorua Volcanic District', *Transactions of the Royal Society of New Zealand*, 64, pp. 1–44.

Massey, Captain C.H., 1918. 'Vulcanicity of Rabaul District', *Rabaul Record*, 3, no. 1, pp. 8–10.

——, 1923. 'Notes on the Physiography of Eastern New Guinea and Surrounding Island Groups: With Special Reference to the Volcanic Features of the Rabaul District of New Britain', *Proceedings of the Royal Society of Queensland*, 35, pp. 85–108.

McCarthy, J.K., 1971. 'When Matupit Blew, It Was Time to Go', *Post Courier* (Port Moresby), 20 July, p. 5.

Neumann, K., 1996. *Rabaul Yu Swit Moa Yet: Surviving the 1994 Volcanic Eruption*. Oxford University Press.

Pearce, G., 1937. 'Volcanic Disturbances at Rabaul', Commonwealth of Australia, Parliamentary Debates, Session 1937 (1 GEO.VI), 153, pp. 43–44.

Simkin, T. & R.S. Fiske, 1983. *Krakatau 1883: The Volcanic Eruption and its Effects*. Smithsonian Institution Press, Washington D.C.

Simkin, T., & L. Siebert, 1994. *Volcanoes of the World: A Regional Directory, Gazetteer, and Chronology of Volcanism during the Last 10,000 Years*. Geoscience Press, Tucson.

Stamm, J., 1930. 'Wie die Fischreusen mich beinahe um meinen ersten Tauftag in Nakanai brachten', *Hiltruper Monatshefte*, 47, pp. 85–91.

Stanley, E.R., 1923. 'Report on the Salient Geological Features and Natural Resources of the New Guinea Territory including Notes of Dialectics and Ethnology'. Report on the Territory of New Guinea, Commonwealth of Australia Parliamentary Paper 18 of 1923, Appendix B.

Stehn, Ch. E., 1929. 'The Geology and Volcanism of the Krakatau Group', *Proceedings of the Fourth Pacific Science Congress, Batavia*, pp. 1–55.

Stehn, Ch.E. & W.G. Woolnough, 1937. 'Report on Vulcanological and Seismological Investigations at Rabaul', Commonwealth of Australia Parliamentary Paper 84 of 1937, pp. 149–58.

Thomas, G., 1937. 'Our Volcanic Issue', *Rabaul Times*, no. 633, Friday 4 June.

Verbeek, R.D.M., 1885. *Krakatau*. Batavia.

Wilkinson, R., 1996. *Rocks to Riches: The Story of Australia's National Geological Survey*. Allen & Unwin, St Leonards.

Williams, H., 1941. 'Calderas and their Origin', *Bulletin of the Department of Geological Sciences*, 25, pp. 239–46.

7. Eruptions during the Pacific War and Postwar Recovery: 1941–1950

It was a most awe inspiring sight to see this great mass of smoke, a greyish orange and purple colour just billowing overherad [sic] and sort of rolling over itself and accompanied by the sound of the roar of hundreds of mighty furnaces.

Dennis Taylor (1943)

Fisher and Renewed Activity from Tavurvur

Concerns expressed in early 1941 about future volcanic activity at Rabaul were to a large extent set aside by the greater anxiety of a likely war in the Pacific. The Second World War had broken out in Europe in September 1939, and Japan in 1941 was expanding militarily in the western Pacific and eastern and southeast Asian regions. Nevertheless, a small but vulnerable Australian garrison, Lark Force, was dispatched to Rabaul in March–April 1941.[1] It was made up largely of soldiers of the 2/22nd Battalion of the 23rd Brigade, 8th Division, 2nd Australian Imperial Force. The battalion was supplemented by smaller units, one of them the New Guinea Volunteer Rifles which was made up of white residents of Rabaul, including volcanologist N.H. Fisher. Consideration was given by the Australian Government to strengthening Rabaul and its magnificent caldera harbour as a major military base in view of the Japanese threat, but this required the resources of the United States, which had not yet entered the war.

Tavurvur and Vulcan volcanoes had not been active since 1937. Fisher, however, had been noticing geothermal changes on parts of Tavurvur since August 1940, and a series of tectonic earthquakes had been taking place in the region — including a particularly powerful one just 30 kilometres west of Rabaul on 14 January 1941— which Fisher related to the increased volcanic unrest.[2] Vigorous gas emissions were evident in growing fumarolic areas on Tavurvur and temperatures by April 1941 were in excess of 300 °C. These are all signs that a volcano may be about to become active. Gas was being expelled with a 'roar like a train' by 27 May, and then Tavurvur broke out into full explosive eruption on 6 June, the first of many, but intermittent, eruptions that would last until December 1943. Fisher's systematic tracking of the temperature rises at Tavurvur is an excellent example of how simple instrumental monitoring of

---
1 Wigmore (1957), Nelson (1992) and Stone (1994).
2 Fisher (1944, 1976).

Fire Mountains of the Islands

volcanoes — in this case using hand thermometers inserted into the ground to measure temperatures — can be taken as a warning of impending eruptive activity.

Figure 49. Tavurvur volcano is shown producing a small vulcanian eruption in this photograph, which is believed to have been taken on 25 June 1941. Similar clouds were emitted from Tavurvur until after the Japanese invasion of 23 January 1942.

Source: E.A. Hawnt. Johnson & Threlfall (1985, figure on p. 142). Geoscience Australia (M2447-11A).

The eruptions during 1941 produced spectacular night-time displays of incandescent strombolian-vulcanian type. Dust and ash that periodically swept over Rabaul town during the south-east season were a considerable nuisance to Rabaul residents, but the town was not affected significantly by the new eruptions, which were generally small.[3] They were, nevertheless, a further reminder of how unsuitable Rabaul town was as a centre of Australian administration, and indeed as a site for any expansion as a major military base. Discussions followed between the Australian Government in Canberra and the now knighted Administrator, Sir Walter McNicoll. A proposal to transfer the Territory's capital to Lae on the New Guinea mainland was approved by the Australian Cabinet in the first week of September.[4] The Australian flag was lowered on Namanula Hill in late November when the Administrator and his wife moved to the mainland and the new capital. Shifting all of the Administration's bureaucracy from Rabaul, however, was no simple task and was expected to take several months to complete, particularly as wartime budgets were austere. The full transfer in any case was interrupted by the Pacific War.

Japan attacked Honolulu, Hawaii, on 7 December 1941[5] and Japanese forces advanced southwards towards the Territory of New Guinea in January 1942. Rabaul was bombed in early January and then again on the 20 January in advance of a land invasion by the Japanese Nankai Shitai, or South Seas Force, early on the morning of 23 January.[6] The under-resourced Lark Force was soon overwhelmed, the Europeans dispersed or captured, and the Japanese took possession of Rabaul even as Tavurvur continued its period of eruptive activity. Fisher, Clem Knight and others escaped capture by walking southwards to the south coast of New Britain, commandeering a small boat, and sailing to Australia-occupied Papua. Australian volcano monitoring at Rabaul was thus abandoned, at least for the time being.

Kizawa and the Sulphur Creek Observatory

The Japanese navy immediately began developing Rabaul as a military base for its planned advances further south, including the proposed capture of Port Moresby, initially by sea, but when that failed, by sending land troops southwards over the Owen Stanley Range from the Gona area on the north coast of Papua. The navy took advantage of the large, deep, caldera harbour and the commercial wharfs of Rabaul town. Simpson Harbour became a major

---

3   Johnson & Threlfall (1985).
4   See National Archives of Australia (1941) for official correspondence.
5   8 December, New Guinea time.
6   Wigmore (1957), Nelson (1992), and Stone (1994).

fleet anchorage, including a submarine base and seaplane mooring, as well as a deep-water port for major shipping, all protected by six nearby airfields and anti-aircraft batteries.

The Japanese military authorities soon recognised the volcanic risk to its operations in and out of the harbour at Rabaul, particularly from Tavurvur, the eruptions of which had welcomed the occupying force.[7] Japan has a long documented history of domestic volcanic disasters and, by 1942, it also had a long history of instrumental monitoring of its active volcanoes at home. The response of the Japanese military to dealing with the volcanic threat at Rabaul was therefore much swifter than had been that of the Australians and Germans before 1937. A request was sent to Tokyo for the secondment of a volcanologist to serve in Rabaul, and Dr Takashi Kizawa, a civilian seismologist with the Central Meteorological Observatory in Japan, arrived in Rabaul in May 1942 to begin monitoring of the volcanoes.

Kizawa spent his first few weeks at Rabaul familiarising himself with the volcanoes and attempting to find any information about them that might have been left behind by the Australians.[8] He received an order on 10 June to report by 10 August on the feasibility of forecasting eruptions and likely volcanic disasters at Rabaul. Kizawa recommended that a volcano observatory be established on the northern side of Sulphur Creek, less than three kilometres from Tavurvur, but was hesitant in promising a reliable eruption-prediction capability. Nevertheless, a large building for the observatory, together with a new wooden gate and the sign K Corps Volcanological Research Institute, in Japanese, was erected at Sulphur Creek in very short time. The observatory was equipped with two seismographs — a Japanese-made two-component Omori and a German three-component Wiechart — together with two tromometers. Kizawa later redesigned the Omori to measure not just ground vibrations from earthquakes but also changes in the slope of the ground at Sulphur Creek, by slowing down the speed of the drum recorder such that it made one revolution only every seven days. This allowed him to interpret the tilting of the seismograph itself and so conclude that the tilt of the ground at the observatory was being caused by Tavurvur volcano.[9]

The Japanese advanced rapidly during early 1942 to occupy virtually all of the Australian Mandated Territory of New Guinea as well as the main islands of the

---

7 The Japanese produced film footage of Tavurvur in eruption, evidently to satisfy the interest of the Japanese public. Clips of the footage were included in a video produced by Film Australia (1993).
8 Nitta (1980).
9 Kizawa (1951, 1961) and Kusaka (1976). Two-component seismographs, such as the Omori, produce twin traces by measuring an earthquake's ground vibrations on two horizontal parts pre-set at 90 degrees to each other. A three-component seismograph such as the Wiechert has an additional third part that measures the earthquake in the third dimension. Differences between the different traces are indicative of the character of any particular earthquake — its size, closeness, direction of travel, and so forth.

7. Eruptions during the Pacific War and Postwar Recovery: 1941–1950

British Solomon Islands Protectorate, but they met resistance later in the year from the Allied forces at the Battle of the Coral Sea, the landings at Milne Bay, and on the Kokoda Trail just west of Lamington volcano. The Japanese were more successful during the Battle of Savo Island in the Solomons on 8 August 1942 when they inflicted considerable losses on the American and Australian navies, including the sinking of the Australian cruiser HMAS *Canberra*. The inactive Savo volcano lay squarely in the channel battleground between Florida Island and Guadacanal, where the strategically important Henderson Airfield was being defended by the Americans.[10]

Figure 50. Seismologist Takashi Kizawa is here working with seismographic equipment at the Sulphur Creek Observatory, Rabaul, probably in 1943.

Source: Kusaka (1976, one of several unnumbered photographs between pp. 160 & 161). Digital copy provided by Y. Nishimura, Hokkaido University.

Kizawa did not witness any eruptions from Tavurvur during his first 18 months at Rabaul, but from 11 October 1943 onwards he recorded a series of earthquakes that culminated in a strong, sharp earthquake beneath Rabaul on 16 October, followed by aftershocks, and also detected the changes in slope of the ground at the observatory using the Omori instrument. He thought that the earthquakes and ground tilt were a reflection of changes within Tavurvur itself and that all the energy in the volcano had not been expended by the earthquake on

---

10   McCarthy (1959).

the 16th. This, he concluded, was corroborated when Tavurvur broke out into explosive activity on 24 November at the start of a period of activity that lasted about a month.[11] The eruptions were small, the winds of the north-west monsoon drove the ash mainly away from the town, and the Japanese forces were not significantly affected by them. These were the last volcanic eruptions to take place at Rabaul for another 51 years. Kizawa's work was the first to indicate that a combination of seismology and ground-tilt measurements might be useful in forecasting eruptions at Rabaul, complementing Fisher's demonstration of the importance of ground-temperature monitoring.

Japanese troop numbers at Rabaul reached almost 100,000 in 1943, and intensive bombing of the Japanese base by Allied aircraft began in October of that year.[12] Ground shaking from bomb impacts complicated the seismograph traces being collected by Kizawa, and instrumental observations became even more difficult when the bombing reached its peak in February 1944. Rabaul became encircled militarily in April, and the Japanese forces were blocked and bypassed as the Allies swept northwards towards the Philippines. The observatory at Sulphur Creek was in a conspicuous position for Allied bombers approaching Rabaul from the south, scientific monitoring there became impossible and the observatory was eventually destroyed by air strikes. Bombs were also dropped by Allied aircraft into the craters of Tavurvur in the false hope of triggering a volcanic eruption. This was witnessed by Kizawa himself who, rather than being bemused, was impressed by the inventiveness and determination of the Allies.[13]

The Japanese who were trapped at Rabaul endured the rest of the Second World War in an extensive and ingeniously constructed network of tunnels dug into the pumice deposits — particularly the ignimbrites — of Rabaul volcano, and which they used successfully for protection against the ongoing air attacks. The town, and surface installations in general, were being systematically razed by the Allied bombing raids, and so digging underground facilities was the only sensible strategy for Japanese survival. More than 500 kilometres of tunnels had been constructed by the end of the war, ranging from simple dugouts and caves to labyrinths of timber-supported roofs and walls for use by virtually all of the Japanese resources — from personnel, to barges, to gun emplacements.[14] Particularly significant features volcanologically were the caves of a Japanese submarine base that were dug into the cliffs on St Georges Channel south of Cape Tavui, and linked to a concealed railway serving Rabaul.[15] The submarines were able to gain close and unseen access to the coastal caves by means of deep

---

11 Kizawa (1951).
12 United States Strategic Bombing Survey (1946) and Stone (1994).
13 T. Kizawa (personal communication, 1981).
14 United States Strategic Bombing Survey (1946).
15 Stone (1994).

7. Eruptions during the Pacific War and Postwar Recovery: 1941–1950

water close to the shoreline. The precipitous sea floor here is now known to be the upper part of the western wall of the 10-kilometre-wide Tavui submarine caldera which, however, would not be identified as such until 1985.

Figure 51. Rabaul is here being attacked from the south by Allied bombers on 2 November 1943. The Japanese volcanological observatory can just be made out on the far bank of Sulphur Creek, the narrow inlet of water in the foreground, as shown by the arrow.

Source: Australian War Memorial (AWM 100146).

Kizawa also retreated underground. The destroyed Sulphur Creek Observatory was useless for any further volcano-monitoring work and, in February 1945, Kizawa constructed the beginning of an underground observatory near Latlat village in the caldera cliff overlooking Vulcan, where he thought that there was a smaller risk of being bombed.[16] A simple seismograph was built by assembling pieces from the instruments at Sulphur Creek. Kizawa made plans to extend the tunnel and live there, but was prevented from doing so by food shortages. The Second World War ended in August 1945 and Kizawa attempted to continue his scientific work at Vulcan, but in October he was interned in Liguan Group Camp

---

16   Kizawa (1961).

prior to repatriation to Japan the following year. Kizawa posted messages on the door to the Vulcan observatory, which contained the still-operating seismograph and other instrument pieces, before leaving it for the last time:

> For the coming generation, these machines have wrought each function to the civilised world and progress of the scientific world during the war ... I offer them to the people of the future ... The machines in this room should be presented to the Australian Corps ...[17]

Eruptions at Goropu Volcano, Papua

An unexpected and still little-known volcanic eruption surprised Australian military personnel who were stationed in the Tufi area of south-eastern Papua in 1943–1944. Four columns of 'smoke' and ash were seen by coastal villagers in October 1943 in the hills south of Tufi and Victory volcano, as well as other unexpected clouds in November.[18] Earthquakes had been felt in the Tufi area during the previous two years. Rumours spread amongst villagers that the Japanese had an aerodrome in the hills, as suggested by aircraft diverting over the smoking area. Japanese forces in mainland New Guinea by this time, however, had been forced northwards back over the Owen Stanley Range in their failed bid to take Port Moresby. Fighting was continuing in the ranges and coastal areas of New Guinea, but not in south-eastern Papua.

Australian civil administration had ceased in February 1942 and the Australian New Guinea Administrative Unit, ANGAU, controlled the Australian territory, which was now being increasingly abandoned by the Japanese, including the Tufi area. The 'smoke' in the hills south of Tufi was in fact from volcanic vents in a gently sloping area known locally as Waiowa at the northern foot of the Goropu Mountains in the Owen Stanley Range, from where eruptions had never been reported previously and where no volcano was known to exist. The aircraft were not Japanese and their pilots were flying over the volcanic area simply out of curiosity.

A much larger eruption took place on 27 December 1943, producing a 5,000-metre-high cloud of volcanic ash that caused great concern at Wanigela on the coast to the north-north-east:

> People were screaming and running in all directions and kids and babies were crying and wailing, [and] the place was in a panic. ... It was a most awe inspiring sight ... No wonder the people were frightened. Jack and I were scarred [sic] stiff.[19]

---

17 Kizawa (1961), p. 3.
18 Marsh (1944).
19 Taylor (1943), p. 1.

There were additional eruptions on 13 February and 23 July 1944 and a fourth and final one is thought to have taken place on 31 August. The exact number of eruptions during the entire eruptive period, and any significant differences between them, however, are unknown because no systematic visual observations were made and the eruptions were not studied at the time by either volcanologists or geologists.[20] Occasional ash falls that affected Port Moresby, Lae, and Tufi were light, and there were no reported deaths or significant damage to coastal villages and gardens closest to the active area. Nervous villagers nevertheless evacuated after the December eruptions to the larger settlements of Tufi and Wanigela, where there were still some Europeans, and from where the villagers were later supplied with emergency provisions by the ANGAU administration. This represents, even though on a small scale, the first example of any colonial authority in the region showing primary concern for the safety of Melanesians under threat from an active volcano, by supporting them with emergency supplies after they had left their homes and gardens. This contrasts with the case of Rabaul in 1937, when the welfare of the European citizens of Rabaul town was the primary concern.

ANGAU officers also visited the volcanic area at Goropu from time to time, and some photographs were taken from the air by others on different occasions — including shortly after the war — so the general character of the eruption can be assessed from these sources, particularly the ANGAU field reports. The Goropu area today is shrouded by vegetation, except for a 200-metre-wide crater lake or explosion pit — technically known as a **maar**. The crater looks, from the air, like a water-filled hole in the ground although in fact it occupies the summit of a low-angle cone representing the build-up of new volcanic material around it. The crater or pit is one of several vents that were active at Goropu, or Waiowa, volcano at different times, although the exact number is unclear, and it appears to have been the only crater active at the end of the eruptive period in 1944.

Maars form typically as a result of **hydrovolcanic** eruptions where small amounts of magma rise to the surface and encounter water trapped below the ground. The explosions that result from contact between hot magma and groundwater are particularly violent and cause both rapid break-up of the magma itself as well as the violent expulsion upwards and outwards of different materials from the vents. Some of the material may be non-volcanic, as in the case of Goropu where pieces of ancient metamorphic rocks and fragments of an old sea floor forming the geological core of the Owen Stanley Range, were flung out. Hydrovolcanic eruptions also form water vapour-rich clouds that rise in great columns, as well as lateral or directional 'blasting' of the immediate area.

---

20   See, however, Baker (1946).

Figure 52. Villagers from Uiaku stand on and near large volcanic bombs in the devastated area at Goropu volcano, probably on 4 December 1943. Trees have been stripped of branches and foliage, an active crater can just be seen in the middle background, and a depression or old riverbed in the foreground, has silted up.

Source: D.R. Marsh.

A former ANGAU officer, Lieutenant D.R. Marsh, noted that after each eruption witnessed by him in the crater area 'the build-up of dust around the crater wasn't very high, but it blew itself off each time and sort of almost flattened it out until perhaps [after] the last one or two explosions, it built up a bit'.[21] Other evidence for this hydrovolcanic blasting at Goropu volcano include stripping of vegetation from trees, snapping of trees trunks, the blowing down of trees, pitting of tree trunks by volcanic material 'as though with a heavy machine gun',[22] bubbling of mud in the active craters, mud rains, surface flows of mud, changes in surface-water drainages, and silted streams. The extent to which the volcano 'dried out' towards the end of the eruptive period is unknown and so there is some uncertainty whether hydrovolcanic eruptions continued up to the end of the active period. No lava flows emerged from Goropu, however, although there were unconfirmed reports of crater glow being seen from the air during the day.

---

21 D.R. Marsh (personal communication, 2005).
22 Marsh (1944), p. 2.

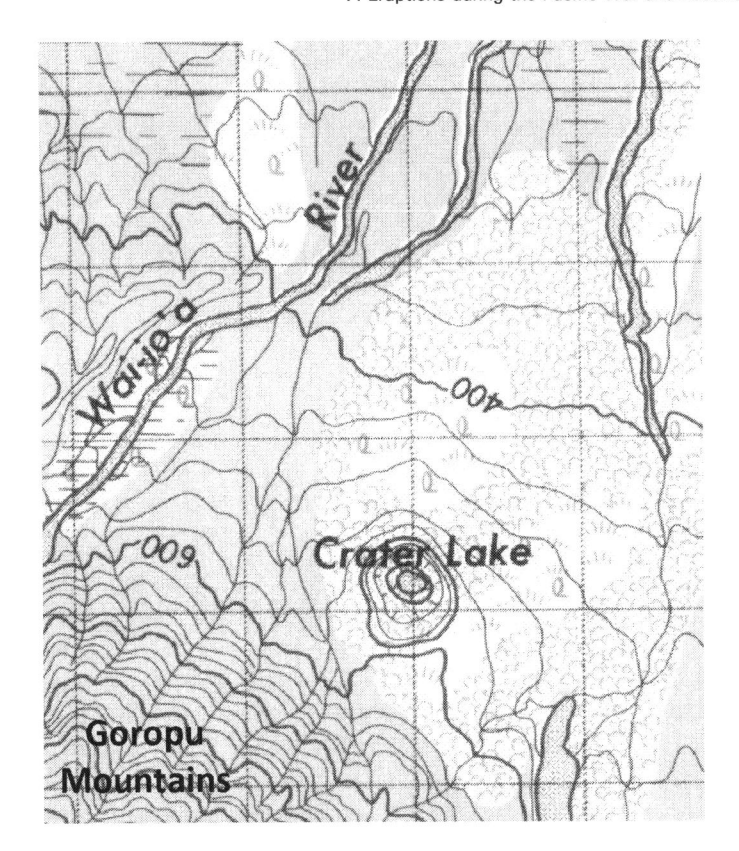

7. Eruptions during the Pacific War and Postwar Recovery: 1941–1950

Figure 53. Goropu volcano is shown here in this detail from a 1973 topographic map, together with its single crater lake and low-angle cone. The grid has spacings of one kilometre.

Source: Royal Australian Survey Corps, Sheet 8778 (edition 1), Series T 693.

The new volcanic area that formed at Goropu in 1943–1944 is separated from the much larger Victory volcano to the north, which was previously active in the late nineteenth century and which some Europeans suspected initially must have been the source of the new eruptive activity. There are, however, small areas of young volcanic rocks and hot springs that are strung out to the west of Goropu[23] as far as the Managalase Plateau immediately south-east of Mount Lamington, where there are many small volcanic cones, craters and maars. Most of these other areas, however, were not discovered until after the war, but together — Goropu included — they may represent a discontinuous volcanic field or series of 'fields' stretched out along the slopes of the Owen Stanley Range. Volcanic fields are characterised by numerous — hundreds in some cases — but generally small, so-called 'monogenetic' volcanoes — that is,

---

23 Smith (1981) referred to the youthful volcanic nature also of the small Sessagara Hills, including Mount Maisin, just a few kilometres to the north-east of Goropu volcano.

volcanoes like Goropu which form during short eruptive periods and may never again become active. Conduits that delivered magma to the surface become solidified and new magma batches break out elsewhere in the 'field' forming a new eruptive centre.

Goropu is not the only historically active volcano along the northern fall of the Owen Stanley Range. A small, **basalt** cinder cone called Kururi on the Managalase Plateau is thought to have been active sometime in the first half of the twentieth century, although even less is known about this eruption than the one at Goropu. An Australian geologist mapping the Managalase area in 1963 heard stories from elderly men at Kuriri village who recalled witnessing 'fire and smoke', perhaps between the two world wars.[24]

Hiroshima, Surges and Postwar Recovery

The 1941–1943 eruptions at Tavurvur, Rabaul, and the explosive activity at Goropu in 1943–1944 dominate the known history of eruptive activity during the Pacific War, and comparatively little is known about eruptive activity at other volcanoes. This deficiency may represent another example of attention being directed away from the reporting of volcanic eruptions because of the major global distraction of a world war, just as seems to have happened during, and indeed continuing from, the Great Depression.[25] There appear to be no records of wartime eruptions even at Manam, although the volcano may well have continued its activity, presumably at a fairly low level. Some minor activity took place at Ulawun,[26] and lava must, as usual, have been flowing on Bagana, judging by a comparison of wartime aerial photographs.[27] Furthermore, a newly formed volcanic island in the caldera lake at Long Island was seen on aerial photographs taken in 1943. The island had been absent when Fisher saw the lake in 1938, but villagers — after further eruptions at Long in 1953 — reported there had been minor eruptions in the lake in 1933, 1938 and 1943.[28] The reporting for these exceptions, however, is not comprehensive.

The Pacific War came to an end on 15 August 1945 following the atomic bombings of Hiroshima and Nagasaki, Japan, on 6 and 9 August. Newsreels and photographs of the high-rising, eruption-like, 'mushroom' clouds were dramatic imagery, marking symbolically the beginning of the atomic and nuclear eras, including the further testing of nuclear weapons and the beginning of the

---

24   Ruxton (1966), Ruxton et al. (1967) and B.P. Ruxton (personal communication, 2006).
25   Simkin & Seibert (1994).
26   Cooke (1981).
27   Bultitude (1979).
28   Best (1956).

7. Eruptions during the Pacific War and Postwar Recovery: 1941–1950

Cold War. The United States began atmospheric tests in the shallow waters on Bikini Atoll in 1946 and close-up observations were made there of the character of the explosions. Striking features of the test explosions were not only the radioactive clouds mushrooming into the sky but also a ring-shaped basal cloud that swept outwards at an estimated initial velocity of more than 50 metres per second. Atomic-test scientists coined the term *base surge* for these rapidly expanding and surface-hugging clouds and attributed them to development of an increasingly widened explosion crater and progressive destruction of the explosion-crater rim. Volcanologists would later adopt this concept of 'base surge' in identifying similar, flat, disc-like clouds in hydrovolcanic eruptions taking place just below the surfaces of seas or lakes.[29] Groundwater-driven surges almost certainly caused much of the damage at Goropu in 1943–1944, although they were not recognised as such at the time.

Figure 54. An example of a base surge is seen in this photograph taken in May 1955 from the western side of the caldera rim on Long Island, about five kilometres away from the active volcano. The surge is radiating out from the volcanic vent across the surface of Lake Wisdom. Tolokiwa Island can be seen in the far distance, beyond the eastern rim of the caldera.

Source: G.A.M. Taylor. Geoscience Australia (GA-9336).

ANGAU passed Australian administration of the two, formerly separate, prewar territories to a single civil-government authority of a new Territory of Papua and New Guinea in the months after the formal Japanese surrender in New Guinea

---

29 Moore (1967).

on 13 September 1945 and after approval by the newly formed United Nations.[30] The Australian military had demonstrated that a combined administration was possible. The new Administrator of the Territory appointed under the Australian Labor Government in 1945, was Colonel J.K. Murray, previously principal of the Queensland Agricultural College. Murray was based in the new capital of Port Moresby in Papua, rather than at Lae which had been the prewar choice of a capital for the Mandated Territory. The new capital certainly would not be at Rabaul town, which had been razed by Allied bombing. The Australian military in 1945 had begun erecting its temporary accommodation at Rabaul, but the town remained a shambles for many years. One Administration officer in 1947 wrote:

> All Administration buildings are temporary and insecure against theft, and the living conditions of the staff is [sic] of the worst .... The Chinese Community of about 1500 souls are living in hovels which never would be tolerated in ordinary times .... The conditions existing in Rabaul is [sic] most depressing. The sanitary arrangements are deplorable, the old Army type of pit latrines are still being used and the water supply is being obtained from wells .... Flies and mosquitoes are becoming very troublesome. On 1st January Army withdrew their Malaria Control and Sanitary squads from the township and I fear for the health of the community. ... The foreshores of the harbour are littered with wrecks of all kinds.[31]

One reason for this situation was lack of funding in the immediate postwar environment for infrastructure reconstruction in the town, but another more significant one was ongoing and contentious discussion on whether Rabaul should be established in the same place at all[32] — the same debate that had taken place after the May 1937 eruption. N.H. Fisher gave his opinion that the most sensible approach — bearing in mind the volcanic threat and the fact that so little remained of the original town anyway — would be to build afresh somewhere outside the caldera.[33] Murray was of the same opinion, favouring the longer term development of Kokopo, the former German administrative centre, but little could be done to advance a final decision in Canberra. European and Chinese business interests gradually returned in Rabaul, developing into a lobby group which favoured remaining there, despite the depressed condition of the town.

Murray, aware of the contents of the report by Ch.E. Stehn and W.G. Woolnough, and no doubt influenced by his own scientific background, also supported immediate re-establishment of volcano monitoring at Rabaul, but extracting

---

30   Downs (1980).
31   McDonald (1947).
32   Johns (1996).
33   Fisher (1946).

approval from Canberra was difficult on this matter, too. Earthquakes, however, were felt in the town during 1949 and particularly on 17 October, again raising fears of the risk to Rabaul from volcanic and earthquake activity. Percy C. Spender had been appointed the Australian Minister of External Territories late in 1949 under the new Liberal Government of R.J. Menzies, but Murray — whose own favourable appointment had been made under Labor — had an uneasy relationship not only with Spender, but later with Spender's successor, Paul Hasluck, in 1951. Murray's frustration by February 1950 is indeed palpable:

> In view of the fact that the Commonwealth has not seen fit to appoint a Vulcanologist for duty at Rabaul, I consider it to be especially necessary that some suitable person should make a weekly inspection of all the vulcanological centres mentioned in the list of phenomena [presented in the Stehn-and-Woolnough report].[34]

The main reason for delay in appointing a volcanologist was the non-availability of suitably trained geoscientists immediately after the war. This shortage was in fact being addressed by the Australian Government through the Commonwealth Reconstruction Training Scheme for ex-service personnel, and Murray, perhaps also in part as a result of his own persistent lobbying, did not have to wait too long. Australian geologist G.A.M. 'Tony' Taylor, then 32 years of age, was selected for the position of the Territory's volcanologist at Rabaul. Taylor came to the run-down town in April 1950 to begin what would be a highly successful career in volcanology — propelled initially by his work during the Lamington eruption of 1951 — but which would come to an end prematurely while undertaking fieldwork on Manam volcano in August 1972.

Changing the Volcanological Leadership

German-born Stehn, Fisher's co-investigator in 1937 of Rabaul's volcanic nature, had been transported from the Netherlands Indies to British India at the outbreak of the war with Japan. He had died in internment in India on 17 May 1945 shortly before war's end.[35] Fisher, however, had returned to Rabaul in 1946, mainly to assess the postwar condition of the observatory that he had established in 1940. Fisher found Kizawa's tunnel behind Vulcan, removed the Japanese seismograph, and sent it to Australia. Fisher told me in 2006 how he had driven down to Sydney from Canberra to pick up the seismograph from the Sydney wharfs, and the instrument was then forwarded to Melbourne for repairs and maintenance. The old seismograph can be seen today in the recording room at RVO headquarters where it is referred to rather nostalgically as 'the old Omori'.

---

34 Murray (1950).
35 Bemmelen (1949).

Fisher, also in 1946, was appointed Chief Geologist in the Commonwealth of Australia's Bureau of Mineral Resources, Geology and Geophysics (BMR) — Australia's first national geological survey — which had been created in April of that year.[36] This promotion, and Taylor's appointment, meant that Fisher would no longer have the same hands-on involvement in volcanology that he had previously. His prime responsibilities from now on would be overseeing the systematic geological mapping of the Australian continent and the new Territory of Papua and New Guinea, together with the postwar search for mineral resources. He would, however, maintain a supervising involvement with volcanology in the Territory in his new capacity and indeed a strong interest in the subject for the rest of his life.

The British too were interested in the search for mineral resources in the immediate postwar environment. The former British Empire had disappeared altogether and been replaced by a new Commonwealth of Nations. Even the name 'British' had been dropped from the title of the Commonwealth in 1949. British geologist John C. Grover came to Honiara in the British Solomon Islands Protectorate in April 1950 — the same month as Taylor's arrival in Rabaul.[37] Grover's career there would cover 17 years during which a focus on the young volcanoes of the protectorate would be included only as part of his wider duties as senior geologist in geological mapping and investigation of resource potential. His volcanological interests would involve not only Savo but also a submarine volcano, Kavachi, in the New Georgia Group south-east of Bougainville Island.

Eruptive activity at Kavachi had been noticed in 1939 by villagers from nearby Vanguna and Nggatokae islands, although there had been earlier stories of undated 'fire on the water', and there was activity too in 1942.[38] The volcano was given its name in 1966 after Kavachi, a legendary sea god of the Marovo district in the New Georgia Group. The name is a shortened form of *Rejo te Kavachi*, meaning 'Kavachi's oven'. Further volcanic activity at Kavachi was reported for late 1950, which continued — evidently intermittently — until 1953.

This timing of the eruptions at Kavachi in 1950–1953 overlaps with that at Lamington volcano in Papua in 1951–1952, again raising — together with eruptions from other volcanoes — the possibility of a 'time cluster' of eruptive activity in Near Oceania during the 1950s.[39] The Lamington eruption, however, had considerably greater impact than did those at Kavachi. It produced a major disaster including a death toll that has not been exceeded since by any other subsequent natural geophysical event in Australian-administered territory.

---

36 Wilkinson (1996).
37 Grover (1955a).
38 Grover (1955b) and Johnson & Tuni (1987).
39 See, for example, Michael (1969).

7. Eruptions during the Pacific War and Postwar Recovery: 1941–1950

Figure 55. The reconditioned Omori seismograph is seen here in operation at Rapindik, Rabaul, in the early 1950s. Rapindik is near the north-western shore of Greet Harbour, even closer to Tavurvur volcano than was the Sulphur Creek Observatory where the two-component instrument was first used by Kizawa in 1942.

Source: M.A. Reynolds. Geoscience Australia (no registered number).

References

Baker, G., 1946. 'Preliminary Note on Volcanic Eruptions in the Goropu Mountains, Southeastern Papua, during the Period December, 1943, to August, 1944', *Journal of Geology*, 54, pp. 19–31.

Bemmelen, R.W. van, 1949. 'Charles Edgar Stehn (1884–1945), Nécrologie', *Bulletin Volcanologique*, 8, pp. 133–37.

Best, J.G., 1956. 'Investigations of Recent Volcanic Activity in the Territory of New Guinea'. *Proceedings of 8th Pan Pacific Science Congress, Manila, 1953*; 2: Geology, Geophysics and Meteorology, pp. 18–204.

Bultitude, R.J., 1979. *Bagana Volcano, Bougainville Island: Geology, Petrology, and Summary of Eruptive History between 1875 and 1975*. Geological Survey of Papua New Guinea Memoir, 6.

Cooke, R.J.S., 1981. 'Notes on the Activity of Ulawun Volcano, 1700–1958: Results of a Literature Survey', in R.W. Johnson (ed.), *Cooke-Ravian Volume of Volcanological Papers*. Geological Survey of Papua New Guinea Memoir, 10, pp. 147–51.

Downs, I., 1980. *The Australian Trusteeship: Papua New Guinea 1945–75*. Australian Government Publishing Service, Canberra.

Film Australia, 1993. *Waiting for the Big Bang*. Film Australia, Lindfield (video).

Fisher, N.H., 1944. 'The Gazelle Peninsula, New Britain, Earthquake of January 14, 1941', *Bulletin of the Seismological Society of America*, 34, pp. 1–12.

——, 1946. 'Administrative Centre for the Rabaul District'. Department of Supply and Shipping, Mineral Resources Survey Branch Report 1946/32.

——, 1976. '1941–42 Eruption of Tavurvur Volcano, Rabaul, Papua New Guinea', in R.W. Johnson (ed.), *Volcanism in Australasia*. Elsevier, Amsterdam, pp. 201–10.

Grover, J.C., 1955a. 'Summary of the Activities of the Interim Geological Survey from April 1950 to January 1953', in *Geology, Mineral Deposits and Prospects of Mining Development in the British Solomon Islands Protectorate*. Interim Geological Survey of the British Solomon Islands Memoir, 1, Appendix B, pp. 105–08.

——, 1955b. 'Seismic Activity and Vulcanism', in *Geology, Mineral Deposits and Prospects of Mining Development in the British Solomon Islands Protectorate*. Interim Geological Survey of the British Solomon Islands Memoir, 1, pp. 21–24.

Johns, E., 1996. 'To Move Rabaul', *Journal of Pacific History*, 31, pp. 92–103.

Johnson, R.W. & N.A. Threlfall, 1985. *Volcano Town: The 1937–43 Rabaul Eruptions*. Robert Brown and Associates, Bathurst.

Johnson, R.W. & D. Tuni, 1987. 'Kavachi, an Active Forearc Volcano in the Western Solomon Islands: Reported Eruptions between 1950 and 1982', in

B. Taylor & N.F. Exon (eds), *Marine Geology, Geophysics, and Geochemistry of the Woodlark Basin–Solomon Islands*. Circum-Pacific Council for Energy and Mineral Resources Earth Science Series, 7, pp. 89–112.

Kizawa, K., 1951. 'Volcanic Tremors and Tilting of the Ground', first report, *Kenshin Jihō*, 15, no. 2, pp. 18–34 (in Japanese).

——, 1961. Unpublished letter to G.A.M. Taylor, 23 August 1961.

Kusaka, J., 1976. *Rabaul Sensen Ijyo Nashi* [*All Quiet on the Rabaul Front*]. Kowado, Tokyo (in Japanese).

Marsh, D.R., 1944. 'Separate Report on Volcanic Disturbances Inland Collingwood Bay Tufi District'. National Archives of Australia, Canberra, Series A518/1, Item 432711, Folios 56–59.

McCarthy, D., 1959. *South-West Pacific Area — First Year: Kokoda to Wau. Australia in the War of 1939–1945*, Series 1 (Army), 5. Australian War Memorial, Canberra.

McDonald, J.H., 1947. Report on Rabaul. Letter to the Government Secretary, Port Moresby, 24 February 1947. PNG National Archives, Port Moresby, SN 677, AN 244, File GH 1-5-3, folios 11–12.

Michael, M., 1969. 'Volcanic Pulses in the New Guinea–Solomons Region', *Papua and New Guinea Scientific Society Transactions*, 10, pp. 8–13.

Moore, J.G., 1967. 'Base Surge in Recent Volcanic Eruptions', *Bulletin Volcanologique*, 30, pp. 337–63.

Murray, J.K., 1950. Volcanic risk at Rabaul. Letter to the Government Secretary, Port Moresby, 14 February 1950. PNG National Archives, Port Moresby, SN 388, AN 247, File CA 32/5/5A, folios 15–16.

National Archives of Australia, 1941. Transfer of Administration to Lae, 1941–1946. Correspondence files, multiple number series with alphabetical prefix, A518. Item AK800/1/3/PT4.

Nelson, H., 1992. 'The Troops, the Town and the Battle: Rabaul 1942', *Journal of Pacific History*, 27, pp. 198–16.

Nitta, J., 1980. 'Kazan Gun [Volcano Group]', in *Hyogen. Hijyou no Burizaado* [Ice Field: Sad Blizzard] (in Japanese), pp. 55–116. Shincho-sha, Tokyo.

Ruxton, B.P., 1966. 'A Late Pleistocene to Recent Rhyodacite-Trachybasalt-Basaltic Latite Volcanic Association in North-east Papua', *Bulletin Volcanologique*, 23, pp. 347–74.

Ruxton, B.P., H.A. Hantjens, K. Paijmans & J.C. Saunders, 1967. *Lands of the Safia–Pongani Area, Territory of Papua and New Guinea*. Land Research Series, 17. Commonwealth Scientific and Industrial Research Organization, Australia, Melbourne.

Simkin, T. & L. Siebert, 1994. *Volcanoes of the World: A Regional Directory, Gazetteer, and Chronology of Volcanism during the Last 10,000 Years*. Geoscience Press, Tucson.

Smith, I.E.M., 1981. 'Young Volcanoes in Eastern Papua', in R.W. Johnson (ed.), *Cooke-Ravian Volume of Volcanological Papers*. Geological Survey of Papua New Guinea Memoir, 10, pp. 257–65.

Stone, P., 1994. *Hostages to Freedom: The Fall of Rabaul*. Ocean Enterprises, Yarram.

Taylor, D., 1943. Untitled letter to an unidentified recipient, 29 December (assumed to be 1943). Anglican Archives, New Guinea Collection, University of Papua New Guinea, Port Moresby. AA Box 57, File 7.

United States Strategic Bombing Survey (Pacific), 1946. *The Allied Campaign Against Rabaul*. US Naval Analysis Division, Marshalls–Gilberts–New Britain Party.

Wigmore, L., 1957. *The Japanese Thrust. Australia in the War of 1939–1945*, Series 1 (Army), 4. Australian War Memorial, Canberra.

Wilkinson, R., 1996. *Rocks to Riches: The Story of Australia's National Geological Survey*. Allen & Unwin, St Leonards.

8. Disaster at Lamington: 1951–1952

> ... in the tropics decomposition sets in quickly and thousands of rotting corpses were scattered throughout the devastated area mostly covered in ash. Many hundreds more were spread out along the road from Higaturu where they had been attending church services on the Sunday morning ... A couple of hundred more were huddled together inside a church. Large numbers of the bodies had split open with intestines spilling out. The stench was appalling.
>
> Des Martin (2013)

Higaturu and the Orokaiva

We have a volcano! This is the exclamatory statement that Margaret or 'Peggy' de Bibra wrote down after breakfast on the morning of the Lamington catastrophe on Sunday 21 January 1951.[1] Miss de Bibra was the principal of Martyrs' Memorial School at the Anglican Christian Mission at Sangara on the northern flank of Mount Lamington, less than two kilometres downslope from the Territory's government headquarters for the Northern District at Higaturu. Sangara and Higaturu were only 10–12 kilometres from the summit of the mountain, which rose as an impressive and scenic backdrop to almost 1,800 metres above sea level.

The Higaturu-Sangara area was, in many ways, located advantageously. It had been developed by Europeans in the higher country away from the oppressive heat of the coast and had, like Rabaul, rich volcanic soils. A wide variety of vegetables and fruit was grown there, and rubber, coffee and cocoa were harvested on commercial plantations managed by Europeans. Soil richness was also the reason why so many Orokaiva villages and hamlets had already spread throughout the area — including a concentration around Isivita, where the Anglicans had a mission station a few kilometres south-west of Higaturu.

There were, however, darker memories of wartime conflict in the area. Australian authorities had abandoned much of the Northern District for several months in 1942 when the Japanese military landed in the Gona area. The landing marked the beginning of the Kokoda campaign for the Allies, culminating in savage fighting at Buna and Gona. Anglican missionaries had been betrayed in 1942 by some of the local Orokaiva, and six of the Anglicans were murdered by

---
1 White (1991), p. 45.

the Japanese. Three corpses were recovered and later buried at Sangara at the Martyrs' Cemetery, the name of which memorialised the dead missionaries. The returning Australian military captured the Orokaiva betrayers, as well as Orokaiva accused of other crimes, and publicly hanged 22 of them at Higaturu in early 1943.[2]

The Papuan people who speak the Orokaiva language live mainly in the northern foothills of Mount Lamington and on the adjoining plains. Their language is the one most commonly used by people of the Binandere language group and is spoken widely in the more densely populated parts of the Northern District. Group relationships within Orokaiva-Binandere territory and with adjacent outside groups such as in the Managalese area south-east of Mount Lamington, were managed through traditional exchange and reciprocity arrangements.[3] Yega people on the coast to the north-east of the mountain traditionally considered the inland, 'mountain' Orokaiva as enemies, and they regarded Mount Lamington fearfully, even before the 1951 eruption, as the home of the departed spirits of the Orokaiva, later blaming the powerful spirits of the mountain for the disastrous eruption itself.[4]

De Bibra included the following pieces in the remainder of her unfinished draft article:

> Mt Lamington ... lies behind us and consists of four or five sugar-loaf peaks ... We have always loved her for her beauty and nearness ... [but now] she has changed from fairy queen to a wicked witch, and the gossamer scarves of mist [seen in the early mornings] have turned into smoky outpourings of some bubbling cauldron ...
>
> For days we had tremors ... and the face of Lamington became scarred with great patches of bare earth, caused by landslides. Then one morning — January 18th [the previous Thursday] — after a night of continuous tremors, smoke appeared [that later] ... came pouring out in great thick puffs high into the sky, wreathing and curling in awe-inspiring cauliflower shapes ...
>
> What do the people think? ... What will it mean? ... How will it affect the faith of new Christians? ... Will you think of the people here, particularly the Managalas people, and those near our stations of Sewa and Sehaperete? Pray for them and for us, that out of this good may come, and as the dead mount came to[5]

---

2 Grahamslaw (1971) and White (1991).
3 Schwimmer (1973).
4 Benson (1955).
5 White (1991), pp. 45–47.

8. Disaster at Lamington: 1951–1952

Her final sentence was never completed.

Volcanologist G.A.M. 'Tony' Taylor arrived at the Lamington disaster area by air on the day after the catastrophe to begin an intensive study of the eruption and its volcanic deposits, and involving fieldwork that lasted almost two years. This was followed by a period of report writing in Canberra, resulting in the publication in 1958 of Bureau of Mineral Resources Bulletin 38 which today is recognised as a landmark in the history of volcanological studies.[6] Taylor compared the Lamington eruption to those at Mont Pelée and Soufriére in 1902 and to subsequent eruptions at these two Caribbean volcanoes, and called the 1951 Lamington eruption peléean in type. Bulletin 38, however, carries no hint of the disaster-management controversies of the catastrophe at Lamington.

Taylor achieved fame not only amongst the international volcanological fraternity, who read and learnt from Bulletin 38, but also amongst the general public in both the Territory of Papua and New Guinea and in Australia. This is because Taylor was awarded the George Cross for courage as a result of his work during the dangerous first few months following the 21 January catastrophe. He thus became a high-profile scientist–hero, despite being a private, serious-minded, if not shy person. Taylor was assisted at Lamington by Leslie Topue, a villager from the Rabaul area, who had recently joined the Rabaul Volcanological Observatory. Topue would be awarded the British Empire Medal. Taylor was also assisted in the disaster area initially by Bureau of Mineral Resources (BMR) geologist John G. Best who was, however, soon sent on to supervise the operations at Rabaul Volcanological Observatory (RVO) during Taylor's long absence from Rabaul.

Build-up to Catastrophe

Mount Lamington was not known to be a volcano, and there were no traditional stories or legends about previous eruptions. Both of these statements were true as far as the people living on the northern flank of Mount Lamington in 1951 were concerned. They were true, also, for Administration officials in Port Moresby and even for volcanologist N.H. Fisher when he later wrote the foreword to Bulletin 38. Both statements, however, need qualification given that previously 'buried' pre-1951 information came to light and became more significant after the catastrophic eruption.

Australian geologists before 1951 had recognised Mount Lamington and the Hydrographers Range to its east as both volcanic in nature and Quaternary in age — that is, geologically very youthful — and a Netherlands Indies geologist,

---

6   Taylor (1958).

R.W. van Bemmelen, had even referred to Mount Lamington itself as being an 'active' volcano.[7] The young volcanic features of Lamington volcano, including a north-facing crater, youthful lava domes — the 'sugar-loaf peaks' mentioned by de Bibra — and a thermal area, are visible on aerial photographs taken in 1947.[8] Furthermore, an Administration patrol officer visited and identified the volcanic crater on Lamington in 1948 and recounted, before the 1951 eruption, a traditional story that a lake had once existed at the summit but its waters had burst out northwards, devastating several Higaturu villages and causing great loss of life.[9]

The sugar-loaf peaks of Mount Lamington feature in powerful legends about the mountain, which the Orokaiva regarded as the centre of their universe, a place where death, strife, warfare, fires, marriage and other cultural elements originated. The Orokaiva called the mountain Sumbiripa Kanekari, meaning 'the separation of Sumbiripa', alluding to a time when the mountain had opened and split into separate crags.[10] The Orokaiva man Sumbiripa, or Sumbirip, and his wife had been hunting on the mountain but they became separated on different peaks. Sumbiripa died, the first Orokaiva to do so, and became the spirit master of the mountain, living inside it with other Orokaiva people who died after him. This is not a clear-cut volcano or eruption legend as such, but the story does hint at earlier, witnessed, eruptive activity.

Europeans wanting to visit the summit of Mount Lamington could not always find Orokaiva people willing to guide them because of its special spiritual significance, but a group of missionary women managed to climb the volcano in the early 1930s. Their apprehensive guides had warned them in advance that the earth there shook and that a roaring noise could be heard at the top. The missionaries confirmed — before the 1951 eruption — these claims, as there was

> a great roar at the summit, the roar of a mighty waterfall in the chasm below, but the boys began to talk together in whispers. 'Sister, listen' said one, 'That big noise not river.'[11]

None of this information was generally available when Lamington started precursory activity six days before the paroxysmal outburst of 21 January 1951. There may have been even earlier warning signs, and a seismograph possibly could have picked up any earlier build-up of earthquake activity, but the mountain had no such instruments. Increases in tremors, landslides and gas emissions were noted from Monday 15 January until the time of the initial ash

---

7   Stanley (1924), Bemmelen (1939) and Montgomery et al. (1950).
8   Taylor (1958).
9   Murphy (1951).
10 Schwimmer (1973, 1977). 'Sumbiri' (without the suffix 'pa') was the name of a tribal warrior from the Angereufu Clan of the Songe Tribe, and his wife was called Suja (Maclaren Hiari, personal communication 2013).
11 Tomblin (1951), pp. 132–33.

eruption of the following Thursday, when the resident District Commissioner at Higaturu, Cecil F. Cowley, informed Port Moresby of the 11 am outburst and that 'there was no need for alarm'. The Administrator, J.K Murray, was travelling in other parts of the Territory and Justice F.B. 'Monte' Phillips was the acting Administrator in Port Moresby at the time. Cadet Patrol Officer Athol J. Earl wrote the following to his parents from Higaturu on 18 January:

> Things have been happening here the last two days. We started off yesterday with earth quakes [sic], they were not very bad but frequent, every three to five minutes [t]his went on all day and on the Lamington Mountains behind us we could see great land slides. The quakes kept up all last night and this morning great rumblings commenced. At about ten o'clock it blew its top and we now have a volcano just behind us. Great masses of smoke have been belching out ever since and the lava [sic] can be seen running down the mountain side. We have looked at it through a telescope and you can see rocks, and so forth being tossed into the air. The native[s] from all around here deserted with all their belongings, however, I notice they had started to come back tonight. Earth quakes are still continuing and great rumbling is going on … [12]

Phillips flew the next day, Friday, to Popondetta where there was a tiny settlement — including a Buntings general store and a new government agriculture station alongside the airstrip — more than 20 kilometres north-east of Mount Lamington. Phillips was met there by Cowley who surprised the judge by saying he had been expecting a volcanologist on the flight, having earlier requested the Administration in Port Moresby to provide one.[13] The European women present, including Mrs Cowley, were anxious about the state of the mountain but her husband, the District Commissioner, was 'cool and confident, and tried to calm his wife's nervousness', wrote Phillips. The people at the airstrip could see an eruption cloud rising lazily from Lamington to the south-west, and Phillips concluded that the volcanic pressure was being relieved quite safely. He pronounced this judgment on the basis of his personal geophysical experience at Rabaul in 1937. Phillips declared 'there was no immediate danger to human life at Higaturu' and flew back to Port Moresby. Cowley, still without the volcanological advice he had requested, had little option but to concur with the acting Administrator's opinion. The decision was supported, too, by the missionary-in-charge at Sangara, Reverend Dennis J. Taylor, who had witnessed the Goropu eruption of 1943–1944 from Wanigela, and who thought that the Goropu volcanic activity was much stronger than the current, weak, eruptions at Lamington.

---

12 Unpublished letter dated 18 January 1951. A copy of the letter was kindly provided by Earl's niece, P. Earl (personal communication, 2011).
13 Phillips (1951), pp. 5, 6.

The Administration and Mission were the political and ruling elite and many of the Orokaiva looked to the Europeans for guidance on the matter of the eruption and its dangers. The decisions made by Phillips and Cowley were distributed by village constables and mission staff, and they influenced people not to evacuate the area. There were, however, spontaneous evacuations from some Orokaiva villages.[14] The Phillips decision, nevertheless, 'was a case where bad science was more dangerous than sound superstition', wrote one commentator later.[15] The Administration's volcanologist, Tony Taylor, was in Rabaul and at this time had not been consulted about either the volcanic eruptions or a need for any evacuation, although John Best recalled many years later that Taylor was aware of the volcanic happenings at Lamington from hearing radio news broadcasts that week.[16] More than one explanation has been proposed for why Taylor was unable to leave Rabaul to investigate the early eruptions at Lamington, but the non-availability that week of suitable air transport from Rabaul seems to be the main reason for the delayed departure. Best, however, hinted at bureaucratic delays in Rabaul, which might have been avoided had the uncertain situation at Lamington that week been recognised in Rabaul as critical.

A black volcanic cloud advanced northwards over Higaturu, Sangara and many Orokaiva settlements on the morning of 21 January, and there are dramatic — at times horrific —records of the individual and in some cases courageous experiences of survivors, rescuers and evacuees. Pilots and passengers of in-flight aircraft probably had the best overall view of the colossal cloud that rose rapidly from Mount Lamington to at least 15 kilometres. A Qantas Dragon aircraft flying from Lae was just about to land at Popondetta when the 'entire mountainside blew out, and Higaturu station was blotted out. The terrible cloud rushed towards them ... [and they] sped back at full speed to Lae'.[17] News of the eruption first reached Port Moresby by means of a radio message from a Qantas DC3 en route to Rabaul from Port Moresby. The aircraft was north-west of the mountain when its captain saw

> a dark mass of ash shoot up from the crater and rise, within two minutes, to 40,000 feet, forming a huge expanding mushroom-shaped summit. The base of the column expanded rapidly as if the 'whole countryside were erupting' ...[18]

---

14   Taylor (1958) and Didymus (1974).
15   Schwimmer (1969), p. 12.
16   Best (1988).
17   Sinclair (1986), p. 118. D.H. Urquart was one of the passengers on the Dragon aircraft and he exhausted a roll of 35 mm film in photographing the eruption from the air. The undeveloped film was posted to the *Sydney Morning Herald*, which syndicated it worldwide on Urquart's behalf (Graham, 1974). Subsequent strenuous efforts by J.R. Horne (personal communication, 2012) to find the film have been unsuccessful, but some of the Urquart photographs were published by different newspapers at the time.
18   Taylor (1958), p. 24.

The Mission outstation and villages at Isivita were perilously close to the sharply defined edge of an area of complete devastation on the flank of the volcano. Reverend Robert G. Porter and others experienced a fearful blackout and substantial fallout of volcanic debris from the eruption cloud. Some injured people from the devastated area managed to reach the Mission on foot and received medical attention from the nursing sister, Pat Durdin, but they were in terrible condition:

> The entire floor [at the Mission] was covered with people in utter agony. Some had almost the whole of their skin burnt off. It hung from their hands like discarded gloves, and their agonising cries were awful to hear …

Deaths followed, graves had to be dug, and 18 died that night at Isivita:

> We carried them all out and laid them on the lawn at the end of the church. As we lifted several of them we could feel the burnt flesh coming away on our hands. It was a terrible sight to see those eighteen poor charred bodies laid in a row.[19]

Figure 56. The climactic eruption of 21 January 1951 was photographed from an in-flight Qantas DC3. The base of the collapsing eruption cloud is expanding and pyroclastic flows are cascading down the flanks of the volcano (right).

Source: Taylor (1958, Figure 11). Photographer Captain A. Jacobson, Qantas Empire Airways. Geoscience Australia (GB-1886).

---

19 Porter (1951) pp. 26–27.

The arterial road linking Gona at the coast to Kokoda in the mountains ran across the northern flank of Mount Lamington. It was outside the devastated area, and on the Sunday it was an important lifeline for the injured, evacuees and the first rescuers, although fallout of volcanic debris and minimal visibility greatly hampered movement along the road, particularly close to the volcano. A small group of mainly European men organised in dangerous circumstances a response to the chaos along the road between the plantations and other settlements. Trucks were used to shuttle the injured and dying to Popondetta. One group entered a village on the western edge of the devastated area and saw the extent of the human tragedy there. The radio at Higaturu headquarters was silent, and the usual weekday 'sked', or radio schedule, was closed each Sunday anyway, but the manager of the Awala rubber plantation, C.E.'Clen' Searle, later that day managed to establish radio contact with Port Moresby, even though strong electrical disturbances interfered with effective radio transmission when the volcano was in strong eruption.

Rescuers from Port Moresby and Lae were unable to reach the Lamington area during that Sunday. Night fell, adding fear and greater uncertainty to those attending to the requirements of the injured, dying and dead. And there was further terror when, at about 8.30 pm, a second violent eruption took place from the volcano. The night-time eruption was loud, producing a great deal of volcanic fallout, and the cloud may have risen at least as high as the one in the morning.[20] The cloud spread across the Owen Stanley Range and ash fell on Port Moresby, closing the airport there to traffic the next morning.

Relief and Recovery

The Territory Administration in Port Moresby was more fully informed of the extent of the catastrophe at Lamington by the morning of Monday 22 January, and a major relief-and-recovery effort was organised to deal with the Northern District disaster. Relief efforts were speedy and effective. The first few days were chaotic, however, before some order could be established, and many Administration personnel became involved in relief work. Aircraft would play a major role in both the relief and recovery phases of the disaster, using airdrops and airstrips in the area to deliver food, tents, medical and other supplies, in amounts that at times exceeded actual requirements.

Murray had left Rabaul by air to Port Moresby early on the morning of Monday 22 January, and had volcanologist Taylor on board with him. Murray was in the air when he heard about the catastrophe at Lamington, and his aircraft was

---

20   Taylor (1958).

diverted to Lae because of the closure of Port Moresby airport. The District Commissioner at Lae, H.L.R. 'Horrie' Niall, who had heard about the disaster from Port Moresby, had already arranged for the Administration vessel *Huon* to sail from Lae on the Sunday evening. The *Huon* had Administration and medical personnel and emergency supplies on board, and it travelled overnight to Cape Killerton on the coast near Popondetta. The personnel included Niall himself as well as patrol officers J.D. 'Des' Martin and R.W. 'Bob' Blaikie, together with Doctor Max Sverklys, an Australian nurse Sister 'Rusty' Maclean, and about six Papua New Guinean police. The party did not know what to expect, but after landing at the beach at Cape Killerton on the Monday morning and then driving to Popondetta they soon started to appreciate the impact and horror of the disaster. They passed on the road a jeep and trailer heading to Gona from Popondetta and carrying the body of Reverend Dennis Taylor who had died overnight at Popondetta from burns after first escaping from the worst of the eruption. The party reached Popondetta later that Monday and saw the extent of burns on the survivors who had been brought in to the small settlement. Sverklys and McLean joined other Europeans who had already been attending to the victims still alive at Popondetta, but the number of injured was relatively few compared to the much greater number of people killed by the eruption. Indeed, the doctor and sister would have relatively little work to do after the first day or two.

Murray, Taylor, the Director of Public Health Dr John T. Gunther, the acting Director of District Services and Native Affairs Ivan Champion, and others flew into the area on Monday 22 January to assess and then implement what had to be done. The Administration set up a forward rescue-and-relief and evacuation centre at Popondetta, and the injured started to be airlifted that morning to base hospitals in Port Moresby and Lae. European women and children were evacuated too, by air and by road to Kokoda, including the Australian anthropologist Marie Reay who had been researching Orokaiva leadership in the area.

The sight of the volcanic destruction astounded and horrified the first people who managed to negotiate the side road that climbed through the devastated area southwards to Higaturu-Sangara from the arterial Gona-Kokoda road. The destruction seemed complete — as if a bomb had exploded, blasting everything away. Vegetation had been stripped, trees felled, and remnant trunks split and abraded. Most buildings had been obliterated. Ash covered everything, creating a bleak and monotonous 'moonscape'. Most horrifying of all were the burnt human corpses, including hundreds littered along the access road itself, the fleeing people apparently having been felled as they attempted their escape. Champion directed Martin and a team of Papua New Guinean police early on Tuesday 23 January to clear the road into Higaturu in order to retrieve moneys

Fire Mountains of the Islands

and documents in the safe at Administration headquarters using spare keys that had been brought over from Port Moresby. Martin and the police tried to move to one side the bodies that lay in the road, but the scale of task, including actual burial of the already decomposed bodies, was overwhelming and the work was clearly futile. Martin reported the situation back to Champion and, years later, wrote of his experiences amongst rotting and ruptured corpses and the overwhelming stench:

> Initially we tried to shovel bodies off the road into drainage ditches with four of us together using shovels to do so. The masses of bodies along the road actually made it difficult to move around without stepping on one. In those days the native police had bare feet and what with ruptured bodies and exuding body fluids the police were slipping and sliding about … . In retrospect it was really was the stuff of nightmares.[21]

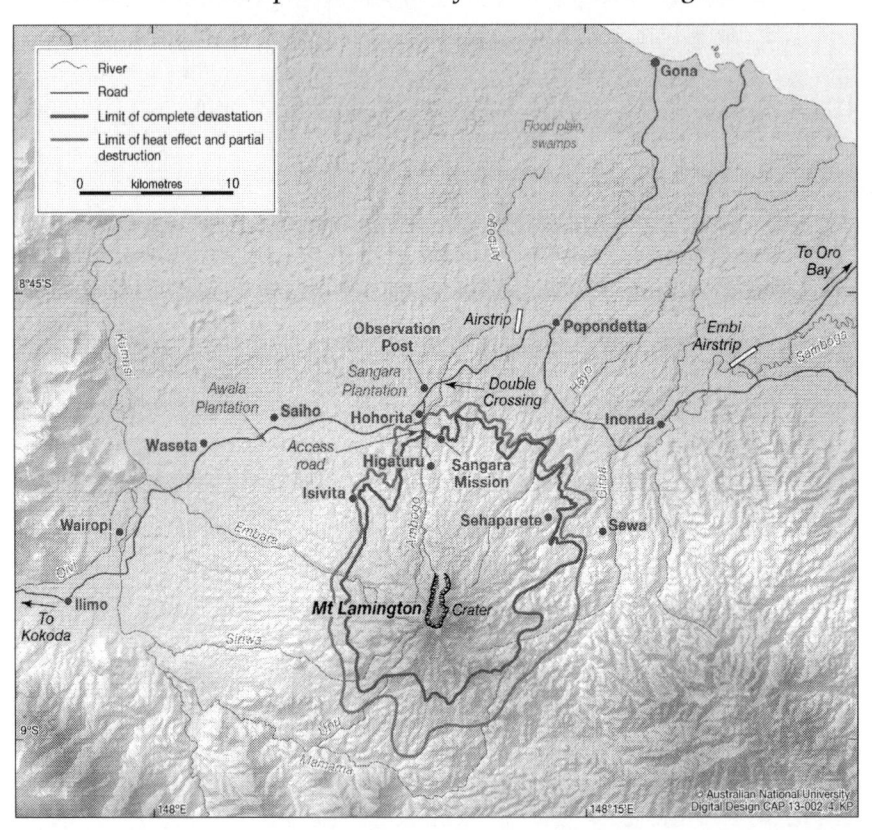

Figure 57. The limit of the devastated area on Mount Lamington, including Higaturu and Sangara, is surrounded by a narrower zone of partial destruction.

Source: Adapted from Taylor (1958, Figure 52).

---

21 Martin (2013), p. 44.

Acting Government Secretary Claude Champion, brother of Ivan, later visited Higaturu on the same day, Tuesday 23 January, and the following day wrote:

> After reaching Higaturu we made a detailed survey, [but] were unable to move collapsed buildings. Several Europeans and natives were found in [the] vicinity of the District Commissioner's residence — highly decomposed condition. It was difficult to distinguish between Europeans and natives as most of the clothing was apparently blown off. Without doubt, we have definitely identified C. Cowley who was sitting in his landrover [the body of his young son, Earl, was nearby] ... [A] Works and Housing jeep was found suspended ten feet from the ground on a tree stump.[22]

Putrefying bodies were strewn around in their thousands, in numbers that far exceeded the number of injured. People within the nearly 200 square kilometres of total devastation were killed, those outside it survived, and the 100–200 injured people were mainly from a narrow 'transitional' zone in between. Identifying the dead was difficult because of an intense body lividity which removed the racial distinctions of skin colour. The authorities realised within two or three days that 35 Europeans were dead or missing but several months of inquiry were needed before the non-European death toll was known. Cadet Patrol Officer Athol J. Earl, for example, was on the list of the missing Europeans but his remains at Higaturu were not identified as his, and his parents informed officially, until February 1952. An estimate of 4,000 dead had been made initially. The final death toll however — made available in October — was 2,907 indigenous people, together with an acknowledgment that a more accurate number would never be known.[23] The great majority of the almost 3,000 dead were Orokaiva from the Sangara area.

Priority for the Administration, however, was not counting the dead. There were thousands of surviving people whose villages had been destroyed but who happened to have been away from them at the time of the Sunday morning eruption, or whose houses and gardens had been seriously damaged by volcanic fallout beyond the devastated area. Survivors in general moved down slope, away from the volcano's active crater, to places of refuge. People in the west descended to the arterial road and congregated at places such as Waseta. Those in the north moved down to Sangara Plantation on the road, displaced people reached Popondetta and the Cape Killerton coastline, and refugees in the east sought shelter at Inonda, near the wartime US airstrip at Embi. A refugee camp was soon established at Wairopi on the western bank of the Kumusi River in the west, where about 4,000 people gathered. Plans to establish another camp at Cape Killerton were abandoned because of inadequate amounts of potable water and the health threat of the swampy conditions there. Evacuees, therefore, eventually were shipped round to the coast east of Lamington to a camp at Oro Bay.

---

22 Champion (1951).
23 Murray (1951).

Figure 58. Many victims were seen on the road to Higaturu. This photograph was probably taken within two or three days of the eruption on 21 January 1951.

Source: Taylor (1958, Figure 76). Geoscience Australia (M-1745).

The refugee camps grew to include medical centres and schools. Medical work gave emphasis to sanitation, prevention of epidemics, and general camp health services. Whooping cough had been diagnosed in the area before the eruption, so a program of inoculation was started to prevent its further spread. Quick burial of the decomposing dead would have reduced the risk of more deadly diseases breaking out, but rapid burial of so great a number was impossible. Not all of the bodies could be entombed, and some 'burials' represented just piles of ash over decomposing bodies. In addition, some cadavers suffered the scavenging effects of hungry dogs and pigs, rendering the grim task macabre. Gunther said that the Department of Public Health succeeded in preventing major disease outbreaks.[24] Nevertheless, deaths at Oro Bay — mainly of children — resulted from a mix of dysentery, pneumonia and whooping cough, disheartening local medical staff.[25]

---

24 Gunther (1951–1952).
25 Biggs (1987).

8. Disaster at Lamington: 1951–1952

Figure 59. The pyroclastic surge that swept across Higaturu, from left to right in this photograph, destroyed vegetation and hurled a jeep onto tree stumps. The end of a bent telegraph pole can be seen on the extreme left. Leslie Topue is in the centre of the photograph.

Source: G.A.M. Taylor (likely photographer). Taylor (1958, Figure 69). Geoscience Australia (GB-1060).

Taylor played a key role during the relief and recovery phases at Lamington. He advised Administration personnel on the state of the volcano and, in effect, had full authority to advise on movements in and out of the devastated and prohibited area. BMR geophysicist W.J. 'Bill' Langron assisted Taylor by operating a seismograph that began recording on 8 February at an observation post at Sangara Plantation. Many explosive eruptions were still taking place, and levels of anxiety were high amongst the relief teams and survivors, who feared further major eruptions like, or even more powerful than, that of 21 January. The devastated area was at times declared off limits by Taylor because of this uncertainty and fear, thus again hindering any thoughts amongst Administration staff of burying the dead. Taylor was admired for his professional commitment and stoicism, and he surprised some colleagues when he would lie flat on his back on the ground to feel tremors or use a glass of water to watch tremor-induced ripples.

Figure 60. A Roman Catholic priest, Father Justin Lockie, left, conducts a requiem mass on the bonnet of a jeep on the road to Higaturu on 24 January 1951. Those attending the mass included senior Territory officials — from left to right in the front row: Ivan Champion, his hat in hand; the Administrator J.K. Murray wearing sunglasses; Dr John Gunther; and the uniformed J.S. Grimshaw, Commissioner of Police. The tall man in the centre-rear wearing a surgical mask around his neck is Patrol Officer Des Martin who had just returned from work in the devastated area. He had doused the surgical mask with disinfectant in order to cope better with the stench.

Source: Anonymous (1951, unnumbered photograph on p. 19). Geoscience Australia (GB-2782, copy only).

The physical and psychological impact on people involved in the relief and early recovery was extreme — what today would be called 'post-traumatic stress disorder' or PTSD. Martin and Blaikie, for example, were recognised as being physically and emotionally exhausted after almost three weeks, having been involved in the relief effort from the beginning. The two patrol officers were relieved from their duties at Popondetta and they returned to their permanent postings in Morobe.[26] They never returned to the relief area at Lamington, but their experiences — and those of many other people — were unforgettable for the rest of their lives.

---

26 R. W. Blaikie (personal communication, 2008) and Martin (2013).

8. Disaster at Lamington: 1951–1952

Figure 61. Tony Taylor at Popondetta Airstrip on 5 February 1951 is about to start an aerial inspection of Lamington volcano — seen in eruption in the background — using a Qantas de Havilland DH84 Dragon aircraft.

Source: Geoscience Australia (GB-2770).

Taylor conducted extensive fieldwork on the ground and made many hazardous aerial inspections of the active crater area in small aircraft flown by adventurous pilots from the Department of Civil Aviation and, later, Qantas Empire Airways. He also received from Canberra copies of the volcanological reports on the 1902 peléean eruptions in the Carribean, and these guided much of his thinking about the kind of eruptions that might be expected. Taylor was on hand for particularly large eruptions on 6 and 18 February, and especially one on 5 March which turned out to be the last of the major explosive activity at the volcano. Ongoing development of a large, bulbous, lava 'dome' in the breached crater characterised much of the remaining eruptive period. Lahars, or volcanic mudflows, became a real threat when rains, unhindered by the absence of vegetation on the flanks of the volcano, formed hot torrents of water-borne ash, rocks, and boulders down major waterways, including the Kumusi River. The evacuation camp at Wairopi was threatened by lahars in early February and had to be transferred to Ilimo, further west and away from the riverbank.

Figure 62. This aerial photograph of Mount Lamington was taken from the north on 8 February 1951. The devastated area dominates the foreground and extends back to the breached crater of the volcano, where the early stages of growth of the active, vapour-emitting, lava dome can be seen. Visible dome growth was especially rapid between 2 and 9 February. The dark areas are mudflows of originally hot ash on the north-east flanks of the volcano in the headwaters of the Ambogo River. Old lava domes form the rugged summit of the mountain.

Source: G.A.M. Taylor (likely photographer). Taylor (1958, Figure 118). Geoscience Australia (GA-9938).

A major settlement including a hospital, was built for the displaced Orokaiva at Saiho, near Awala Plantation on the Gona-Kokoda road. It represented part of an overall resettlement strategy that the Administration had to work out in conjunction with the displaced Orokaiva themselves. Sydney Elliott-Smith had been appointed as District Commissioner to replace Cowley, who had been killed in the eruption, but development of Saiho was the day-to-day responsibility of Assistant District Officer F.P.C. 'Fred' Kaad. Resettlement strategies were not straightforward as many factors influenced final decisions on where the displaced Orokaiva would live, how large their new villages would be, how close they should be to the arterial road, which peoples would live together, whose land would be used, and what type of economic-development initiatives would be promoted in the district. A fundamental starting point for both the Administration and Anglican Mission, however, was a decision not to resettle the devastated area — Higaturu and Sangara would not be rebuilt. The volcanic threat was simply too great, not only in 1951, but also in the years ahead. Administration officer H.T. 'Harry' Plant, who had had anthropological training, undertook a detailed survey of clan relationships as a basis for resettling people of the Isivita villages, and two academic anthropologists, Cyril Belshaw and Felix Keesing, were invited to the area to make independent assessments.[27]

---

27 Plant (1951), Belshaw (1951) and Keesing (1952).

8. Disaster at Lamington: 1951–1952

Figure 63. Tony Taylor, in front, and Patrol Officer W. 'Bill' Crellin approach the active lava dome on the first visit on foot into the newly enlarged crater of Lamington volcano on 11 February 1951. They were accompanied by N.H. Fisher and Leslie Topue. The photograph was taken by Fisher who visited Taylor, his subordinate, at Lamington during February. Only minor explosive activity is taking place — on the rear side of the lava dome — but a much larger eruption took place later that day after the group had left the area.

Source: Taylor (1958, Figure 90). Geoscience Australia (GA-9940).

The Anglican Mission at Sangara sustained grievous losses.[28] Many communicants, missionaries and teachers had died in the eruption of 21 January — a Sunday, when Christians gathered for worship. Reverend Taylor and his family had perished, Taylor himself enduring horrific burns for several hours before being taken to Popondetta and dying there. De Bibra did not finish writing her article because — as surmised later by her missionary colleague, Sister Nancy White — 'It must have been at that moment that the explosion occurred, to make everyone run down the mountain'.[29] De Bibra had arranged for Papuan teachers from different parts of the district to come to Sangara that school holiday weekend in order to prepare school work for the coming year, and they had perished too.

---

28 Strong (1951).
29 White (1991), p. 47.

David Hand, then 32 years old, had been ordained a bishop the previous year, assisting Bishop Philip N.W. Strong who was based down along the north coast of Papua at Anglican Mission headquarters, Dogura. Hand was in Australia at the time of the disaster and returned to the Lamington area which he knew well. The energetic bishop played a prominent role in the recovery process at Lamington, not only of the Anglican Mission itself but also in the resettled village areas, leading memorial and funeral services and supervising the building of a church at the new settlement of Hohorita. The Mission before the eruption had already held ceremonies of consolation, reconciliation and memorialisation after the horrors of the Second World War, but now the Lamington eruption had devastated the Orokaiva, and the Mission, adding to the work of postwar recovery.

Figure 64. Two pyroclastic flows from Lamington volcano on 5 March 1951 ran beyond the limits of the devastated area of 21 January. The one shown here flowed down the Ambogo valley and threatened Sangara Plantation, where Tony Taylor had his observation post, but it eventually swung away to the north-east (left).

Source: G.A.M. Taylor. Taylor (1958, Figure 42). Geoscience Australia (GA-8197).

Hand seems to have had a forceful personality — for some people an abrasive one — and he did not in general establish good relations with Administration staff. Hand claimed after the building of Saiho had started, that the land there had been promised previously to the Mission. Rumours then reached senior Administration staff that the bishop had told villagers the Saiho land belonged to God who would destroy Saiho because of the Administration's takeover. These stories did not impress the Administration. Relations with the bishop remained strained, Gunther characterising the bishop's style of Christianity as 'militant'.[30] Furthermore, other rumours circulated regarding the bishop and a message having been relayed to parishioners of the Mission that the eruption had been an example of the 'Wrath of God'.

Seeking Explanation and Meaning

Tony Taylor soon decided on a volcanological explanation for the Lamington disaster of 21 January 1951. It had been caused by passage of a *nuée ardente* — or, in more modern terminology, a 'block-and-ash' type of pyroclastic flow. The eruption that morning did not develop a high, umbrella-type, plinian eruption column with a 'stalk', as had happened at Vesuvius in AD 79, but rather had formed a large vulcanian cloud. Much of the cloud's great mass was unable to sustain its upwards momentum for longer than a few minutes. It collapsed, spreading outwards as a pyroclastic flow, especially down the northern flank of the volcano, and devastating the countryside and populated areas. The pyroclastic flow stopped quite abruptly. Hot ash clouds then rose, drawing in the cooler surrounding air in surface winds that felt like a 'drawback' of the pyroclastic flow itself to people on the ground.

Block-and-ash flows are deadly because of their speed, temperature and gas content. They are composed of three main parts, all of which Taylor identified from his field investigations of the volcanic deposits at Lamington, supplemented by eyewitness accounts. First, is a dense, incandescent, basal layer that contains most of the energy of the pyroclastic flow. This layer moves down depressions such as valleys, like an avalanche hugging the ground, and it buries whatever lies in its path. A second part of the block-and-ash flow is represented by impressive ash clouds that roil upwards from the surface of the ground-hugging parts of the flow. These thermally driven clouds can be caught by the wind and drift off, dropping their ash on other parts of the volcano, and they conceal from the side-looking observer the fast-moving parts of the flow on the ground.

---

30   Gunther (1951).

The third part of a block-and-ash flow is the *pyroclastic surge*. Taylor called this the 'ash-hurricane' component because of the similar speeds to those of the colder winds of meteorological hurricanes. A surge is able to move over hills and ridges, unlike the valley-hugging avalanche part. It is less dense and deposits less ash, but it is lethal to the unfortunate people who are caught in it and inhale its hot gases and dust. No autopsies were carried out on the dead at Lamington because 'putrefaction was too advanced when the medical services were free from their urgent obligations to the living', but the general conclusion was that asphyxia and rapid damage to respiratory systems were important factors in the causes of death.[31] Cadaveric spasm was suspected from the rigidity of many bodies. The rupturing of stomachs and extrusion of intestines seen at St Pierre in 1902 was also found at Lamington.

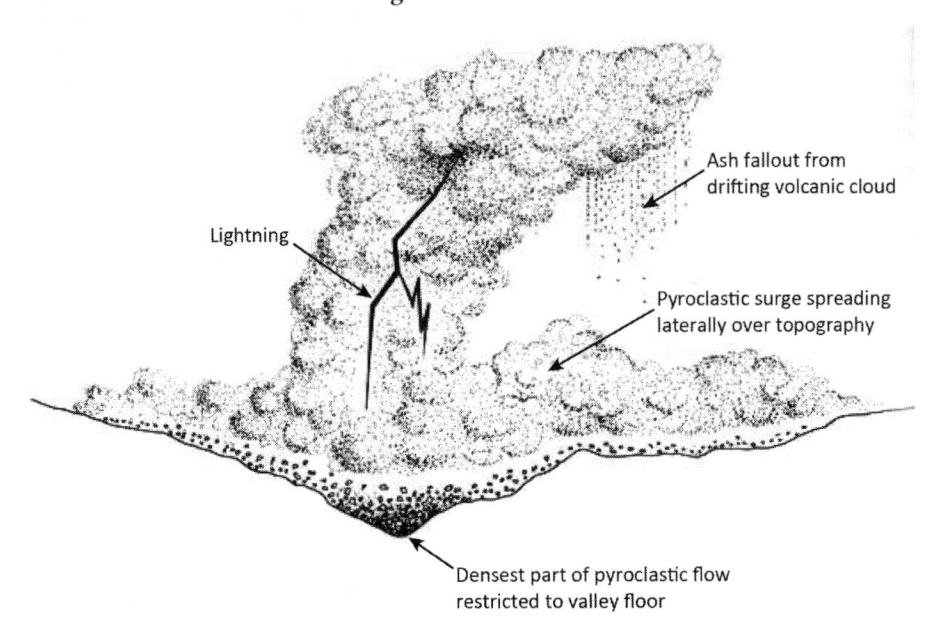

Figure 65. The different parts of a pyroclastic flow being emplaced during a peléean eruption are shown in this diagram, which is based in part on Taylor's landmark study of the 1951 eruption at Lamington. The viewer must imagine facing into the direct path of the fast-encroaching flow.

Source: Adapted from Francis (1993, Figure 12.9). Reproduced with the permission of Oxford University Press.

Many comparisons would be made later by both eyewitnesses and journalists in the Australian media between the Lamington eruption cloud and the atomic-bomb blasts at Hiroshima and Nagasaki less than six years previously. The devastation of the two Japanese cities had been caused by high-energy atmospheric waves that radiated across the cities in seconds. In contrast, the

---

31 Taylor (1958), p. 49.

dark pyroclastic flow that struck Higaturu-Sangara and the surrounding villages took minutes to travel downslope. People in the devastated area are thought to have seen the flow coming and they may have had time to think about escape. Taylor concluded that the speed of the pyroclastic flow on 21 January was between about 100 and 350 kilometres per hour and that the temperatures of the surge — which were insufficient to cause the charring or ignition of wood at Higaturu — may have been about 200 °C for one or two minutes.[32]

Explanations of a different kind were being sought from the Administration in the few weeks after the disaster by the general public and the media and by Percy Spender, the Minister for External Territories.[33] Why did Phillips decide not to evacuate Higaturu-Sangara? Was the Administration aware of the early warning signs that had been seen and felt by people before the devastating eruption? Why did volcanologist Taylor not arrive at the scene until after thousands of people had been killed? These and other questions were being asked and the necessity of an official inquiry was urged. Phillips, in particular, was under considerable pressure and he gave a full report of his actions to the Administrator, and thence to the Minister in Canberra.[34] His detailed report was written with the exactitude and thoroughness of a lawyer, and one can only imagine what emotions he must have endured as a result of his fateful decision, on Friday 19 January, not to evacuate. No official inquiry was considered necessary by the Australian Government.

The rapid onset of the eruption after Phillips' visit on the Friday meant almost certainly that an effective evacuation of the entire area, including Higaturu-Sangara, could not have been undertaken in time, especially over the next day, a Saturday, and without any idea about which areas would be safe from the effects of the unexpected eruption that took place the following Sunday morning. Furthermore, there are considerable doubts whether an arrival by Taylor days earlier would have made any difference. Taylor may have had to undertake many days, at least, of volcanological fieldwork in order to collect information sufficient for making any reasonable assessment of the situation. He would also have had inadequate prior knowledge, if any, of the volcanic geology of the area

---

32 The volcanological interpretations presented here are those of Taylor (1958), who was strongly influenced by the descriptions of the Mount Pelée eruption of 1902. More recently, however, volcanologists who have studied the 1980 eruption at Mount St Helens, United States, have been struck by its similarities to the Lamington eruption of 21 January 1951. The material deposited by a 'surge'-like cloud that swept across the northern flank of Mount St Helens has been called a 'blast deposit', and the blast itself has been interpreted as the lateral release of pressure from a body of magma that had accumulated beneath the northern flank of Mount St Helens, pushing it outwards. The magma pressure was finally released, sideways, on 18 May 1980 when the flank gave way, also producing a debris-avalanche deposit. Overseas volcanologists who have visited Mount Lamington interpret the 'ash-hurricane' or 'surge' materials there as a blast deposit and have also identified debris-avalanche deposits. The results of these field studies (see, for example, Hoblitt, 1982) have not yet been published in the peer-reviewed literature.
33 National Archives of Australia (1951).
34 Phillips (1951).

or of this type of volcano in general. Taylor, had he arrived earlier, would have been killed by the eruption if he had based himself at Higaturu-Sangara or had undertaken fieldwork on the northern side of the mountain.

People in both the Territory and Australia who were seeking a full explanation — if not someone to be held accountable — may to some extent have had their attention diverted from recognising the success of the Administration's disaster relief effort at Lamington. Important factors in this achievement were strong leadership by senior officials, who had known the Territory since before the war; experience in the management and deployment of resources during and after the war; stringent selection and training of a new generation of able, postwar, patrol officers for the Territory; and, the availability of aircraft and nearby airstrips for the delivery of supplies. Important support roles were provided by Melanesians as police, medical assistants and mission workers. The success of the disaster relief, however, is in stark contrast to the lack of volcanic-disaster prevention and preparedness at Lamington before the catastrophe.

Explanation and meaning were explored after the disaster at a deeper metaphysical or religious level by the Orokaiva.[35] Their conclusion was that they must have broken a covenant with a higher being and that, accordingly, the disaster was their punishment. Anthropologists Cyril Belshaw and Felix Keesing found that initially, in 1951, many Orokaiva considered the disaster to be a result of the wrath of the Christian God. This, the Orokaiva thought, was because of their disobedience towards God and the directions of the Anglican Mission, or because they had not supported sufficiently the Administration's efforts in economic development, or had not done enough for the Allies in the war. Another explanation — which may have derived from the Yega coastal people — was that the eruption was a payback for the wartime hangings by the Europeans at Higaturu. One of the 35 Europeans killed at Lamington in 1951 was W.R. Humphries, Director of Native Labour in the Administration who, during war service, had been involved in arranging the hangings of the Orokaiva at Higaturu in 1943.

The Administration levelled criticism at Hand that the Mission itself had been verbally promulgating the Wrath of God explanation amongst the Orokaiva, and they even alleged that the bishop had said he would cause another eruption from the mountain unless he was obeyed.[36] Hand strenuously denied these claims of blackmail by the Mission.[37] Furthermore, former employees of the Mission in 1951 say today that they cannot accept that Hand, a committed Christian, would ever initiate such rumours. Hand did, however, confirm in the

---

35 Belshaw (1951), Keesing (1952) and Schwimmer (1969, 1973, 1977).
36 Schwimmer (1977).
37 Hand (2002).

mid-1960s the prevalence of the Wrath of God explanation found by Keesing and Belshaw in 1951. There is additional uncertainty about the extent to which the Orokaiva themselves were testing, or even exploiting, perceived weaknesses in the relationship between the Administration and Mission, their joint colonial masters.

Many Orokaiva by the mid-1960s, in any case, believed that Sumbiripa, the spirit who lived at the top of Lamington, had caused the disaster.[38] The peace of the mountain had been interrupted by disrespectful acts, such as grenades being let off during the war or hunting on the mountain with guns. One anthropologist concluded that Orokaiva explanations of the Lamington disaster, whether based on the anger of Sumbiripa, or of the Christian God, or even of the government, were consistent with a basic world view of the fundamental importance of exchange-and-reciprocity arrangements in all aspects of life. The Orokaiva, in any of these cases, would have believed that they had violated an agreement and, accordingly, the volcanic disaster was a punishment.

Aftermath

Regrowth of vegetation in the devastated area was rapid, particularly after the explosive eruptive activity diminished at the volcano.[39] A curious, bright orange-yellow bloom of the fungus *Neurospora* appeared in the area following a light shower of rain three days after the 21 January eruption, and it lasted several weeks. Its spores had germinated in the high temperatures of the pyroclastic surges. The fresh ash encouraged growth rather than its retardation, and tuber plants especially — taro, yam and sweet potato — without competition from other plants, began shooting within weeks from the buried soils and from the new humus formed from the destroyed vegetation trapped beneath the ash. Grasses and other secondary growth then took over, and gradually the volcano became shrouded in its more familiar deep green. The slowly cooling lava dome, however, remained bare for many more years. Mount Lamington today is shrouded in vegetation, as it was before the 1951 disaster.

New Orokaiva settlements and land-use patterns were established in ways determined by the effects of the eruption on the land and by the new approaches of the people themselves, but the disaster was only one factor that influenced the future of the Orokaiva. The Administration was committed to postwar strategies of economic development, and Orokaiva people had travelled to other parts the Territory, returning home with new ideas. These are only two of the factors that

---

38 Schwimmer (1969).
39 Taylor (1958).

influenced what was a complex pattern of social change.[40] Popondetta became the new centre for district administration. There were changes, too, in the Territory Administration. Paul M.C. Hasluck took over as Australian Minister for Territories in the re-elected Liberal Government led by R.J. Menzies in May 1951, and he visited the Lamington area that year, flying over the crater with Taylor during an aerial inspection.[41] Hasluck appointed Donald M. Cleland as the Assistant Administrator in the Territory, Murray was dismissed in May 1952, and Cleland then took over as the acting Administrator.

Five more European bodies were found in January 1952 at the deserted ruins of Higaturu, where the regrowth was now metres high.[42] Burial of these bodies at Higaturu was considered inappropriate and a decision was made to have them, and the disinterred remains of some other Europeans, reburied at a new cemetery at Popondetta. The reburials became an occasion for a ceremonial opening of the Mount Lamington Memorial Cemetery in a plot of ground whose pathways were arranged like the Christian cross of crucifixion.[43] The cemetery included a central plaque in memory of all those who lost their lives in the volcanic disaster, and the memorial was opened — and the plaque unveiled — by Hasluck on 24 November 1952. Individual graves were in the upper two quadrants of the cemetery and were mainly those of Europeans. The colonial authorities had there memorialised most of their individual dead but had not been able to individualise the memory of the greater number of Orokavia who had perished.

Fourteen people, mainly European, received medals in recognition of their work during the stressful and dangerous disaster-relief phase of the eruption, and five of them were presented with their awards by Cleland, after the ceremony at Popondetta on 24 November 1952.[44] These included Tony Taylor GC and Leslie Topue BEM. No staff of the Department of District Services and Native Affairs, including patrol officers, were given awards. There were evidently too many worthy candidates, thus making the task of preference and prioritisation difficult, but perhaps also a view prevailed that Murray and his officers had done no more than their duty during the Lamington emergency.[45] One other amongst the worthy candidates was S.A. Lonergan, Government Secretary in Port Moresby, about whom, on the day of Lonergan's retirement in 1959, John Gunther said to the assembled Legislative Assembly:

> One day … someone will write the full history of the Administration's success in meeting the terrible tragedy of Mt. Lamington. I believe the

---

40 Schwimmer (1969).
41 Hasluck (1976).
42 Murray (1952).
43 Australia Department of Territories (1953).
44 Central Chancery of the Orders of the Knighthood (1952).
45 Downs (1980).

8. Disaster at Lamington: 1951–1952

handling of the situation was a magnificent success ... There is no doubt that success can only be achieved if the tools from which people could improvise were made available. In getting the tools and the needs of the field party to them Mr. Lonergan probably contributed more than any other individual towards that great achievement.[46]

Memorialisation is one aspect of the aftermath of any major disaster. Individual grief management and trauma recovery are others. Kindness, sympathy, consideration and prayer abounded after the disaster, and grieving must have been supported privately within individual families and friendship groups. Some traditional wailing could be heard in the camps in the evenings, but Sister Pat Durdin recalled that 'we could not but be impressed by the overall attitude of acceptance [by the Orokaiva, and their] readiness to respond to the immediate demands of the situation'.[47]

Figure 66. Colonial authorities arranged for the ceremonial opening of the Mount Lamington Memorial Cemetery on 24 November 1952 and ensured that everyone knew their place in the official proceedings. The original, single, panoramic photograph has here been split into two overlapping parts.

Source: Australia Department of Territories (1953, first of nine unnumbered plates). Photograph attributed to Papuan Prints, Port Moresby. National Library of Australia.

---

46 Dr Gunther in Legislative Council Debates (1959), p. 577.
47 P. Durdin (personal communication, 2007) — now Sister Patience in holy orders.

Figure 67. Leslie Topue is presented with his British Empire Medal by the Territory Administrator, Brigadier D. M. Cleland, at Popondetta on 24 November 1952.

Source: Photograph supplied by A. Speer.

There were no formal grief-counselling services in the 1950s, and so the number of people who suffered psychological trauma will never be known. Yet individual cases have come to light that hint at the emotional intensity of those who experienced the Lamington tragedy. The anthropologist Marie Reay, for example, is said to have been hospitalised for a long period in 1952 having suffered a mental breakdown on account of her experiences during the eruption.[48] Mrs Ray Kendall, who came to the area as a mission teacher in 1952, told me of the anger in some of the schoolboys in her classroom at that time, only later realising that this was a stage in the grieving process.[49] She also told

---

48 Glick & Beckett (2005).
49 R. Kendall (personal communication, 2006).

the tragic story of an Orokaiva woman who had gone down to the creek to collect water on the morning of 21 January 1951 and had returned to her home to find her husband and children dead. The woman later committed suicide.

More than 60 years have passed since the Lamington disaster and the number of people who experienced, and suffered, its consequences dwindles year by year. The tragedy is unlikely to be forgotten by the present generation of Orokaiva, yet inevitably its immediacy is fading. The green-clad mountain that can be seen from Popondetta, the modern capital of Oro Province today, has not been in eruption since 1951, but few can doubt that Mount Lamington must continue to be regarded as an active and potentially dangerous volcano. There are today population pressures in the Popondetta-Lamington district. Many people have moved back into the area previously devastated by the 1951 eruption.

References

Anonymous, 1951. 'Disaster on Mount Lamington: Two Thousand Feet Blown off a New Guinea Mountain', *Sphere*, 10 February, pp. 190–91.

Australia Department of Territories, 1953. Official Record of the Unveiling of the Mount Lamington Memorial, Popondetta Cemetery, Papua. Canberra.

Belshaw, C.S., 1951. 'Social Consequences of the Mount Lamington Eruption', *Oceania*, 21, pp. 241–52.

Bemmelen, R.W. van, 1939. 'The Geotectonic Structure of New Guinea', *De Ingenieur in Nederlandsch-Indie*, 6, no. 2, pp. 17–27.

Benson, J., 1955. 'The Bapa Saga and the Brothers Ambo', *Anglican* (Sydney), 4 March (no. 134), p. 6.

Best, J.G., 1988. 'A Chronology of Events as They Involved G.A.M. Taylor, Volcanologist, Territory of Papua New Guinea, during the Period January 15–21 1951'. Appendix to letter, dated 1 November, to the Australian Minister for Resources and Energy, Senator Peter Cook.

Biggs, B., 1987. *From Papua with Love*. Australian Board of Missions, Sydney.

Central Chancery of the Orders of the Knighthood, 1952. Supplement to *The London Gazette*, 18 April, pp. 2165–66.

Champion, C., 1951. Telephone message from Government Secretary, Port Moresby, to Department of External Territories, Canberra, 24 January 1951. National Archives of Australia, Series A518 (A518/1), Control Symbol AV918/1 Part 1.

Didymus, H. (trans.), 1974. 'The Disastrous Catastrophe that Shook My Land and My People. The Eruption of Mt. Lamington'. *Territory of Papua New Guinea Public Works Department Newsletter*, March, pp. 5–7.

Downs, I., 1980. *The Australian Trusteeship Papua New Guinea 1945–75*. Department of Home Affairs, Commonwealth of Australia.

Francis, P., 1993. *Volcanoes: A Planetary Perspective*. Oxford University Press.

Glick, P.B. & J. Beckett, 2005. 'Marie Reay, 1922–2004', *Australian Journal of Anthropology*, 16, pp. 394–96.

Graham, G.K., 1974. Letter to Mr J.R. Horne dated 1 April. Department of Agriculture, Stock and Fisheries, Port Moresby, File 1-8-6.

Grahamslaw, T., 1971. 'Grim Retribution for Papuans who Backed the Losing Side', *Pacific Islands Monthly*, 42, no. 5, pp. 41–45, 105–18.

Gunther, J.T., 1951. Letter to J.K. Murray, 30 July 1951. National Archives of Papua New Guinea. Anglican Mission Confidential Matters. File GH47-19, SN 677, AN405, BN 887.

——, 1951–1952. Untitled report. Territory of Papua Annual Report for the period 1st July, 1950, to 30th June, 1951. Parliament of the Commonwealth of Australia, Canberra, pp. 35–36.

Hand, D., 2002. *Modawa: Papua New Guinea and Me 1946–2002*. SalPress, Port Moresby.

Hasluck, P., 1976. *A Time for Building: Australian Administration in Papua and New Guinea 1951–1963*. Melbourne University Press.

Hoblitt, R.P., 1982. 'Reconnaissance of the Area Devastated by the January 21, 1951 Eruption of Mount Lamington, Papua: August 10–18, 1982'. Unpublished administrative report, United States Geological Survey.

Keesing, F.M., 1952. 'The Papuan Orokaiva vs Mt. Lamington: Cultural Shock and its Aftermath', *Human Organisation*, 2, no. 1, pp. 16–22.

Legislative Council Debates, 1959. Retirement of Mr S.A. Lonergan. Territory of Papua and New Guinea Legislative Council, 23 March 1959, 4, no. 5, pp. 576–79.

Martin, J.D., 2013. 'Mount Lamington'. Una Voce, 2, pp. 42–45.

Montgomery, J.M., M.F. Glaessner, & N. Osbourne, 1950. 'Outline of the Geology of Australian New Guinea', in T.W.E. David (ed. W.R. Browne), *Geology of the Commonwealth of Australia*. Arnold, London, 1, pp. 662–85.

Murphy, J.J., 1951. Mt. Lamington volcano ascent [on 28 May 1948]. Unpublished memorandum dated 8 February to Officer in Charge, Administration Field Group, Popondetta.

Murray, J.K., 1951. Mount Lamington disaster death roll (native) — list of surviving dependents. Memorandum to Secretary, Australian Department of Territories, Canberra, 29 October 1951. National Archives of Australia, Series A518 (A518/1), Control Symbol AV918/1H.

———, 1952. Higaturu victims: Europeans. Memorandum to Secretary, Australian Department of Territories, Canberra, 9 February 1952. National Archives of Australia, Series A518 (A518/1), Control Symbol AV918/1B.

National Archives of Australia, 1951. Volcanic eruption — Mount Lamington. Correspondence file. Series A518 (A518A), Control Symbol AV918/1 Part 3.

Phillips, F.B., 1951. Mt. Lamington eruption. Memorandum to the Administrator, Government House, Port Moresby, 3 February. National Archives of Australia, Series A518 (A518A), Control Symbol AV918/1 Part 3.

Plant, H.T., 1951. Re-establishment of the Isivita villages. Memorandum to the Director of District Services and Native Affairs, Port Moresby, 24 March 1951. Manuscript copy lodged with the National Research Institute, Port Moresby.

Porter, R.G., 1951. 'Eye-witness' Record', *Australian Board of Missions Review*, 34, pp. 24–27.

Schwimmer, E., 1969. *Cultural Consequences of a Volcanic Eruption Experienced by the Mount Lamington Orokaiva. Comparative Study of Cultural Change and Stability in Displaced Communities in the Pacific*, Report 9, Department of Anthropology, University of Oregon, Eugene.

———, 1973. *Exchange in the Social Structure of the Orokaiva: Traditional and Emergent Ideologies in the Northern District of Papua*. St Martin's Press, New York.

———, 1977. 'What did the Eruption Mean?', in M.D. Lieber (ed.), *Exiles and Migrants in Oceania*. University Press of Hawaii, Honolulu, pp. 296–341.

Sinclair, J., 1986. *Balus: The Aeroplane in Papua New Guinea*. 1: *The Early Years*. Robert Brown & Associates, Bathurst.

Stanley, E.R., 1924. *The Geology of Papua*. Government Printer, Melbourne.

Strong, P., 1951. 'A Message to the Wider Church from the Bishop of New Guinea', *Australian Board of Missions Review*, 34, pp. 40–50.

Taylor, G.A.M., 1958. 'The 1951 Eruption of Mount Lamington, Papua'. Bureau of Mineral Resources, Canberra, Bulletin 38.

Tomblin, J.W.S., 1951. *Awakening: A History of the New Guinea Mission*. New Guinea Mission, London.

White, N.H., 1991. *Sharing the Climb*. Oxford University Press, Melbourne.

9. Tony Taylor and an Eruption Time Cluster: 1951–1966

... within a few minutes the roar of a full-scale eruption could be heard all over the island [Manam]. Activity fluctuated but was always on a gigantic scale ... The main crater began roaring like an enormous blow lamp and huge blocks were ejected as far as the forest margins on the northern slopes. So great was the pressure of this emission that ... barometric readings on the island were impossible.

G.A.M. Taylor (1960)

Eruptions of 1951–1957

The Lamington disaster of January 1951 had an immediate impact on public and Administration attitudes towards volcanic threats in the Territory, including fears of other volcanoes soon breaking out in concert with Lamington. Such immediate responses might seem in hindsight to be based on an understandable nervousness following the Lamington tragedy, but good reasons emerged for such public anxiety. This is because the seven-year period from 1951 to 1957 turned out to be another example of a 'pulse' or volcanic time cluster of eruptions in Near Oceania, one in which eight or nine different volcanoes became active. Four of these volcanoes — active in 1953–1956 — were in the western part of the Bismarck Volcanic Arc alone.

The eruption at Lamington also had longer term consequences. Its terrible death toll was regarded as a clear reason why instrumental early warning of eruptions should be carried out by the Rabaul Volcanological Observatory (RVO) throughout the Territory — and by the authorities in the British Solomon Islands Protectorate too — and it provided the rationale for the development of RVO capability in particular. Volcanologist G.A.M. 'Tony' Taylor was now a respected authority with influence and reputation signified by the award of his George Cross. His career had been launched spectacularly by the Lamington disaster, and he was now well positioned to develop the Territory's volcanological service.

Lamington volcano, by the end of 1952, was no longer presenting a threat. Taylor returned to Canberra first on recreation leave, and then began to write Bureau of Mineral Resources (BMR) Bulletin 38. He did not resume full-time duty in the Territory again until 1957. BMR geologist John G. Best was in charge of RVO and was joined by M.A. 'Max' Reynolds in late 1954. They were the first of many expatriates from Australia and other countries to work at RVO over the following decades. Furthermore, N.H. Fisher together with John Grover in 1956 — and later Taylor in 1959 — visited the geothermally active Savo Island in the British Solomon Islands Protectorate, noting the similarity of its geology to that of Mount Lamington, and thus raising questions about future safety on the volcanic island.[1] Similar questions of safety were raised about other 'sleeper' volcanoes in the region.

Table 3. Volcanoes in Eruption in Near Oceania from 1951 to 1957

Kavachi 1950–1953. Eruptions took place during this period and in 1957–1966.[a]
Lamington 1951–1952. Growth of the 1951 lava dome, after the catastrophic eruption of 21 January 1951, continued into at least 1952.[b]
Bagana 1951 and onwards. Bagana had already been in significant eruption in 1950, but activity recurred in early 1951 when numerous block-and-ash type pyroclastic flows were observed.[c]
Long 1953–1955. The first eruption from Long Island to be documented by an observer was of the active volcano on the floor of Lake Wisdom breaking through the lake surface in May 1953.[d]
Tuluman 1953–1957. A protracted period of mainly submarine explosive volcanism immediately south of Lou Island eventually created small islands that included lava flows of obsidian.[e]
Langila 1954–1956. Increases in volcanic-gas emissions from the volcano at the western end of New Britain in 1952 were investigated prior to explosive activity starting there in 1954.[f]
Bam 1954–1955. Explosive activity was minor and the island was evacuated temporarily in late 1954, but with tragic results.[g]
Manam 1956–1966. Major eruptive activity took place during the 1957–1960 period, including 1958, after the population of the island had been evacuated.[h]
Unnamed submarine volcano near Karkar in 1951. Information about the activity is sparse, but a marine disturbance and dead fish were noted on 24 November 1951, 35 kilometres north-north-east of Karkar in a position reported in 1945 to be that of a 'mud volcano'.[i]

a. Johnson & Tuni (1987).
b. McKee (1976).
c. Best (1956b).
d. Best (1956a).
e. Reynolds et al. (1980).
f. Taylor et al. (1957).
g. Taylor (1955a).
h. Taylor (1960).
i. Fisher (1957).

---

1   Grover (1958) and Taylor (1965).

9. Tony Taylor and an Eruption Time Cluster: 1951–1966

Figure 68. Increased emission of gases at Langila volcano, western New Britain, in 1952 prompted investigations which confirmed that future eruptive activity was likely. Tony Taylor is here seen descending into the active crater at Langila. Low-grade vulcanian eruptions did indeed start at Langila — in 1954 — but without any serious effects on nearby communities.

Source: J.G. Best (likely photographer). Geoscience Australia (GA-9348.tif).

Taylor, Best and Reynolds, and Fisher before them, were Australian geologists trained in how to identify rocks of different types and ages, how to map rock distributions, and how to assess their mineral and petroleum potential. Transforming themselves into volcanologists, assessing the eruptive potential of active volcanoes, and advising the authorities on whether to evacuate at-risk communities were new skills that they each had to learn soon after arriving in the Territory. These challenges were significant, bearing in mind the number of active volcanoes to be monitored, the general deficiency in monitoring instrumentation at that time, and the fact that so many volcanic eruptions needed to be investigated during the 1950s time cluster. The general approach was based on the need to investigate each report of volcanic activity as soon as it was received in Rabaul, to keep instruments operating, and to maintain effective communication links with Administration officials in the volcanically active districts of the Territory of Papua and New Guinea. Investigations of the volcanoes and their activity in the 1950s were aimed at understanding how the volcanoes 'worked' in a scientific sense, but the focus was oriented foremost towards a practical use — that is, on identifying and interpreting for public safety purposes the precursory signs of impending eruptions.

Experiments in Prediction

A foremost challenge for RVO staff in the early 1950s was being able to proffer realistic and scientifically based recommendations to authorities who were faced with signs of an impending eruption, or with a volcano that had already started activity. A general framework was required for understanding volcanoes and their activity and this is what Taylor began to address.

Taylor developed a three-part hypothesis for volcanic outbreaks and eruption prediction against which observational data could be assessed and interpretations tested. The first part of what Taylor called an 'experimental' approach was assessing the condition of each volcano itself, including the behaviour of the underlying magma. Visual observations of crater activity were critical. So, too, were the data obtained from seismographs, tiltmeters, thermometric instruments, and volcanic-gas tests, although instrumentally based techniques for volcano monitoring and interpretation of the recorded data were all primitive by present-day standards. Taylor also developed a particular fascination with volcanic sound, in part because he thought it might indicate the nature of crater activity where the tops of the active volcanoes were covered in weather cloud, but also where particular types or trends in sound might presage an eruption.

The second part of the hypothesis represented Taylor's conviction that volcanoes at low latitudes, such as in New Guinea, were influenced by the changing positions of the Sun and Moon relative to the Earth at different times of the year and month. Taylor took the advice of the well-known American volcanologist Frank Perret who, before the Second World War, had recommended access to an ephemeris of tide times before planning fieldwork on active volcanoes. Taylor had noticed such 'earth tide' relationships during the 1951 Lamington eruption where the gravitational pull of the Sun and Moon seemed to have had a triggering effect, particularly during the early and highly explosive phase.[2] He later pointed out other, similar correlations during the 1951–1957 period, including the use of solstice and equinoctial times to remarkable effect in forecasting major explosive activity at Manam volcano.[3]

The third part of Taylor's hypothesis was that any upsurge of tectonic earthquake activity near a volcano might lead to an eruption, as Fisher had proposed for the 1941 eruption at Tavurvur, Rabaul. Taylor himself became convinced of the usefulness of this relationship between 'regional tectonic-stress release' and subsequent volcanic activity after investigating eruptive activity at Ambrym volcano in the New Hebrides.[4] He believed that tectonic earthquakes in 1949–1950 had triggered the 1951–1957 volcanic upsurge and proposed, too, that the

---

2   Taylor (1958a).
3   Taylor (1960).
4   Taylor (1952, 1956).

severe earthquake and tsunami damage reported by Miklouho-Maclay on the Rai coast in the mid-1870s was the tectonic trigger for the subsequent pulse of eruptive activity.[5]

The earthquake data available to Taylor was, on present-day standards, severely limited for testing these ideas, as he could use only relatively imprecise data on large earthquakes recorded at worldwide recording stations and supplied from the Riverview Observatory in Australia. Taylor therefore urged the establishment in the Territory of volcanological observatories on Manam Island and at Esa'ala on Dawson Strait in the D'Entrecasteaux Islands, where regional earthquake recording could also be undertaken. These two observatories were planned to complement the earthquake recording already being done at Rabaul, and at Lamington, and to provide a capability for the detection of precursors of possible eruptions at Manam and from volcanoes in Dawson Strait.[6]

There were three reasons why Taylor recommended establishment of a full-scale observatory at Esa'ala. He was concerned, first, about reports in 1953–1955 of numerous, generally small, local earthquakes being felt in the Dawson Strait area — some taking place soon after others in so-called 'seismic storms'. Second, he was aware of the geological youthfulness of the volcanoes of Dobu, Oiau and Lamonai in Dawson Strait. And, third, there were ongoing and worrying reports about changes to the character of geothermal fields in the area, especially the large ones at Deidei and Iamelele on Fergusson Island. Taylor thought that the Dawson Strait earthquakes were part of a pattern of regional tectonic-stress release in eastern Papua, which, following the volcanic eruptions at Goropu in 1943–1944 and then Lamington in 1951, might mean new explosive outbursts taking place in the Dawson Strait area.[7] The signs of unrest at Esa'ala in the 1950s, however, became of less importance when no volcanic eruptions took place and as the volcanic threat at other volcanoes began to dominate the time of RVO volcanologists during 1953–1957. Those Esa'ala earthquakes would nevertheless resonate historically in 1969 when new earthquake activity and the fear of a volcanic outbreak in Dawson Strait would cause the evacuation of thousands of people.

Long Island Evacuation of 1953–1954

Most of the 1951–1957 volcanic activity affected only under-populated areas, but three of the volcanoes — Long, Bam and Manam — in the western part of the Bismarck Volcanic Arc — caused disaster management problems for local communities and involved decisions to evacuate by the Territory Administration. Long Island was the first of these.

---

5   Taylor (1955c, 1960).
6   Taylor (1966).
7   Taylor (1955b).

A Lutheran missionary on Umboi Island reported to RVO on Saturday 9 May 1953 that an eruption appeared to be taking place in the vicinity of Long Island to the north-west of Umboi. Best was then in charge of RVO, and that afternoon in Rabaul he requested that an aircraft of Qantas Empire Airways divert over Long to investigate the eruption. The pilot reported back to RVO and to aviation authorities that 'a small horseshoe island' in the centre of Lake Wisdom was 'hurtling jets of mud and steam every ten minutes to a height of three to five hundred feet'.[8] News of the eruption, however, did not reach the District Commissioner, C.D. Bates, at Madang until the following morning, Sunday 10 May. Bates immediately organised a flight over the lake, saw the eruptive activity for himself, and concluded that 'present indications were a major eruption was imminent'. He then put in place arrangements for a seaborne evacuation of the whole island. Long Island villagers live on the coast, rather than on the shores of the lake, and they evidently had been unaware that an eruption was taking place. All 377 islanders were, with their concurrence, taken on board three vessels that left Long Island before dawn on Wednesday 13 May. Furthermore, arrangements began to be made by the District Commissioner for the permanent settlement of the evacuees at Saidor on the mainland, near to Long Island.

No volcanological advice concerning the evacuation was sought from RVO, and neither was the District Commissioner's head of department in Port Moresby involved in the prompt decision to evacuate the Long population permanently. Any such rapid decision to evacuate presumably would have had to be balanced against the longer term effects that the social dislocation may have on the lives of evacuees, which are put 'on hold' in a new and often stressful environment. Evacuees are away from their homeland, their own traditional lands and may, against their own will, have to become strongly dependent on, and subservient to, the requirements and resources of the people hosting them. Furthermore, there are financial costs involved for the governing authorities. There seems little doubt, however, that the District Commissioner put the safety of the islanders ahead of any such difficulties likely to be faced by the Long evacuees, and the Administration, on the mainland.

Best undertook an aerial inspection of the active volcano in Lake Wisdom on 12 May, and that day discussed the situation with the authorities in Madang, but he was not in a position to make any useful, evidence-based recommendations regarding the likely development of the Long Island eruptions. Best was the only volcanologist in the Territory at that time. Long Island was a long way from Rabaul, the island could be reached only by sea, and RVO had neither instrumentation nor staff on the island. Also, Best had insufficient background knowledge on the nature of the volcano, and monitoring the dangerously active centre itself presented insurmountable practical difficulties — that is, in the

---

8   Bates (1953), p. 1. See also Best (1955, 1956a).

middle of a lake situated several kilometres from the lake shore. Best had to rely almost entirely on observations made during aerial inspections to determine how the eruption was playing out. He was, however, aware from fieldwork on Long Island with Taylor in 1952 that a major caldera-forming eruption had taken place at Long within recent centuries.[9] Best saw that the current eruptions in the lake were minor in comparison to these earlier events, but he exercised caution and thus, initially at least, supported the need for the evacuation.

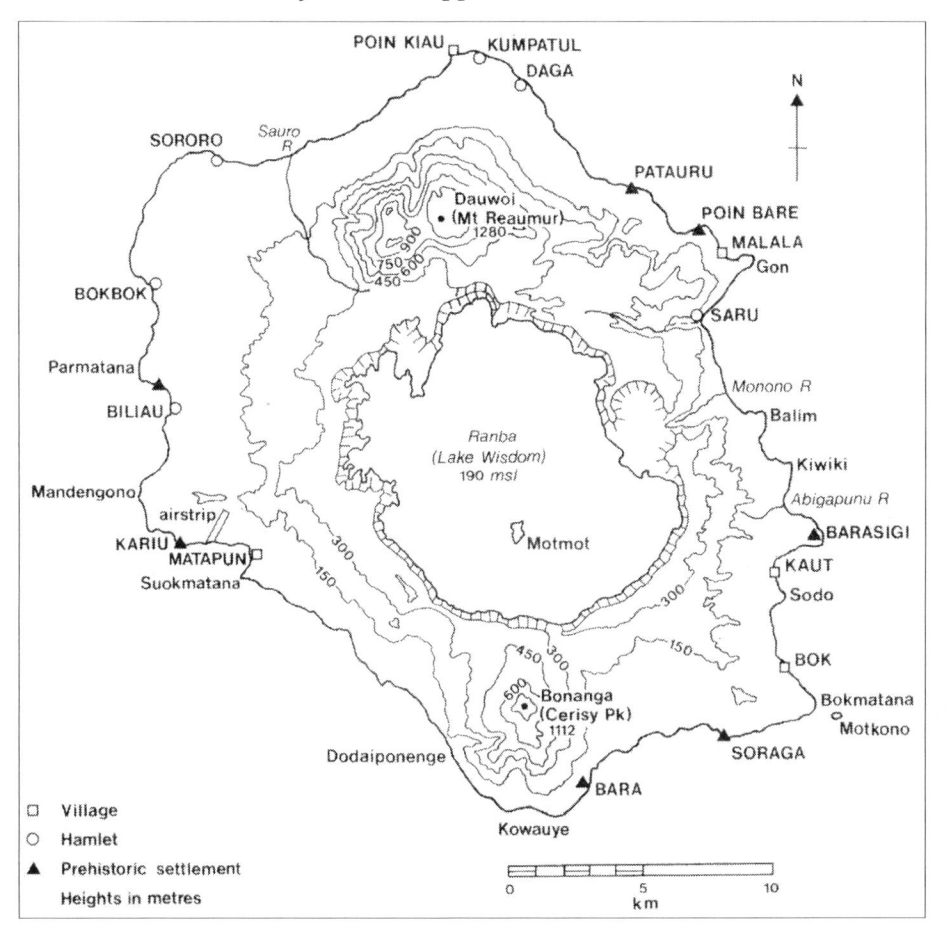

Figure 69. Villages and prehistoric settlements on Long Island are restricted to the coast. No people live on the shores of the freshwater caldera lake where the active lake-floor volcano periodically forms the island of Motmot, the local name for 'island'. Dauwoi and Bonanga are extinct stratovolcanoes at either end of Long Island.

Source: Ball & Hughes (1982, Figure 2). Reproduced with the permission of the authors.

---

9   Taylor (1953).

Figure 70. This explosive volcanic activity in Lake Wisdom, Long Island, was photographed during an aerial inspection on 24 May 1953. Strong winds are forming a curtain of dark ash and white water vapour away from an island that is building up out of the lake water. The type of eruption can now be identified as *surtseyan*.

Source: J.G. Best. This photograph is one of several taken during the same aerial inspection. Geoscience Australia (no registered number for this photograph, but see others in the GA series MRG-2 to 13).

Mention of previous major eruptions at Long Island understandably caused a great deal of concern and uncertainty in the minds of Administration staff about ever returning the evacuated people to the island — even in June 1953, during which no further eruptive activity was reported and, in July, when Best visited the island and confirmed that the eruptions apparently had stopped.[10] Opinions about the situation were provided by Taylor and Fisher, both in Canberra, and a view was soon reached in RVO that a recommendation to return the Longs was warranted. Some senior staff in the Administration, however, thought that the volcanological advice should be disregarded and that permanent relocation to the mainland was still the safest option. For example, the Assistant Government Secretary, Claude Champion, on 5 October wrote: 'After the terrible disaster of Mount Lamington, I am of the opinion that the natives should remain on the mainland, and not return to Long Island'.[11] Champion had seen, firsthand, the horror of the devastation and human toll at Higaturu on 23 January 1951, and

---

10 See National Archives of Papua New Guinea (1953–1954) for Administration correspondence.
11 Champion (1953), p. 2.

this experience clearly influenced his strong opinion. Nevertheless, ongoing discussions continued for weeks, those in authority gradually moved away from the idea of permanent relocation, and the Administrator, Brigadier D.M. Cleland, by end of December 1953 had authorised the return of the displaced people to Long Island. The islanders were taken home between 9 and 17 March 1954, ten months after the benign lake eruptions had first been observed, and a few weeks after what may have been a final outburst from the lake, at least for this eruptive period, in January 1954.

Memories of the 1953–1954 evacuation remained as the most dramatic event of the lives of Long Island people for at least a generation, even though the eruptions had not affected them directly.[12] The volcanic eruptions from the lake at Long Island in 1953 were in fact too distant and too small to be of any danger to the sea coastal villages, and similar eruptions in 1955, 1968, 1973 and 1976, for example, never again triggered the need for evacuation of the Long Island people.

Bam Tragedy of 1954–1955

Almost 430 people in the mid-1950s lived in three villages on a narrow coastal shelf on the northern tip of Bam Island. Their lives seemed volcanically precarious to any visitor, as Bam is only 3.2 kilometres from north to south, and the volcano rises steeply out of the sea to an active summit crater at 640 metres above sea level, less than two kilometres from the villages. Furthermore, the island has no natural harbours or safe anchorages for larger vessels. Bam people were, nevertheless, committed to life on the island and did not doubt its sustainability. They were expert canoeists and fishermen, and had traditional ties with neighbouring communities, particularly on the nearby islands of Kadovar and Blup Blup. They grew coconuts and fed on the abundant marine life of the surrounding waters that were enriched by outflows from the nearby Sepik River. Bam Island evidently was also relatively free from diseases such as malaria and tuberculosis, which meant, however, that the Bams would have been vulnerable to them elsewhere if exposed to infection.

Bam people were familiar with the normally mild eruptions from the volcano. They also had stories of stronger eruptive activity — in the 1860–1870s, for example, when the islanders escaped to safety on Kadovar and Blup Blup.[13] Other eruptive activity had been noted by the Bam people in about 1909, the early 1920s, and the late 1940s. They had observed, too, that volcanic explosions seemed to coincide with the change in seasons or with eruptive activity at Manam. Europeans had recorded eruptions at Bam before 1954, although the

---

12 Ball & Hughes (1982).
13 Cooke & Johnson (1981).

descriptions in most cases are somewhat brief and unclear. Miklouho-Maclay saw some sort of activity there on 13 November 1877, in one account stating — without elaboration — that both Bam and Manam were 'in full eruption'.[14]

News of renewed volcanic eruptions on Bam was received at RVO on 4 November 1954 and the activity was investigated by Best a few days later.[15] He noted that the eruptions had been mild and had not affected the villages, but that lava blocks had been expelled and were hot enough to have scorched grass near the summit crater. Best concluded that the immediate threat was not significant, but that the longer term risk of stronger and potentially disastrous volcanic activity was too great. He therefore recommended to the Territory Administration the somewhat drastic action of permanent evacuation of the island and the resettlement of the people elsewhere.

RVO volcanologist Reynolds visited the island a few days later, but he could find no good reasons to justify an evacuation.[16] The islanders themselves seemed unperturbed by the behaviour of the volcano and told him that the eruptions had been normal. Reynolds indeed was more concerned about new reports of a possible resurgence of volcanic activity at Long Island and left Bam for Long to investigate. Taylor, in Canberra, however supported an evacuation of Bam on the basis of his assessment that a period of volcanic 'regional unrest' was underway and that further eruptions at Bam might be expected.[17] Other options to enforced evacuation, he said, were to establish a European observer and volcano-monitoring equipment on the island, and to provide additional boats for the islanders in the event of a large eruption.

The Administrator, on 25 November 1954, authorised evacuation of Bam Island, having taken all of the volcanological assessments into account; and the removal of the islanders was completed by 2 December under the overall direction of the District Commissioner at Madang, F.E. Bensted. The Bams were taken to the mainland and were allocated some land at Dagoi, a few kilometres east of Bogia, for a new village and gardens where they could become farmers. There was a swamp nearby that needed draining.

Life for the Bams at Dagoi began to deteriorate in the weeks after their evacuation from Bam in 1954. Their decline on the mainland was reported by the Port Moresby news media in February 1955, and comments were made about the 'disgraceful' and unhygienic conditions at Dagoi, about two deaths from dysentery, and about the absence of supervision of anti-malarial medication.[18] The Bams were evidently vulnerable to mainland diseases and had problems

---

14   Miklouho-Maclay (1884–1885), p. 965.
15   Best (1954) and Taylor (1955a).
16   Reynolds (2005).
17   Lambert (1954).
18   Anonymous (1955).

adapting to a new diet. Furthermore, cultural ties with their ancestral island home were too strong for them simply to start a new life in a strange, if not alien environment, and serious psychological effects began to be noticed. The number of deaths at Dagoi continued to rise. An evacuation based on a scientific assessment of the volcano and recommended with the best of intentions, was turning into a tragedy.

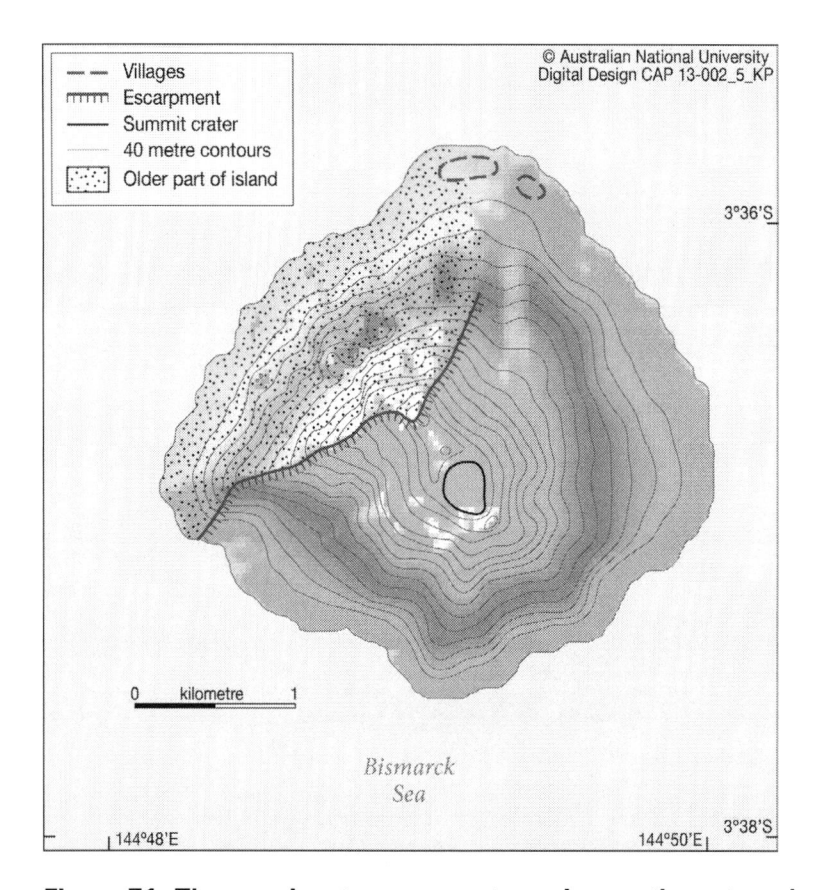

Figure 71. The prominent escarpment running north-eastwards across Bam Island and down towards the villages was probably formed by a prehistoric gravitational collapse or collapses of the volcano towards the south-east. Later eruptions from the central crater then built up the volcano alongside the older, north-western part of the island.

Source: Adapted from Cooke & Johnson (1981, Figure 3).

Figure 72. Bam Island from the south-west in 1970. The peak on the right is at the south-eastern end of the summit crater, and the peak on the left is the summit of the collapse escarpment.

Source: R.W. Johnson. Geoscience Australia (GA-5783).

Figure 73. Dejected islanders being evacuated from Bam Island on 30 November 1954.

Source: Photograph supplied by M.A. Reynolds.

The Director of Public Health, John T. Gunther, was critical and terse in a memorandum, dated 14 February 1955 that he sent to the Administrator through the Government Secretary. Gunther pointed out that his headquarters in Port Moresby did not become aware of the evacuation until the second week of December and that, had they known, they would have been able to manage the evacuation and resettlement more effectively and humanely than had actually been the case. He and his staff, after all, had experience managing the health aspects of the relief and recovery phases of the 1951 Lamington eruption. Gunther wrote:

> Resettlement needs good and constant leadership. If it is to succeed such leadership must ensure, if possible, that the people are gently led into their new environment and that sanitation is satisfactory, disease is prevented, food is proper and adequate, and that occupation and recreation maintains mental health against the longings for a lost home. We did none of these things until it was too late. We had made no [Health] staff available at the beginning to meet the impacts, and, as so often occurs, we are trying to correct errors and undo harm.

He added:

> Vulcanologists are in a most invidious position when asked to assess the need for evacuation. If they are conservative a great tragedy might occur, thus they must tend, if there is any doubt, to recommend the safest precautions ... I believe their position is very like a medical practitioner's ... In surgery there are risks. Every surgeon must measure the risks ...
>
> Bam Island was like that. The history of its behaviour was scanty and a constant watch could not be kept. The islanders apparently agreed to go because too many of the pros for going (e.g. Mt. Lamington) were stressed to them.[19]

Gunther recommended that the Bams be allowed to return to the island, subject to another visit and assessment of the situation at Bam by a volcanologist. Others were of the same opinion. For example, Bishop A.A. Noser of the Vicariate Apostolic of Alexhaven wrote to the Administrator on 16 February 1955 concerning the physical and mental state of the Bams on the mainland:

> I have serious doubts about the final outcome of the present arrangement. Dissatisfaction can be very contagious and burst into conflagration at a most inopportune moment. Personally I think it is best to let them return home.[20]

---

19 Gunther (1955).
20 Noser (1955).

The Bams themselves in February also expressed their dissatisfaction and frustration, if not repressed anger, in a letter originally written in Tok Pisin:

> Why have we left our home, the home of our forefathers ... . We are dissatisfied with the mainland; we are constantly sick — the hospital is full of men, women and children ... . We have lost many things ... . Eight children, one young woman, two men and one old woman have died ... . The natives of Bogia ridicule us — they are angry with us and object to us. Native police are angry with us and call us thieves ... . We are people without a homeland. Why? Release us and let us go back to our island.[21]

Twenty-four Bam islanders died at Dagoi. Cleland eventually, in the second week of May 1955, issued firm instructions outlining the requirements that would need to be met for the return of the Bams to their island.[22] This included immediate installation by RVO of appropriate volcano-monitoring instruments on Bam, the training of a local observer in the use of the equipment, radio communications with Rabaul, and the procurement of additional canoes to ensure the success of any future complete evacuation of the island. The Bams returned to their island in June 1955 and monitoring instruments were installed later the same month. The Administration had, therefore, backed off from any consideration to abandon the island permanently, as recommended by Best, to a position that recognised the self-sufficient character of the islanders when living in their homeland, coupled with the need for ongoing volcano monitoring. The Bam tragedy had been salutary in illustrating the need for government administrations to assess the relative degrees of self-reliance and vulnerability of 'at-risk' communities before enforcing evacuations from restless volcanoes.

Eruptions at Bam volcano continued intermittently for another five years after 1955, but these were mainly mild gas and vapour explosions and weak, soundless, ash ejections of low-grade vulcanian type that had little impact on the villagers and their lives. The monitoring results did not point to any significant escalation of eruptive activity, and monitoring was withdrawn in January 1957.[23] It was, however, resumed for a short period in 1958 because of fears that Bam volcano might break out again in activity following the major eruptions that had been taking place at nearby Manam.

---

21 Roberts (1955).
22 Cleland (1955).
23 Taylor (1958b).

9. Tony Taylor and an Eruption Time Cluster: 1951–1966

Evacuation of Manam and the 1956–1966 Eruptions

Eruptions at seven or eight volcanoes of the 1951–1957 'cluster' had already started by 1955–1956, but Manam — the region's most frequently active volcano — had remained 'anomalously quiet'[24] since its previous significant eruptions in 1936–1939 and 1946–1947.[25] Thus, the start of new strombolian activity at the volcano in December 1956 that lasted until the following February, was not wholly unexpected. Indeed, a detailed evacuation plan for Manam had been drawn up by district authorities in Madang as early as October 1955, triggered by cautious remarks in volcanological reports by Taylor and Best.[26] Taylor related the volcanic outburst in December 1956 to previous tectonic earthquake activity west of Manam in the nearby Sepik area. More eruptive activity at Manam followed in May–July 1957, and pyroclastic flows were observed in October 1957, alerting Taylor to the danger of this kind of volcanic hazard to the population on the island.

An observation post was established at Waris, or Warisi, village on the east coast of Manam Island in June 1957. It was manned by a local villager, Nelson Wangiga, who had previous observatory experience on Bam Island. Wangiga was provided with a low-magnification earthquake recorder, a single-component bubble tiltmeter, a barometer and a thermometer. Taylor was required to make a prognosis about eruptive activity, but was hard pressed to do so on the basis of the instrumental data, particularly bearing in mind there was no 'baseline' information available for comparative purposes. He nevertheless, in early November 1957, informed the Administration's district headquarters in Madang that 'the volcano was building up to a phase of climactic release probably in December–January 1957–58 or June–July 1958'[27] — that is, during the southern summer or winter solstices.

Further activity culminated in a series of powerful outbursts on 6–8 December 1957, the islanders themselves expressed a desire to leave the island, and an evacuation was initiated by the Administration, starting on 11 December and completed by 13 December. The evacuation was expedited efficiently without significant incident, using the detailed plan and a flotilla of district vessels that plied between Manam and Bogia and which were assisted by good weather and calm seas.[28] More than 3,300 Manams were evacuated and then accommodated in mainland villages stretching 70–80 kilometres along the coast. The efficiency

---

24 Taylor (1960), p. 8.
25 Palfreyman & Cooke (1976).
26 Skinner (1955).
27 Taylor (1960), p. 6.
28 Williams (1957).

Fire Mountains of the Islands

of the Administration response was reminiscent of that characterising the relief phase at Lamington in 1951, and was in marked contrast to the failures resulting from the earlier Bam evacuation.

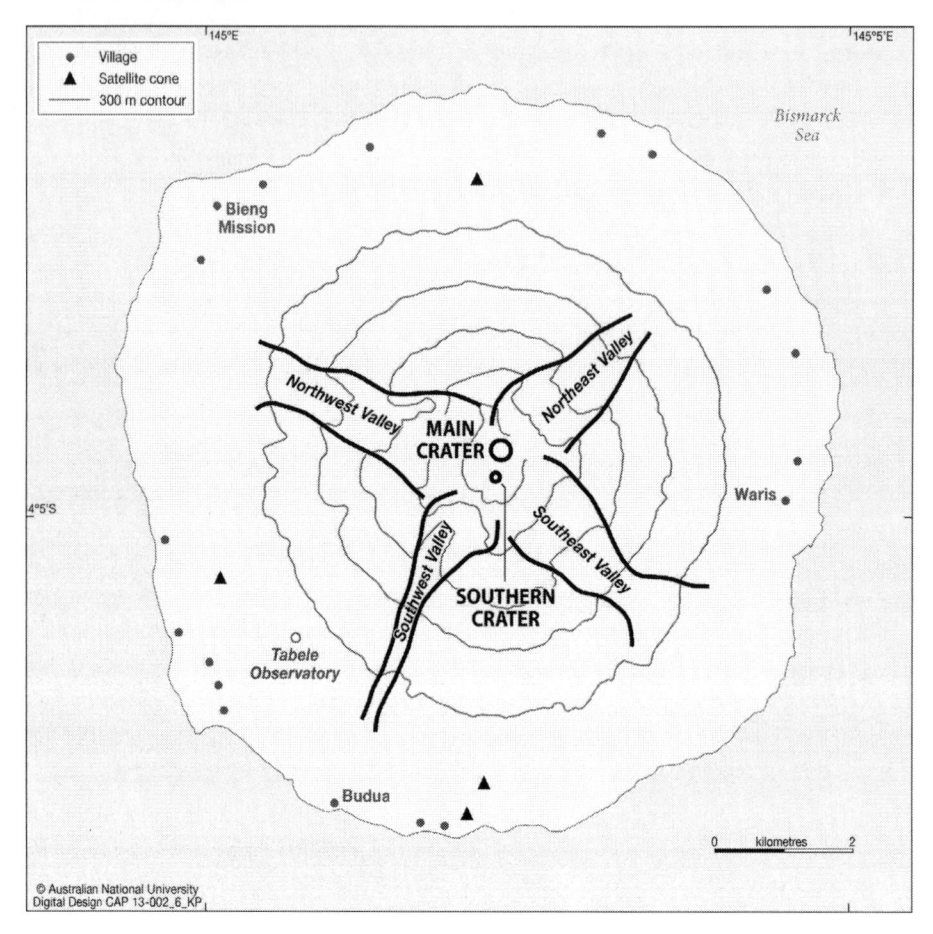

Figure 74. The four 'radial valleys' of Manam Island tend to focus the paths of pyroclastic flows that are emitted from the two summit vents and which discharge down to the coast. Villages are in the locations they occupied in the 1950s. Tabele Observatory was built in 1964.

Source: Adapted from Palfreyman & Cooke (1976, Figure 2).

Manam continued in activity in 1958. Taylor came to the island, installing a Willmore seismograph, and — consistent with Taylor's general forecast — the volcano reached a climax on 25 January 1958, which Taylor described vividly, including the following:

> At 0800 hours nuees ardentes were expelled onto the south-eastern slopes ... . A towering column of vapour and ash rose to more than 30,000 feet and was carried to the west by the high altitude winds. For

the next five hours the column was fed by incessant explosions roaring and rumbling from the summit vents. Near midday the emission reached a new peak of intensity … .

Heavy nuees ardentes … came down the four great valleys dissecting the cone. The nuees from the southern crater were voluminous enough to enter the sea below the south-eastern and south-western valleys. Part of the evacuated village of Budua was completely destroyed and new areas of forest were annihilated.[29]

The fallout of ash and dust from the eruption of 25 January caused village houses to collapse, gardens to be destroyed, and leaves to be stripped from forests across much of the island.[30] Wangiga was still based at Waris maintaining observations as best he could, but he was compelled to abandon his quarters and to occupy one of the remaining houses in the village. Effects from the pyroclastic flows were devastating but, significantly, they were confined mainly to the four main radial valleys. Few coastal villages were sited at the openings to these valleys, the villagers apparently having recognised the vulnerability of such areas. Budua village, however, in the south-west was an exception.

Figure 75. Damage was caused by volcanic ash falling on houses and vegetation in a Manam village in 1958.

Source: G.A.M. Taylor. Geoscience Australia (no reference number).

---

29 Taylor (1960), p. 5.
30 Taylor (1958c).

Administration officials were impressed by the way the mainland villages accepted the 3,341 Manam evacuees, given that the villagers did not know how long their guests would be staying: 'the Bogia Coastal natives have set to and afforded every possible aid to the evacuated people billeted on them. The nearer inland villages have also come to the aid of all by keeping up a continuous supply of building materials for the construction of houses and latrines'.[31] In addition, land was made available to the Manams for temporary gardens, and efforts were made even to find government-owned land where the islanders might settle more permanently. The extent to which provincial government authorities at this time were aware of the true nature of the strong traditions binding the Manams to their island is unclear, including the islanders' views of the cause of these particular eruptions and what could be done to stop them. Nevertheless, an Australian anthropologist had been working in the southern part of Madang Province since 1949, concentrating particularly on the recent rise of different cults of cargoism, and including the key role played by Yali, a famous messiah-like figure from a village on the Rai coast. The anthropologist wrote in 1964:

> In 1953 and 1957, there had been volcanic eruptions on Long and Manam Islands, which [Yali] was said to have caused by invoking the local deities in order to express his hatred of Europeans. In 1957–8, the natives of Manam and the north coast approached him clandestinely to stop the volcano by interceding with the gods once more, and appeared to have offered him a considerable sum of money as payment.[32]

Major eruptions continued at Manam until March 1958 when, on the 4–5 March, there was a powerful and sustained eruption. Such activity prevented any decision on a return of the evacuees to the island, and added great uncertainty to considerations of their future.[33] Evacuations and resettlement involved financial costs, and Administration officials had other, more routine duties to perform in the district, such as patrolling and the collection of taxes. There was concern, too, for the older Manams who, not seeing any prospect of a return to their island, might lose heart and die.[34] Eruptive activity fortunately declined after March and, following a mid-year recurrence of some strombolian activity, preparations were made for the evacuees to return to Manam. This followed favourable advice from Taylor and from W.L. Conroy of the Department of Agriculture who undertook an agricultural and soil survey of the island.[35] A staged return began and much of the island was reoccupied by October 1958.

---

31   Donaldson (1957), p. 1.
32   Lawrence (1964), p. 268.
33   Williams (1958a).
34   Williams (1958b).
35   Conroy (1958).

9. Tony Taylor and an Eruption Time Cluster: 1951–1966

Strombolian volcanic activity took place on Manam each year up to 1966 and, in March 1960, there was yet another phase of greater explosive activity, including pyroclastic flows which descended the North East Valley.[36] Earthquake and tilt measurements gave early warning that enhanced eruptive activity was possible, and Taylor again used the argument of preceding tectonic-earthquake activity and a likely correlation with earth tides. He noted, however, that the gas content of the 1960 eruptions was lower than in 1957–1958 and that, if the villagers stayed away from the four radial valleys, then no evacuation would be required — and, by implication, its attendant costs and disruption would be avoided. Heavy lava outpouring, incessant lava fountains, and brilliant luminous effects were observed from the volcano mid-year when the summit glow was clearly visible at Wewak, 170 kilometres away, but the explosive activity of 1960 had greatly reduced by year's end.

Figure 76. Ash clouds rise from a block-and-ash pyroclastic flow descending the North East Valley on Manam volcano on 17 March 1960.

Source: Taylor (1966, unnumbered photograph on p. 17). Geoscience Australia (M-2044).

Thus ended an eruptive period that was remarkable for the successful application of a volcanological hypothesis by Taylor for evacuation purposes, for the disaster-management lessons learnt by the Administration, and for the absence of deaths that otherwise might have been caused by the major

---

36   Taylor (1963).

explosive eruptions of 1957–1958 and 1960. This work also drew attention to the importance for disaster-management purposes of the four radial valleys on the island. These have the strong effect of channelling pyroclastic flows and lava flows originating from the summit craters.[37]

Wangiga was recommended for an award,[38] but nothing seems to have come of it. The course of Wangiga's subsequent life appears not to have been documented.

Figure 77. Deposits of pyroclastic flows laid down in 1958 in the South West Valley on Manam volcano, are shown in the foreground of this photograph taken in 1963. RVO volcanologist Colin Branch is looking up at the summit where minor explosive activity is taking place. The smaller of the two peaks on the mid-slope, down to the right of the summit, is Iabu Rock.

Source: G.A.M. Taylor. Geoscience Australia (G-5501).

---

37 See also McKee (1981).
38 Taylor (1958d).

9. Tony Taylor and an Eruption Time Cluster: 1951–1966

Tuluman 1953–1957 and the Obsidian Miners of Lou

Eruptions at Tuluman off the southern tip of Lou Island in the St Andrew Strait area, south of Manus Island, in 1953–1957 did not cause any volcanic disasters, but the eruptive activity there[39] is significant for two reasons. First, it represents the type of eruption that had built up Lou Island, where archaeological investigations have yielded some insight on the impact of earlier eruptions on the lives of people who were living there before the arrival of Europeans.[40] Second, the final eruptive phase at Tuluman volcano in 1956–1957 produced, possibly uniquely, sluggish *rhyolite* lava flows that spilled out over newly formed pumice islands, and the lava flows on cooling formed black volcanic obsidian rather than crystallising. There appears to be no other historical eruption anywhere else in the world that has produced obsidian lava flows.

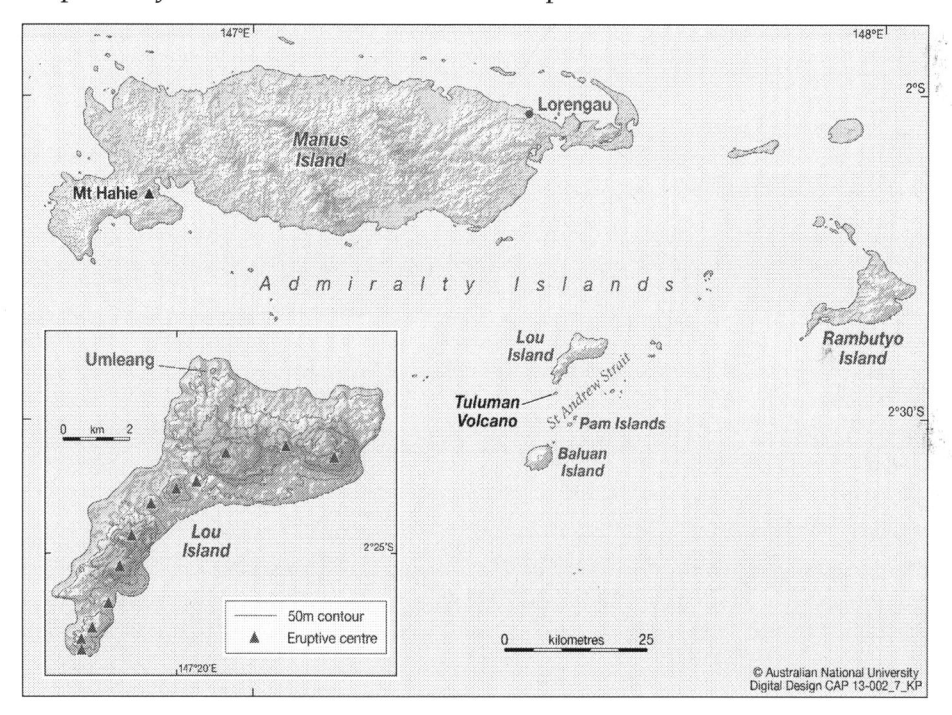

Figure 78. Lou and Pam islands and Mount Hahie are traditional sources of high-quality obsidian in the Admiralty Islands. Tuluman volcano, off the southern tip of Lou Island, produced lava flows of poorer quality obsidian during 1956–1957.

Source: Inset adapted from Johnson & Smith (1974, Figure 3).

---

39 Reynolds et al. (1980).
40 Miklouho-Maclay (1884–1885) observed, from an anchorage on the north coast of Manus Island, a volcanic eruption on 28 March 1883, which may have been from Tuluman or Lou islands.

RVO volcanologists John Best and Max Reynolds were to identify seven separate phases of eruption and eight different eruption centres during the 1953–1957 activity at Tuluman. Their counted number of phases and centres, however, may be imprecise as the two men were unable to make continuous observations over the entire three-and-a-half years of the eruption, because of the need to attend to more important crises at other active volcanoes. The first eruption at Tuluman began in June 1953 a few days after the winter solstice in the Southern Hemisphere — one of several apparent correlations between Tuluman eruptions and earth tides. Much of the Tuluman activity was submarine and, at times, it consisted of blocks of gas-inflated lava which floated up from sea floor vents and exploded on reaching the sea surface.

Lou is arcuate and made up of pumice, ash and obsidian lava flows produced from 12 volcanoes that form the prominent and curved 12-kilometre chain.[41] Tuluman represents the 13th, and most southerly, volcano in the Lou–Tuluman arc, which in a general geological sense seems to be growing anticlockwise along a ring fracture. The pumice and ash layers on Lou Island provide evidence for four main explosive eruptions in the recent past, separated by buried soil horizons that have been radiocarbon dated at about 1,650, 1,920, and 2,100 years ago.[42] These soils in turn contain evidence of early human occupation at times between the eruptions when the island was suitable for horticulture.

People on Lou made use of the island obsidian resource by knapping pieces of high-quality obsidian for the manufacture of blades for knives and spears. Obsidian is a remarkable material, able to be struck into exceptionally sharp edges and therefore valuable for cutting and scraping and, in turn, for use in trade and exchange. The cutting edges are brittle, but obsidian itself is archaeologically durable. Furthermore, obsidian sources are found in only three areas in Near Oceania, one of them being the Admiralty Islands which includes not only Lou but the two nearby Pam Islands, as well as Mount Hahie in western Manus Island.[43] Obsidian from the three main areas, and from sub-sources within each area, can be chemically 'fingerprinted', meaning that trade-and-exchange routes can be traced archaeologically. Lou obsidian has been found in archaeological sites as far west as Malaysia and as far east as the Santa Cruz Islands in Remote Oceania.[44]

---

41 Johnson & Smith (1974).
42 Ambrose (1988). See also Pain (1981).
43 The cultural use of obsidian by people on Manus Island was described by Moseley (1877), the naturalist on board HMS *Challenger*, which anchored off north-western Manus on 3–10 March 1875. Moseley noted obsidian-headed spears or lances, which served as knives, as well as tattooing, which he thought was probably made with obsidian flakes. The Manus people pointed to the mountains of Manus itself as the source of their obsidian — most likely Mount Hahie.
44 See, for example, Summerhayes (2009).

Figure 79. Tuluman Island on 12 January 1957 looking east-southeastward towards Pam Lin Island, in the right background, and showing — in the centre — the dark, new, obsidian flows from Cone 7 joined to the paler island of Cones 2 and 4. The sea to the north-east is stained by volcanic effluent and the remnant islet of Cone 3 is seen in the foreground.

Source: Reynolds et al. (1980, Figure 22). Geoscience Australia (MRG-732).

Artefacts on the surface of the 2,100-year-old soil on Lou Island include obsidian flakes and triangular-sectioned points, charcoal and hearth stones, and similar evidence of reoccupation is also found on the 1,650-year-old soil that was sealed by the latest of the four pumice-and-ash layers.[45] This upper layer evidently concealed obsidian of high quality because people were prepared to dig numerous shafts at Umleang — one 17 metres deep — in order to reach and exploit it.[46] Human remains as evidence of eruption fatalities have yet to be found on the buried-soil surfaces of Lou.

The picture that emerges for Lou Island from the available archaeological evidence is of small settlements of self-sufficient, shoreline peoples who gathered seafood from coastline waters, grew crops, were long-range seafarers, fished in deeper waters, and traded and exchanged high-quality obsidian, possibly over

---

45 Ambrose et al. (1981) and Ambrose (1988).
46 Fullagar & Torrence (1991).

large distances. They evidently coped with repeated volcanic eruptions on their small island — at least, escape in canoes to nearby islands in St Andrew Strait, or further, at times of devastating eruptions seems likely. They were perhaps motivated to return in order to access again the valuable obsidian resource. This general view of self-sufficiency, mobility and resilience has some relevance, albeit contrasting, to what happened on Bam, also a small island when colonial authorities intervened in 1954, evacuated the island population, and caused a disaster rather than preventing one.

References

Ambrose, W.R., 1988. 'An Early Bronze Artefact from Papua New Guinea', *Antiquity*, 62, pp. 483–91.

Ambrose, W.R., J.R. Bird & P. Duerden, 1981. 'The Impermanence of Obsidian Sources in Melanesia', in F. Leach & J. Davidson (eds), *Archaeological Studies of Pacific Stone Resources*. British Archaeological Records, International Series 104, pp. 1–19.

Anonymous, 1955. 'Six Bam Islanders Die in Mainland Home', *South Pacific Post*, Port Moresby, 16 February.

Ball, E.E. & I.M. Hughes, 1982. 'Long Island, Papua New Guinea — People, Resources and Culture', *Records of the Australian Museum*, 34, pp. 463–25.

Bates, C.D., 1953. Summary of Volcanic Disturbance at Lake Wisdom, Long Island, Madang District. National Archives of Papua New Guinea, 1953–54, Re-establishment of Long Island Natives. SN667, AN247, BN 319, File CA 35/7/39, Folios 1–4.

Best, J.G., 1954. 'Preliminary Report Bam Island, Madang Sub-District, New Guinea', Bureau of Mineral Resources, Canberra, Record 1954/59.

——, 1955. 'Volcanic Activity Lake Wisdom, Long Island, T.N.G. May, 1953', Bureau of Mineral Resources, Canberra, Record 1955/78.

——, 1956a. 'Investigations of Recent Volcanic Activity in the Territory of New Guinea'. *Proceedings of 8th Pan Pacific Science Congress, Manila, 1953*; 2: *Geology, Geophysics and Meteorology*, pp. 180–204.

——, 1956b. 'Investigation of Eruptive Activity at Mt. Bagana, Bougainville — March 1952', Bureau of Mineral Resources, Canberra, Record 1956/14.

Champion, C., 1953. Memorandum to Government Secretary, 5 October. National Archives of Papua New Guinea (1953–54). Re-establishment of Long Island Natives. SN667, AN247, Box 319, File CA 35/7/39, Folios 31–32.

Cleland, D.M., 1955. Bam Islanders. Memorandum to Government Secretary, Port Moresby, 10 May. National Archives of Papua New Guinea, Port Moresby, SN 677, AN 247, Box 320, File CA 35/7/51, folios 74–75.

Conroy, W.L., 1958. Survey Report — Manam Island. Memorandum to Director, Department of Agriculture, Stock and Fisheries, from Chief of Division of Agricultural Extension, Port Moresby, 6 October. National Archives of Papua New Guinea, Port Moresby, SN 088, AN 82, File 59-2-16, folios 134–39.

Cooke, R.J.S. & R.W. Johnson, 1981. 'Bam Volcano: Morphology, Geology, and Reported Eruptive History', in R.W. Johnson (ed.), *Cooke-Ravian Volume of Volcanological Papers*. Geological Survey of Papua New Guinea Memoir, 10, pp. 13–21.

Donaldson, P., 1957. Progress Report of Activities re Manam Island Evacuation. Memorandum to District Commissioner, Madang, from Acting Assistant District Officer, 22 December. PNG National Archives, Port Moresby, SN 088, AN 82, File 59-2-16, folios 68–71.

Fisher, N.H., 1957. *Catalogue of the Active Volcanoes and Solfatara Fields of the World*, 5: *Melanesia*. International Volcanological Association, Napoli.

Fullagar, R. & R. Torrence, 1991. 'Obsidian Exploitation at Umleang, Lou Island', in J. Allen & C. Gosden (eds), *Report of the Lapita Homeland Project*. Department of Prehistory, Occasional Papers in Prehistory, The Australian National University, Canberra, 20, pp. 113–43.

Grover, J.C., 1958. 'Savo Volcano — A Potential Danger to its Inhabitants'. *Geological Survey of the British Solomon Islands*, Memoir 2, pp. 102–108.

Gunther, J.T., 1955. Volcano: Bam Island. Memorandum to the Administrator, 14 February. National Archives of Papua New Guinea, Port Moresby, SN 677, AN 247, BN 320, File CA 35/7/51, folios 23–26.

Johnson, R.W. & I.E. Smith, 1974. 'Volcanoes and Rocks of St. Andrew Strait, Papua New Guinea', *Journal of the Geological Society of Australia*, 21, pp. 333–52.

Johnson, R.W. & D. Tuni, 1987. 'Kavachi, An Active Forearc Volcano in the Western Solomon Islands: Reported Eruptions between 1950 and 1982', in

B. Taylor & N.F. Exon (eds), *Marine Geology, Geophysics, and Geochemistry of the Woodlark Basin–Solomon Islands*. Circum-Pacific Council for Energy and Mineral Resources Earth Science Series, 7, pp. 89–112.

Lambert, C.R., 1954. Volcano: Bam Island. Memorandum to the Administrator, 30 November 1954. National Archives of Papua New Guinea, Port Moresby, SN 677, AN 247, BN 320, File CA 35/7/51, folio 12.

Lawrence, P., 1964. *Road Belong Cargo*. Melbourne University Press, Carlton.

McKee, C.O., 1976. 'Investigations at Mount Lamington 1960–75', Geological Survey of Papua New Guinea Report 76/21.

——, 1981. 'Geomorphology, Geology, and Petrology of Manam Volcano', in R.W. Johnson (ed.), *Cooke-Ravian Volume of Volcanological Papers*. Geological Survey of Papua New Guinea Memoir, 10, pp. 23–38.

Miklouho-Maclay, N. de, 1884–1885. 'On Volcanic Activity on the Islands Near the North-East Coast of New Guinea and Evidence of Rising of the Maclay-Coast in New Guinea', *Linnean Society of New South Wales Proceedings*, 9, pp. 963–67.

Moseley, H.N., 1877. 'On the Inhabitants of the Admiralty Islands, &c', *Journal of the Anthropological Institute of Great Britain and Ireland*, 6, pp. 379–29.

National Archives of Papua New Guinea, 1953–1954. Re-establishment of Long Island Natives. SN667, AN247, Box 319, File CA 35/7/39.

Noser, A.A., 1955. Letter to the Administrator, 16 February. National Archives of Papua New Guinea, Port Moresby, SN 677, AN 247, BN 320, File CA 35/7/51, folio 37.

Pain, C.F., 1981. 'Stratigraphy and Chronology of Volcanic-Ash Beds on Lou Island', in R.W. Johnson (ed.), *Cooke-Ravian Volume of Volcanological Papers*. Geological Survey of Papua New Guinea Memoir, 10, pp. 221–25.

Palfreyman, W.D. & R.J.S. Cooke, 1976. 'Eruptive History of Manam Volcano, Papua New Guinea', in R.W. Johnson (ed.), *Volcanism in Australasia*. Elsevier, Amsterdam, pp. 117–31.

Reynolds, M.A., 2005. 'Experiences in Volcanology and Life in the Territory of Papua New Guinea 1953–1957', unpublished memoirs, Geoscience Australia, Canberra.

Reynolds, M.A., J.G. Best & R.W. Johnson, 1980. *1953–57 Eruption of Tuluman Volcano: Rhyolitic Volcanic Activity in the Northern Bismarck Sea*. Geological Survey of Papua New Guinea Memoir, 7.

Roberts, A.A., 1955. Petition from All of the People of Bam Island to the Administrator, Including English Translation. National Archives of Papua New Guinea Port Moresby, SN 677, AN 247, BN 320, File CA 35/7/51, folios 50–52.

Skinner, R.I., 1955. Evacuation Plan — Manam Island, Madang. Report to Director, Department of Native Affairs, Port Moresby, from Madang District Headquarters, 28 October (copy in papers of the late G.A.M. Taylor).

Summerhayes, G.R., 2009. 'Obsidian Network Patterns in Melanesia — Sources, Characterisation and Distribution', *Bulletin of the Indo-Pacific Prehistory Association*, 29, pp. 109–23.

Taylor, G.A., 1952. 'Report on Volcanic Activity on Ambrym Island', Bureau of Mineral Resources, Canberra, Record 1952/4.

——, 1953. 'Notes on Ritter, Sakar, Umboi and Long Island Volcanoes', Bureau of Mineral Resources, Canberra, Record 1953/43.

——, 1955a. 'Report on Bam Volcano and an Inspection of Kadovar and Blup Blup', Bureau of Mineral Resources, Canberra, Record 1955/73.

——, 1955b. 'Notes on Volcanic Activity and Thermal Areas in the D'Entrecasteaux Islands', Bureau of Mineral Resources, Canberra, Record 1955/75.

——, 1955c. 'Tectonic Earthquakes and Recent Volcanic Activity', Bureau of Mineral Resources, Canberra, Record 1955/123.

——, 1956. 'Review of Volcanic Activity in the Territory of Papua-New Guinea, the Solomon and New Hebrides Islands, 1951–53', *Bulletin Volcanologique*, 18, pp. 25–37.

——, 1958a. 'The 1951 Eruption of Mount Lamington, Papua', Bureau of Mineral Resources, Canberra, Bulletin 38.

——, 1958b. Evacuation — Bam Island. Memoranda to District Commissioner, Wewak, 15 April and 26 June. National Archives of Papua New Guinea, Port Moresby, SN 088, AN 82, File 59-2-17, folios 25–26, 31.

——, 1958c. 'Volcanological Report for January 1958. Manam Volcano', Rabaul Volcanological Observatory Monthly Reports, 3 February 1958.

——, 1958d. Untitled Letter to Assistant Administrator, Department of the Administrator, Port Moresby, 1 May. PNG National Archives, Port Moresby, SN 088, AN 82, File 59-2-16, folio 175.

——, 1960. 'An Experiment in Volcanic Prediction', Bureau of Mineral Resources, Canberra, Record 1960/74.

——, 1963. 'Seismic and Tilt Phenomena Preceding a Pelean Type Eruption from a Basaltic Volcano', *Bulletin Volcanologique*, 26, pp. 5–11.

——, 1965. 'Notes on Savo Volcano, 1959', *British Solomon Islands Geological Record*, 2, pp. 168–73.

——, 1966. 'The Surveillance of Volcanoes in the Territory of Papua and New Guinea', *South Pacific Bulletin*, 16, no. 2, pp. 15–20.

Taylor, G.A., J.G. Best & M.A. Reynolds, 1957. 'Eruptive Activity and Associated Phenomena, Langila Volcano, New Britain', Bureau of Mineral Resources, Canberra, Report 26.

Williams, H.L., 1957. Evacuation of Manam Island. Memorandum to Assistant Administrator, Department of the Administrator, Port Moresby, from District Commissioner, Madang, 16 December. PNG National Archives, Port Moresby, SN 088, AN 82, File 59-2-16, folios 60–67.

——, 1958a. Resettlement — Manam Evacuees. Memorandum to Assistant Administrator, Department of the Administrator, Port Moresby, from District Commissioner, Madang, 7 February, pp. 81–84.

——, 1958b. Manam Island. Memorandum to Assistant Administrator, Department of the Administrator, Port Moresby, from District Commissioner, Madang, 6 March. PNG National Archives, Port Moresby, SN 088, AN 82, File 59-2-16, folios 91-93.

10. Plate Tectonics and False Alarms: 1960–1972

From the maw of the hole [at Koranga Crater] there belched an evil, sulphurous stink and a sinister column of smoke emerged. The ground was hot to touch for a radius of fifty feet, and after a couple of days began to open in deep, ominous fissures ...

James Sinclair (1981)

Advances in Science and Technology

Remarkable scientific and technological achievements in the 1960s had a major impact on understanding the nature of the volcanoes in Near Oceania. Satellites were being launched — the pioneering Russian Sputnik as early as 1957 — at the competitive Cold War frontier of space travel, so initiating development of both space-borne, earth-observing platforms and satellite communication systems and therefore opportunities for the future monitoring of volcanoes from space. Mainframe computers were being mainstreamed in scientific establishments generally. Telecommunications technology was advancing, too, including the development of ways to convert earthquake vibrations — which, up until then, had been transferred mechanically to the paper of seismograph drums — to electronic signals that could be transmitted by radio or telephone lines linked to centralised recording rooms. The United States in the early 1960s, also as part of its Cold War initiatives, established the World Wide Standard Seismograph Network (WWSSN) for the more accurate global detection of tectonic earthquakes and, more particularly, nuclear explosions.

G.A.M. 'Tony' Taylor in 1966 affirmed that volcanology was still only a 'Cinderella science [that advanced on] the ashes of catastrophe',[1] yet new approaches in volcanology were beginning to emerge. For example, there was growing research interest in studying the pyroclastic deposits of explosive eruptions and in deducing their mechanisms of formation. A specific example of such specialised research was that undertaken on the pyroclastic materials laid down in 1965 during a hydrovolcanic eruption in a lake at Taal volcano in the Philippines, where the volcanological phenomena of 'base surge' was first recognised and their distinctive features documented.[2] Similarly, hydrovolcanic eruptions in 1963 at the summit of a largely submarine volcano, Surtsey, off

---

1 Taylor (1966), p. 20.
2 Moore (1967).

the south coast of Iceland, were so distinctive and came to be recognised at so many other volcanoes, that the term *surtseyan* was coined for them. Surtseyan eruptions can produce base surges. Their defining characteristic, however, is the extreme fragmentation of the erupting magma, caused by water 'flashing' to steam and shattering the magma, such as photographed at Long Island in 1953. The mix of white vapour clouds and dark sprays of exploding ash at times produces striking 'cock's tail'-like plumes.

Figure 80. The distinctive 'cock's tail' pattern of a small surtseyan explosion from the near-surface vent of the submarine volcano Kavachi, Solomon Islands, can be seen in this photograph taken on 17 or 18 July 1977. Kavachi was in eruption several times during the 1970s.

Source: W.G. Muller, Barrier Reef Cruises, Yepoon.

Much more important for the geosciences in general was the emergence of the revolutionary theory of plate tectonics in the late 1960s — the geoscientific equivalent of the theory of evolution in biology — meaning that attempts to understand volcanic disasters now had a rational geological framework. Earth was portrayed as a dynamic planet covered by a thin and broken 'shell' of tectonic pieces of different sizes that move relative to one another. Converging tectonic plates create 'subduction zones', where one plate carrying water-rich rocks and sediments of the ocean floor, disappears beneath a neighbouring plate along the convergent boundary between the two plates. Earthquakes track the descent of these great tectonic 'slabs' deep into the Earth as so-called Wadati-Benioff zones. Melting of deep rock takes place, and buoyant water-rich magma is returned to the surface in explosive volcanic eruptions, such as seen commonly, and experienced catastrophically, in Near Oceania. Erupted magmas are characteristically *andesite* in compostion. Other tectonic plates separate, or diverge, from one another producing shallow earthquakes as well as new volcanic rocks on the ocean floor in the process of 'sea floor spreading'. A good example of this divergent or 'extensional' type of plate boundary and related sea floor volcanism is in the Woodlark Basin area of the Solomon Sea. My own volcanological research in the region began in 1969 when, with several others, I began collecting hundreds of volcanic-rock samples in an attempt to relate their chemical compositions to the different plate-tectonic settings of the sampled volcanoes.

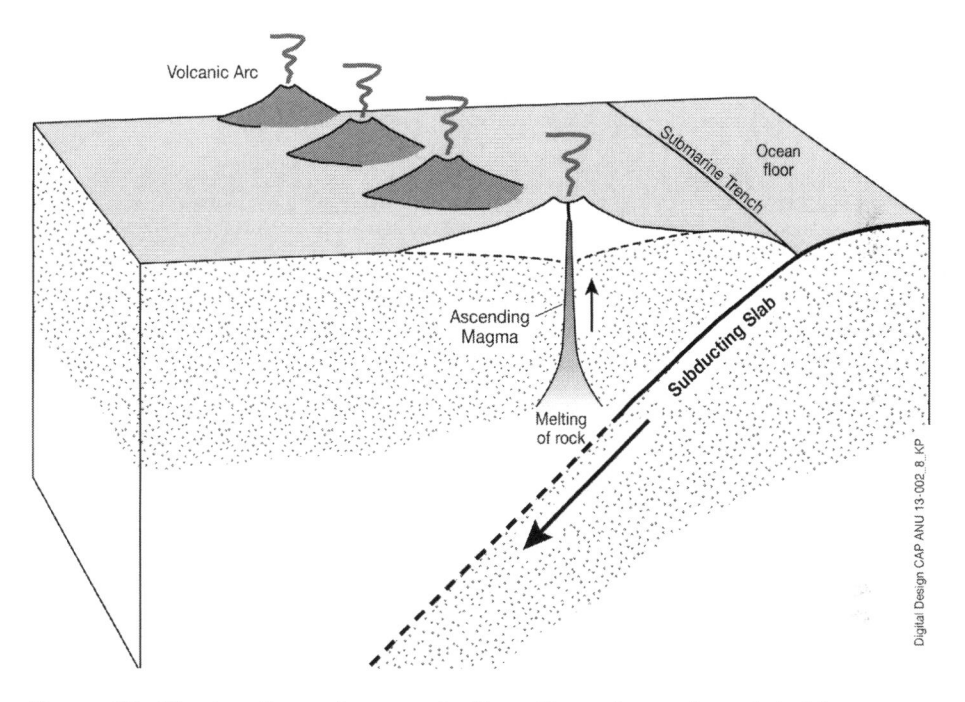

Figure 81. Most active volcanoes in Near Oceania can be related in one way or another to the geological process known as 'subduction'. A tectonic plate shown here on the right plunges down to the left as a 'slab' in front of an overlying plate, forming a submarine trench on the ocean floor. Rock melting takes place above the subducted slab and the magmas so formed rise and erupt from volcanoes that form a line or 'arc' about 100–150 kilometres from the submarine trench.

Source: CartoGIS, The Australian National University.

These external achievements during the 1960s, and other new geoscientific ventures in Near Oceania itself, created a surge of volcanological understanding in the region into the 1970s. Systematic geoscientific mapping by both Bureau of Mineral Resources (BMR) and the Commonwealth Scientific and Industrial

Research Organisation (CSIRO), throughout the Territory of Papua and New Guinea, yielded valuable volcanological information in those areas that included young volcanoes. Furthermore, the search for copper and gold by mineral-exploration companies focused attention on the roots of young volcanoes where 'porphyry copper' and gold mineralisation were thought to be present. A separate Port Moresby Geophysical Observatory (PMGO) began producing instrumental data in 1958. It was established by the BMR Geophysical Section in order to fill a gap in the Australian region in the production of magnetic, ionospheric, gravity and tectonic-earthquake information. WWSSN stations were established in the early 1960s at the new Port Moresby Geophysical Observatory, as well as at RVO and in Honiara.

A network of volcanological observatories was established throughout the Territory of Papua and New Guinea in the 1960s. The observatory at Rabaul was named the 'Central Volcanological Observatory' and was staffed by professionally and technically trained people who also oversaw operation of the smaller and more remote 'outstation' observatories elsewhere in the territory. The headquarters observatory at Rabaul was upgraded under Taylor's direction so that signals from recording stations in the harbour area could be transmitted to the RVO recording room by telephone and radio.[3] Furthermore, major geophysical surveys were conducted in the late 1960s in order to determine Rabaul's deep-crustal structure. A second type of observatory was built in 1963–1965, at considerable capital investment, on Manam Island and at Esa'ala in the D'Entrecasteaux Islands, again under the direction of Taylor. Both of these observatories had deep cellars constructed for the installation of modern seismometers and water-tube tiltmeters.

A third class of lesser observatories that did not have the luxury of quiet cellars was installed at Ulamona Roman Catholic Mission and Sawmill near Ulawun volcano, and near the then-operational Cape Gloucester Airstrip close to Langila volcano. All of these outstation observatories were staffed by part-time 'volcano observers' recruited locally from villages, the Administration, or religious missions. Headquarters staff in Rabaul maintained radio contact with the observers and local Administration staff, but instrumental records from the outstations had to be sent by mail to the central observatory at Rabaul for scientific assessment. Receiving reports of possible volcanic unrest at the other active or potentially volcanoes in the Territory and in British Solomon Islands Protectorate depended on local people, or the pilots of passing aircraft, informing the respective Administration offices and passing on the information to volcanological headquarters.

---

3   Newstead (1969).

Figure 82. Leslie Topue and Ben Talai attend to monitoring equipment in the recording room at RVO in the late 1960s.

Source: RVO photographic collection.

Tony Taylor during the 1960s began taking on more supervisory roles for BMR in Port Moresby and in Canberra where he was raising a family. BMR geoscientists John H. Latter, Colin D. Branch, Gianni W. D'Addario, and W. David Palfreyman successively took over as head of the RVO into the 1970s, although Taylor and N.H. Fisher remained in control of the observatory's strategic directions. A key responsibility of all RVO staff — whoever was in charge — was maintaining effective communication links with Territory Administration staff, as had happened in the 1950s, and especially with the District Commissioners of volcanically active districts and with Australian patrol officers who maintained close ties with more remote communities.

Eruptions took place at several volcanoes in the Territory during the 1960s, but none of these produced disasters or threats that led to the displacement of at-risk communities. For example, RVO staff in May 1966 investigated prominent explosive activity, including the production of pyroclastic flows and lava flows, at Bagana volcano on Bougainville Island.[4] Settlements, however, were far removed from Bagana, even during this significant eruption, and therefore were not threatened. Nevertheless, exploratory drilling by Conzinc Riotinto of Australia Limited was taking place less than 40 kilometres to the south-east. This drilling led to the identification of the massive copper and gold deposits at Panguna, to the eventual commissioning in 1972 of the huge Panguna Mine, and to an ongoing interest in the condition of Bagana volcano from the extraction company, Bougainville Copper Limited, BCL. A seismometer operated by PMGO was established near the mine, which to some extent could also be used to monitor activity at nearby Bagana volcano.

Gas Emissions from Two Highlands Volcanoes

Taylor became involved in a renewed interest in the young volcanoes of highlands New Guinea. Regional geological-mapping surveys on the mainland were being undertaken by government agencies and by both oil and mineral-exploration companies, and the full extent of the volcanoes in the Fly-Highlands province was recognised for the first time.[5] The discovery was made, too, that some of the volcanoes were emitting volcanic gases and, therefore, that they could not necessarily be regarded as extinct.

The first of these potentially active highlands volcanoes to be discovered was Yelia volcano, near Menyamya, in 1962 when a public health official reported to Taylor that sulphur-bearing gases were being emitted from a summit crater on the mountain at the extreme eastern end of the Fly-Highlands volcanic province. Subsequent geological investigations in 1963 by Branch confirmed that the volcano was indeed geologically youthful — and, furthermore, that its volcanic geology was similar to that of Lamington volcano — but the gas emissions were weak and cold.[6] The identification of Yelia as a typical 'sleeper' volcano did not cause any immediate concern at RVO or amongst the local communities.

Taylor's attention was directed towards another gas-emitting highlands volcano in the late 1950s and 1960s — that of Doma Peaks, east of Tari township in the extreme north-western part of the Fly-Highlands volcanic province. An anthropologist working amongst the populous Huli people in the Tari area told

---

4   Bultitude (1979) summarised the eruptive history of Bagana up to 1975.
5   Mackenzie & Johnson (1984).
6   Branch (1967).

Taylor about local legends of a highlands eruption a few generations earlier,[7] and then Patrol Officer Bill Crellin — who had been with Taylor at Lamington in 1951 — informed him that local people were aware of the danger of volcanic eruptions from Doma Peaks. Next, in the mid-1960s, came reports from pilots who noticed the smell of sulphur-bearing gases when flying over the volcano. Taylor and RVO volcanologist R.F. 'Bob' Heming investigated the volcano in 1968, discovered its gas-producing parts, and concluded that it could have potential for further eruptive activity.[8]

The Huli people use the word *bingi*, or *mbingi*, to describe their world view of cyclic destruction and rejuvenation.[9] *Bingi* means, literally, 'time of darkness' and Huli at different times have drawn the attention of Europeans to layers of young volcanic ash in soil profiles of the Tari area, pointing to it, and saying that it formed during *bingi*, when the sun's light was blocked out and darkness prevailed. Early estimates were that the ash was deposited sometime in the 1880s, and that the source volcano was perhaps Krakatau in 1883, or Doma Peaks itself, or some other highlands volcano.[10] This particular ash layer, or **tephra**, is now known to scientists as Tibito Tephra. Questions on the origin of Tibito Tephra were not settled, however, until well into the 1970s.

There is also Huli oral history that a crater lake in the central part of Doma Peaks volcano burst out catastrophically sometime during 1860–1880, producing devastating mudflows westwards towards Tari:

> the hills broke and blocked the river Arua up there on the mountain. A large lake formed … . Then it broke … . There was a huge noise on the mountain and then a loud roar, and the water came down the valley and poured out over the gardens. Rocks, trees, and dirt covered up the gardens and many people were killed … . Sometimes we find ashes and soil from the old houses and gardens.[11]

Good geological evidence exists for these and earlier catastrophic outpourings of huge volumes of sediment and debris from Doma Peaks, but none that the outpourings were necessarily accompanied by volcanic eruptions. The volcano is thought to be weak and gravitationally unstable and to have slumped away westwards on several occasions in the geological past, forming a complex of west-facing escarpments on the volcano itself and great debris fans into the Tari Basin. The date of the latest eruption of magma from Doma Peaks is still unknown.

---

7 Glasse (1963).
8 Taylor (1971).
9 Ballard (1998).
10 Glasse (1963), Watson (1963) and Nelson (1971). See also Blong (1979).
11 Allen & Wood (1980), p. 343.

Volcanic-Disaster Preparations at Wau Township in 1967

Considerable community concern was generated in late May 1967 when gas emissions were reported from new hot ground that appeared suddenly in Koranga Crater, only one kilometre north-west of the township of Wau in Morobe Province. Wau is in the high country, 75 kilometres south-south-west of the present-day provincial capital of Lae, and is a small but long-established township created to serve the dredging operations of alluvial goldminers in the Morobe Goldfield, particularly in the later 1920s and 1930s. Gold was first found there in Koranga Creek, a tributary of the Watut-Bulolo Rivers system, by the legendary prospector Bill 'Shark-eye' Park in 1922.[12] Fisher, as Australian Government geologist, first went to work at Wau in 1934 before becoming involved in volcanological work at Rabaul in 1937.

An Administration officer at Lae, James Sinclair, later recalled the 1967 crisis at Koranga in a book on the wide-ranging work of *kiaps*, including his own role in disaster management at Wau:

> One fine morning the ADO [Assistant District Officer] Wau, Austin Tuohy, reported by telephone the overnight appearance of a strange sink-hole near the Koranga gold workings ... . The Wau townspeople, white and black, were naturally highly concerned about the Koranga sink-hole. A volcanic eruption in the Wau valley, girt as it was by high mountains, would have catastrophic results. The Government Vulcanologist [Tony Taylor] was called in ... [and] Ian Skinner, the Controller of Civil Defence flew immediately to Wau [from Port Moresby]. Des Ashton [the acting Morobe District Commissioner] rocketed up from Lae by car and I accompanied him ...[13]

The appearance of the hot ground and gas emissions at Koranga was notified to the local administration authorities in Wau on 22 May 1967, and instrumental recordings of temperature, seismicity and ground tilting were started on 2 June by Taylor and R.G. 'Bob' Horne, the resident geologist at Wau, assisted at times by RVO volcanologist Heming.[14] Emissions were from a small mound in the open-cut mining area within a crater-like, eastward-facing amphitheatre. They had appeared suddenly after a landslide. The maximum ground temperature measured was 640 °C in early June, and the emitted gases were sulphur dioxide and carbon dioxide — both typical of the gases dissolved in deep-seated magma. Furthermore, Koranga had been known for many years to be an old

---

12   Nelson (1976).
13   Sinclair (1981), p. 249.
14   Pigram et al. (1977).

eruptive centre, and its geologically youthful rocks to be of volcanic origin.[15] Taylor raised the possibility of large-scale volcanic activity and recommended the preparation of an evacuation plan.[16] Sinclair recalled the effort involved in producing it:

> For days we laboured, preparing an emergency evacuation plan. The best escape routes were decided upon, numbers of people involved listed, available vehicles counted, food stocks calculated, medical resources evaluated and detailed instructions written out, covering every conceivable complication and consequence. Planning of this nature is mainly a matter of identifying and assessing probable escalations of danger, and slavish attention to detail to ensure that nothing is overlooked — not vivacious work, and very tiring.[17]

Figure 83. Koranga Crater is seen from the east after the landslide of May 1967. The active mound of hot ground and gas emission is shown by the arrow.

Source: Pigram et al. (1977, Figure 3). Geoscience Australia (GA-9081).

The Wau Evacuation Plan was released on 16 June 1967 and contained the following introductory statement, based on Taylor's scientific advice: 'It is

---

15 See, for example, Fisher & Branch (1981).
16 Taylor (1967).
17 Sinclair (1981), p. 249.

possible that the volcanic vent in the Koranga open cut (Koranga Crater) will increase in activity and that a volcanic eruption of dangerous proportions could follow'.[18] Additional information on the nature of the hot ground and gas emissions was collected in the following days and weeks, however, and doubts began to emerge that they were truly volcanic in origin, and especially when they disappeared altogether after another landslide in the open cut in mid-August 1967.[19] The crisis ended and the emergency evacuation plan was never used.

Not only did the lifespan of the emissions and the active mound at Koranga seem to have been governed by the landslides, but water vapour — a normally abundant component of active volcanic systems — was virtually absent, especially after the mound and nearby landslide material dried out. Furthermore, there was no indication from the seismic, tilt and acoustic observations that a magma body underlay the open cut. Such a body, in any case, would have been expected to have a wider thermal effect on the surface when compared with the actual small size of the active mound. Some geologists later maintained that hydrothermal eruptions may well have taken place at Koranga in 1967 and simply were not witnessed by local people.[20] This seems most unlikely.

The origin of the hot ground and gas emissions at Koranga was probably the result of the oxidation of reactive iron pyrites and carbonaceous materials in the altered rocks of the open cut, such as can cause fires in underground mines.[21] The sulphur-bearing pyrites began burning when the initial landslide allowed access of oxygen, so producing the sulphur dioxide, and the carbonaceous materials ignited to form the carbon dioxide.

There is today some irony in the initial statements made in 1967 that the hot ground and gas emissions at Koranga signified a possible volcanic eruption, if the reactive iron-pyrites interpretation is correct. Karl Sapper, the young German volcanologist who had visited German New Guinea in 1908, wrote the following in 1931, then as a well respected international scientist reviewing centuries-old explanations for volcanic activity:

> The gradual development of Chemistry first suggested to L. de Capoa, in 1683, that volcanic heat originated in chemical processes. After Martin Lyster, in 1693, had called attention to the decomposition of pyrites, and [Nicolas] Lemery in 1700 offered experimental proof (spontaneous

---

18 Skinner (1967), p. 1.
19 Pigram et al. (1977).
20 Sillitoe et al. (1984).
21 Pigram et al. (1977).

combustion of a mix of iron filings and sulphur buried in moist earth) of the possibility of explosions from chemical processes, this opinion prevailed for a long time.[22]

Chemical processes were the likely cause of the crisis at Wau in 1967, but were not a precursor to any actual volcanic eruption. The Koranga crisis was a volcanic 'false alarm'.

Evacuation from Dawson Strait in 1969

Perhaps the most significant crisis in any volcanic area in Near Oceania in the 1960s took place in the Dawson Strait area of the D'Entrecasteaux Islands in eastern Papua. A series or 'swarm' of local earthquakes was felt around Easter time in 1969, particularly by people on Dobu Island, on south-eastern Fergusson Island, and on north-western Normanby Island, where the Administration had its headquarters at Esa'ala.

Monica Russell was a school teacher at Salamo on the south-east coast of Fergusson Island across the strait from Esa'ala, and wrote in her diary for Saturday 5 April 1969: 'What a day! It started off at just after 4 am with a tremor. There were quite a few and a very strong one ... just about 4.45 am Esa'ala recorded 14 in two hours.'[23] Then later that evening: 'Just got into bed when there was a very strong shake — things fell again ... . We had little shakes then on and off all through the night.' Another teacher, Lucille Piper, said many years later that her easel blackboard immediately began to dance at the beginning of a tremor, alerting her and the students to get prepared for an ensuing 'shake'.[24]

Dawson Strait in 1969 was well known as an area of young, potentially active volcanoes, three of them conspicuous landmarks. Dobu Island itself is the summit of a volcano built up from the sea floor in the strait, and Oiau and Lamonai are small volcanic peaks several kilometres to the north of Dobu on Fergusson Island.[25] In addition, the numerous quasi-permanent geothermal areas of the D'Entrecasteaux Islands, including the well-known Dei Dei hot spring area near Oiau, were another reminder of active volcanic processes and prospective volcanic eruptions in the area. There is, however, no documentation of historical eruptive activity at Dobu, Oiau or Lamonai, nor any reliable local stories of any volcanic eruptions having taken place in the more distant past.

---

22 Sapper (1931), p. 19.
23 M. Russell (personal communication, 2009; from a transcript of her diary entries for 6–12 April 1969). Documented information on the Dawson Strait crisis in 1969 is scant, and much of the material presented here is based on personal communications during 2008–2010 with individuals who were there at the time. The late Herman Patia assisted greatly in disinterring information on the crisis.
24 L. Piper (personal communication, 2008).
25 Taylor (1955) and Smith (1976, 1981).

Fire Mountains of the Islands

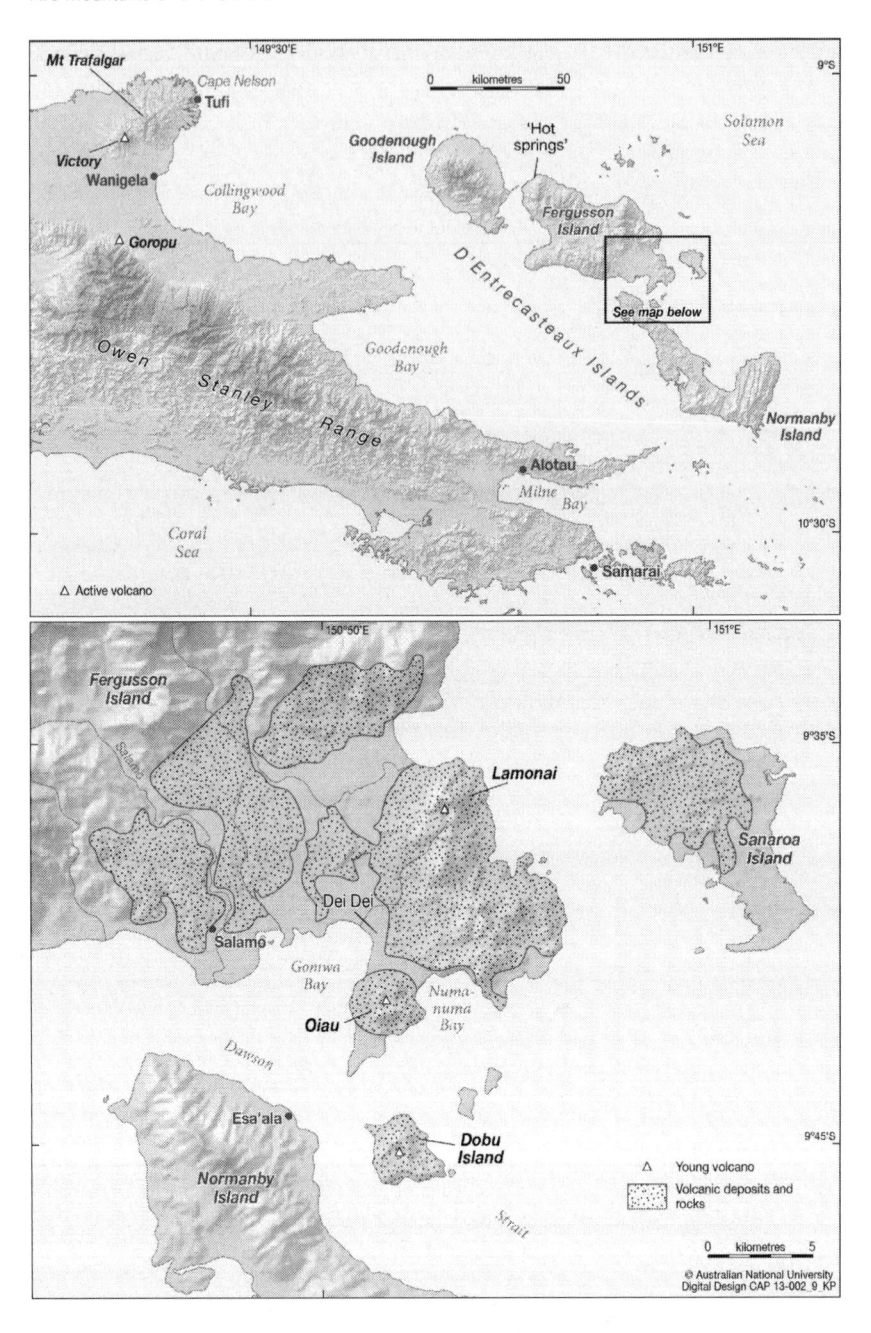

Figure 84. Some of the volcanoes and settlements of the D'Entrecasteaux Islands, the Dawson Strait area, and the south-eastern mainland of New Guinea are shown in these maps. The area marked 'Hot springs' in north-western Fergusson Island is where the crew of the *Basilisk*, captained by John Moresby, landed in 1874.

Source: Adapted from Smith (1976, Figure 1).

April 1969 was not the first time that local earthquakes had been felt in the Esa'ala area. Seismic events in 1953–1957, for example, are particularly noteworthy because, so soon after the volcanic disaster in 1951 at Lamington volcano to the west, on the mainland of New Guinea Island, the earthquakes had 'alarmed local inhabitants and government officers responsible for the welfare of the people.'[26] Taylor, by late 1953, had arranged for a volcanological reporting scheme to be established through which Territory Administration officials in volcanically active districts could report earthquakes and any changes within the volcanic areas to RVO, and numerous felt earthquakes, seismic 'swarms', aftershocks, and 'tremors' were reported by this mechanism from the Esa'ala district and other nearby places between October 1953 and February 1957.[27] The precise location of the earthquakes was uncertain, but many appeared to have taken place beneath the sea floor in Gomwa Bay, immediately west of Oiau volcano. Taylor and RVO volcanologist Reynolds both speculated that these local and shallow earthquakes could be part of a longer sequence of tectonic earthquakes dating from 1939, which had affected eastern Papua as a whole — that is, an extended period of general geophysical unrest which also included the volcanic eruptions at Goropu in 1943 and Lamington in 1951. They feared that the 1953–1957 earthquakes, although probably tectonic in origin, could be precursors to further volcanic activity somewhere in the general area of southeastern Fergusson Island and Dobu Passage. This concern was the reason why Taylor, during the 1951–1957 eruption time cluster throughout the Territory, urged the construction of the volcanological observatory at Esa'ala in order to monitor any potentially threatening volcanic activity. The observatory, including an instrumental cellar housing a seismograph and tiltmeter, was built at Esa'ala in 1964.

The earthquakes of Saturday 5 April 1969, and those in the days following, caused consternation and anxiety, particularly amongst those people who were feeling them most strongly. About 5–7 shocks per day were being felt at intensities I–III on the Modified Mercalli Intensity Scale.[28] The earthquakes seemed to be taking place at or near Dobu volcano, so the possibility of volcanic eruptions was foremost in the minds of many people, including the Administration officers at Esa'ala. An Australian newspaper journalist on Wednesday 9 April reported on the earthquake activity and community fears, mentioning that on the previous day the District Commissioner for Milne Bay, Max Denehey, who was based in Alotau, had announced that a government trawler carrying extra food and medical supplies was on its way to the area as a precautionary measure, and that people on isolated Dobu Island might have

---

26 Reynolds (1957), p. 1.
27 Taylor (1955) and Reynolds (1956, 1957).
28 D'Addario (1969a).

to be evacuated.[29] The Assistant District Commissioner at Esa'ala, J. Edwards, was quoted as saying that the seismic activity was declining. Nevertheless, an evacuation of Dobu Island was well underway by Thursday 10 April, following recommendations made by the authorities at Esa'ala.[30] Some villagers may well have decided to leave their homes before the official declaration and, indeed, a later RVO report contains the statement that 'villagers evacuated spontaneously from the coasts ...'[31] Ian Skinner, Director of Civil Defence in Port Moresby flew over the area on Thursday 10 April and noted about 30 canoes crossing the strait between Dobu and Normanby Island.[32] The evacuation was seen to be orderly and voluntary. People from Dobu joined others moving away from the island and from Esa'ala to other more distant and safer parts of the Normanby Island coast.

Evacuees were also moving westwards and purposefully along the south-eastern coast of Fergusson Island towards the safer Morima area on the west coast, or they stopped off with relatives or friends along the way.[33] These evacuees included people from the Oiau-Lamonai and Dei Dei hot spring areas on Fergusson as well as from Dobu Island. Several hundred people came along the track near the mission, schools, and hospital complex at Salamo, in a long line carrying personal belongings, some with a pig in tow, while teachers were trying to run school classes. Some children at the schools ran out to meet family members amongst the evacuees. Salamo itself was not evacuated, although preparations were made for a quick departure if necessary for school and hospital staff, students, and patients. Arrangements were made with the government authorities in Esa'ala to wait for further directions:

> The signal to leave was to be flares let off at Esa'ala, so two school boys were posted down at the shore at Gomwa village to keep watch each night and if a flare went off they were to run back to the station to get the generator started. This would put the station lights on which would be the signal to move out ...[34]

RVO volcanologist D'Addario visited the Dawson Strait area from 10 to 15 April to assess the nature of the earthquakes and to inspect the condition of

---

29 *Sydney Morning Herald* (1969a).
30 R. Morioga (personal communication, 2009, to Herman Patia at RVO). Ruben Morioga worked for the Administration in April 1969 as a court interpreter. He accompanied patrol officers during their visits to villages throughout the D'Entrecasteaux Islands. Morioga recalled two policemen being instructed by John Absalom at Administration headquarters to travel by dinghy to Dobu Island, south Fergusson Island, and north Normanby Island, and there to inform villagers that they should get ready to move out. Morioga later joined RVO as a part-time observer at the Esa'ala Volcanological Observatory.
31 D'Addario (1969a), p. 5.
32 *Sydney Morning Herald* (1969b).
33 Former school teachers at Salamo who kindly provided information in 2008–2010 are Heather Anderson, John Beasley, Rev. Lucille Piper, Pat Riddell, Sheila Rudofsky and Monica Russell.
34 L. Piper (personal communication, 2008).

the volcanoes.[35] He concluded that there were no indications that any of the volcanoes would break out in eruption. There may have been some changes to the geothermal areas but these were not regarded as volcanologically significant. The earthquakes were reported by D'Addario to have taken place under Gomwa Bay to the west of Oiau and north-west of Dobu and, therefore, not directly under either volcano — or under the more distant Lamonai volcano. The District Commissioner was informed of the results of the volcanological inspection and evacuees began the return to their homes and more normal lives soon after 15 April, although occasional earthquakes continued to be felt. RVO volcanologists continued to visit Esa'ala, and there was a strong local earthquake on 11 July.[36] Furthermore, the district medical officer at Samarai, Dr M. Powell, distributed to health workers throughout the Milne Bay District an outline of action, dated 19 August, that should be followed in the event of a volcanic emergency at Esa'ala.[37] The 1969 'crisis', as far as local people were concerned, had lasted, however, for only two or three weeks in April. The exact number of people who evacuated during this volcanic 'false alarm' is unknown, but it was certainly more than 1,000 and probably more than 2,000.[38]

Origin of the Dawson Strait Earthquakes and a Note on Volcanic False Alarms

What was the cause of the earthquakes in Dawson Strait in 1969? How did they originate if they were not volcanic in origin — that is, caused by the movement of magma beneath the volcanoes? The year 1969 was when seismologists in Near Oceania were beginning to interpret earthquakes in terms of the theory of plate tectonics and to map and publish on the exceptionally complex pattern of plate boundaries in the region.[39] The Woodlark Basin, immediately to the east of the D'Entrecasteaux Islands, was identified in 1970 as a divergent or sea floor spreading type of plate boundary, where plates move apart from each

---

35 D'Addario (1969a, 1969b).
36 D'Addario (1969a, p. 6) wrote briefly that a Willmore portable seismograph was installed near Oiau volcano between 20 and 29 July 1969. Epicentres for weak earthquakes were calculated and were identified as corresponding '… to two areas, one near the Dobu Passage and the other between Cape Dawson [probably Sebulugomwa Point] and Bomwa [Gomwa] Bay. These two areas coincide with the two previous locations indicated during the April [earthquake] storm'. This is the only known reported evidence of the earthquakes having taken place in two separate areas. D'Addario (1969b) also noted that a 'seismic swarm' had taken place in the Esa'ala area on 19 and 20 November 1968 — that is, only five months before the April 1969 earthquakes — but there are no further details about their strength or location. Reynolds (1956) also reported earthquakes in the Gomwa Bay area in 1955.
37 Powell (1969).
38 R. Morioga (personal communication, 2009, to Herman Patia at RVO) suggested that the number of evacuees may have exceeded 3,000.
39 Denham (1969).

other — in this case to the north and south.[40] Geoscientists have since taken a particular research interest in trying to identify what happens to this plate boundary when it is traced westwards towards the D'Entrecasteaux Islands and the south-eastern end of New Guinea Island. The details are still not clear, given that much of this area is covered by sea, but there is a general consensus that the area is being rifted tectonically, and probably complexly, and that it has not, or not yet, produced the kind of sea floor spreading observed in the Woodlark Basin to the east. Some geophysicists in the early 1970s traced a single plate boundary through the general area of the D'Entrecasteaux Islands, including some who proposed specifically that the boundary runs along Dawson Strait between Fergusson and Normanby islands.[41] The earthquakes of 1969, and at other times in Dawson Strait, would therefore seem to be related to the tectonic release of seismic energy along a plate boundary, or at least to a geological structure related to it, rather than directly to a volcanic origin.

Mention of geological 'rifting' in the D'Entrecasteaux Islands perhaps brings to mind the Rift Valleys of East Africa. The comparison is, of course, somewhat inappropriate, given the great size of the magnificent rift escarpments and the vast volcanic regions in, for example, Kenya and Ethiopia. Nevertheless, there is a significant similarity. Rocks from the diminutive D'Entrecasteaux Islands include a type of rhyolite lava, including obsidian, which resembles in some ways the kind of rhyolite found in St Andrew Strait in the Bismarck Sea and from Talasea, New Britain. Petrologists, however, can make a clear chemical distinction between these three rhyolite types. The 'rhyolite' of the D'Entrecasteaux Islands is in fact 'comendite', a rare rock type, which, while not found elsewhere in Near Oceania, exists in places such as East Africa. Its presence in the D'Entrecasteaux Islands is therefore regarded by geologists as being consistent with a rifted or extensional tectonic environment.[42]

There is value in considering further the nature of the 'false alarm' at Dawson Strait in 1969. At least two general types of volcanic false alarm can be distinguished. The first type is where a community has a local volcanological observatory which is well equipped with an array of modern volcano-monitoring equipment, is staffed by experienced scientists and technicians, and is sited on a frequently active volcano where the precursory signs of an impending eruption are well known. Such observatories will also have strong telecommunication links with authorities as well as a staged alert or early warning system, known to the community, which can be used to inform decisions on early evacuation. The staff of these observatories can accrue considerable kudos if they are successful in alerting communities before the advent of a disastrous volcanic eruption.

---

40 Milsom (1970).
41 See, for example, Curtis (1973).
42 Smith (1976).

Conversely, however, the term 'false alarm' can carry a great deal of criticism if such an observatory recommends an evacuation but no eruption follows or, worse, declares there is no reason for community alarm and yet a devastating eruption does take place. Elements of such a situation emerged in Rabaul in 1994, for example.

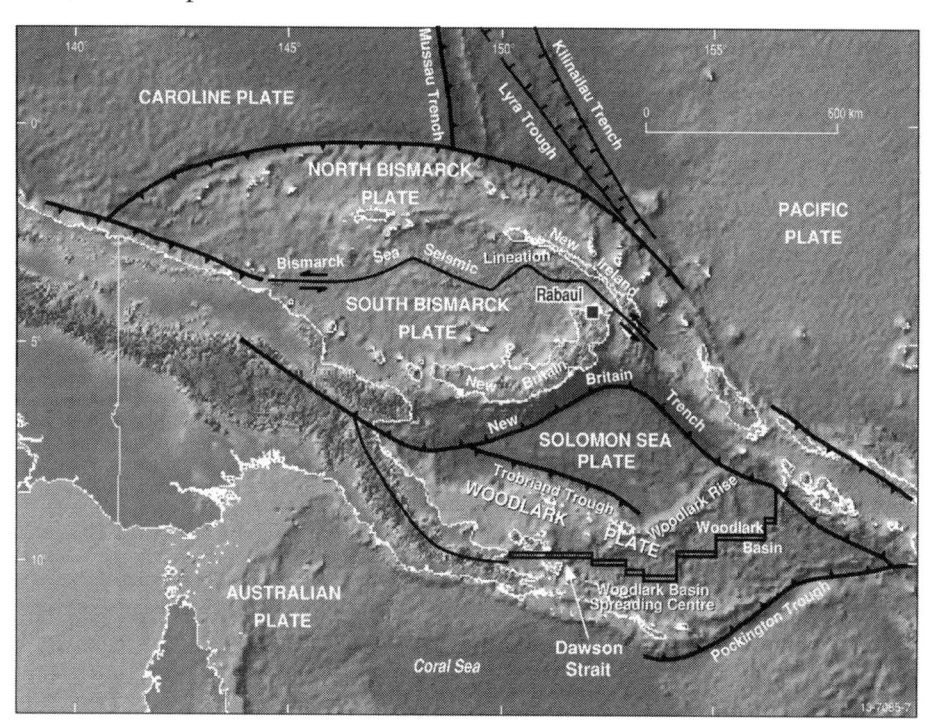

Figure 85. Geoscientists still debate the question of how many plates and plate boundaries exist in the tectonically complex Near Oceania region, but in this map as many as five minor tectonic plates — Caroline, North Bismarck, South Bismarck, Solomon Sea, and Woodlark — are shown sandwiched between the much larger Pacific and Australian tectonic plates. Note how the Woodlark Basin plate boundary or 'spreading centre' is shown running through Dawson Strait in the D'Entrecasteaux Islands.

Source: Finlayson et al. (2003, adapted from Figure 1). Reproduced with the permission of Elsevier.

The second type of false alarm is where a community in a potentially active volcanic area, which is unmonitored instrumentally and has no volcanological observatory, becomes alarmed by apparent geophysical changes such as felt volcanic earthquakes, visible landslides on volcanoes, or the appearance of hot ground and gases, and affected people then inform local authorities. Professionally trained volcanologists are then called in to assess the situation, but the threat dies down, the crisis is called a 'false alarm' and life returns more or less to normal. This was the situation, for example, at Koranga in 1967.

Fire Mountains of the Islands

The false alarm at Dawson Strait in 1969 does not fall neatly into either of these two general categories. The volcanological observatory at Esa'ala was established in 1964 to monitor precursory geophysical signs of possible volcanic activity because of the local earthquakes known to have taken place there. Evacuation in 1969 took place solely because of local fears of the strong April earthquakes, rather than because of any early warning or alert from staff of the Esa'ala Volcanological Observatory. There were no trained scientists at the observatory in 1969, and professional volcanological opinion from Rabaul on the happenings in Dawson Strait came only later, after the worst of the seismic crisis was over and when the seismic records — and those from portable recorders that were deployed later for short periods — could be examined. No full scientific report was ever completed on the Dawson Strait seismic activity in 1969. Indeed, the challenge of relating the tectonic earthquakes in the Dawson Strait area to the sub-surface formation of magma and therefore possible volcanic eruptions has never been addressed properly.

One final conclusion is that the capital outlay in 1964 for a permanent observatory, including an instrument cellar, at Esa'ala has yet to justify its volcanological worth, bearing in mind that no volcanic eruptions have taken place there in the centuries since written records began to be kept in the area. The volcanoes of the Dawson Strait area must still be regarded as having the potential for future volcanic outbreaks, but the key contemporary question is reduced to whether the expense of maintaining and operating the Esa'ala Volcanological Observatory can be justified given the relatively low volcanic risk in the strait, compared with other unmonitored volcanoes in Near Oceania. Volcano-monitoring technologies have improved dramatically since 1969 and an old-style observatory such as that erected at Esa'ala in 1964 would probably never be built today.

Tectonic Earthquakes and the End of the Taylor Era

Four major series of tectonic earthquakes shook New Britain and the north coast of New Guinea over a 15-month period in 1970–1972, and their epicentres were sufficiently close to volcanoes that Taylor was alert to the prospect of new eruptions. I recall him returning to Canberra in late 1970 after the Madang earthquake of 31 October that year, excited — for a normally quiet, cautious, and at times non-communicative man. He said that he had seen changes to the thermal areas in the caldera of nearby Karkar Island,[43] and thought they represented a prelude to new eruptive activity from the volcano. Two

---

43 Taylor (1972).

magnitude-8 earthquakes next took place in the northern Solomon Sea, southeast of Rabaul on 14 and 26 July 1971, causing tsunamis in the harbour.[44] These were followed on 18 January 1972 by another earthquake series in the ranges west of Madang, centred near Josephstaal, and forming a zone that trended north-westwards, almost in line with Manam volcano.[45]

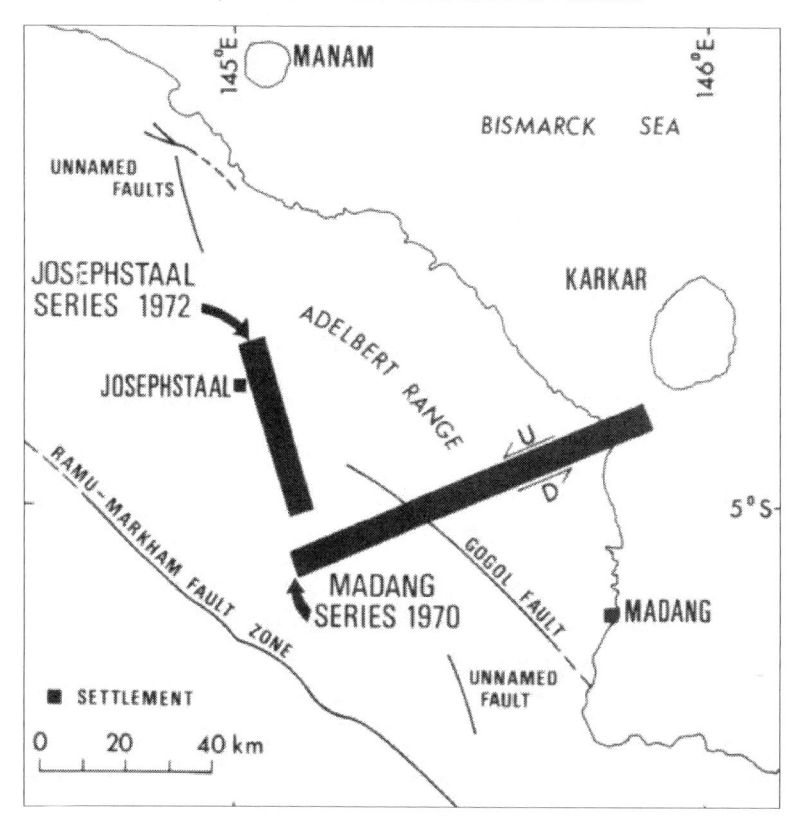

Figure 86. The two black bars represent the general areas of earthquake aftershocks from the Madang and Josephstaal earthquakes. Manam and Karkar volcanoes almost line up with these aftershock zones. There are no known volcanoes between the two volcanic islands.

Source: Cooke et al. (1976, Figure 4). Reproduced with the permission of Elsevier.

Taylor was acting Resident Geologist in Port Moresby in August 1972 when he decided to visit Manam, presumably to see whether there were any changes to the condition of the volcano. He took the steep climb to Yabu Rock on the south-western flank of the volcano on 19 August and, on returning to the coast, collapsed and died, evidently from a heart attack or stroke, aged 54.

---

44   Braddock (1973) and Everingham (1975).
45   Cooke et al. (1976).

Figure 87. G.A.M. Taylor was photographed later in life at a reunion in London of George Cross awardees.

Source: Photograph supplied by M. Hebblethwaite.

Karkar and Manam volcanoes each broke out in full volcanic activity in 1974.

Fisher concluded in an obituary that:

> Taylor had a deep and continuing interest in, almost an obsession with, volcanology, and the main thrust of his work was devoted towards establishing and improving methods of prediction of volcanic eruptions and putting them into practical effect. His work was his main hobby ...[46]

---

46 Fisher (1976), p. xiv.

Taylor's work was both pioneering and impressive, yet his premature death meant he did not complete a planned report on the major eruptions at Manam in 1957–1960. Nor was he able to compile into a definitive account his final assessment of the 'three component' eruption-prediction model that he had been testing over the previous 20 years, and which included the influence of tectonic activity on eruption triggering.

Taylor died the year before Papua New Guinea became self-governing in 1973 and, in a sense, his death marks the end of a 'colonial' era in applied volcanology in Near Oceania. These times had become overarched by great political change in the Pacific region as old colonies and territories became new nation-states. Papua New Guinea obtained its independence in 1975 and Solomon Islands in 1978, each inheriting national boundaries that had been determined as a result of bargaining by the colonial powers of the late nineteenth century. The new nations were characterised as 'developing' or Third World. They were cast into a modern world that, increasingly, was becoming globalised and where the more fortunate 'developed' countries offered financial assistance as a mechanism for sharing global wealth and establishing political influence. Volcano monitoring and disaster management became the responsibility of the new sovereign states, but volcanological leadership in Rabaul and Honiara continued in the hands of seconded, or contracted, expatriate Europeans.

References

Allen, B.J. & A.W. Wood, 1980. 'Legendary Volcanic Eruptions and the Huli, Papua New Guinea', *Journal of the Polynesian Society*, 89, pp. 341–47.

Ballard, C., 1998. 'The Sun by Night: Huli Moral Topography and Myths of a Time of Darkness', in L.R. Goldman & C. Ballard (eds), *Fluid Ontologies: Myth, Ritual and Philosophy in the Highlands of Papua New Guinea*. Bergin and Garvey, Westport, Connecticut, pp. 67–85.

Blong, R.J., 1979. 'Huli Legends and Volcanic Eruptions, Papua New Guinea', *Science*, 10, no. 3, pp. 93–94.

Braddock, R.D., 1973. 'The Solomon Sea Tsunamis of July 1971', in R. Fraser (comp.), *Oceanography of the South Pacific 1972*. New Zealand Commission for UNESCO, Wellington, pp. 9–14.

Branch, C.D., 1967. 'Volcanic Activity at Mount Yelia, New Guinea'. Bureau of Mineral Resources, Canberra, Report 107, pp. 35–39.

Bultitude, R.J., 1979. *Bagana Volcano, Bougainville Island: Geology, Petrology, and Summary of Eruptive Activity between 1875 and 1975*. Geological Survey of Papua New Guinea Memoir, 6.

Cooke, R.J.S. Cooke, C.O. McKee, V.F. Dent & D.A. Wallace, 1976. 'Striking Sequence of Volcanic Eruptions in the Bismarck Volcanic Arc, Papua New Guinea, in 1972–75', in R.W. Johnson (ed.), *Volcanism in Australasia*. Elsevier, Amsterdam, pp. 149–172.

Curtis, J.W., 1973. 'Plate Tectonics and the Papua – New Guinea – Solomon Islands Region', *Journal of the Geological Society of Australia*, 20, pp. 21–36.

D'Addario, G.W., 1969a. 'Interim Progress Report for 1969, Volcanological Section'. Unpublished manuscript, Rabaul Volcanological Observatory.

——, 1969b. Monthly Reports of the Rabaul Volcanological Observatory, April to July.

Denham, D., 1969. 'Distribution of Earthquakes in the New Guinea – Solomon Islands Region', *Journal of Geophysical Research*, 74, pp. 4290–99.

Everingham, I.B., 1975. 'Faulting Associated with the Major North Solomon Sea Earthquakes of 14 and 26 July 1971', *Journal of the Geological Society of Australia*, 22, pp. 61–69.

Finlayson, D.M., O. Gudmundsson, I. Itikarai, Y. Nishimura & H. Shimamura, 2003. 'Rabaul Volcano, Papua New Guinea: Seismic Tomographic Imaging of an Active Caldera', *Journal of Volcanology and Geothermal Research*, 124, pp. 153–71.

Fisher, N.H., 1976. 'Memorial — G.A.M. Taylor', in R.W. Johnson (ed.), *Volcanism in Australasia*. Elsevier, Amsterdam, pp. ix–xiv.

Fisher, N.H. & C.D. Branch, 1981. 'Late Cainozoic Volcanic Deposits of the Morobe Goldfield', in R.W. Johnson (ed.), *Volcanism in Australasia*. Elsevier, Amsterdam, pp. 249–55.

Glasse, R.M., 1963. 'Bingi at Tari', *Journal of the Polynesian Society*, 72, pp. 270–71.

Mackenzie, D.E. & R.W. Johnson, 1984. 'Pleistocene Volcanoes of the Western Papua New Guinea Highlands: Morphology, Geology, Petrography, and Modal and Chemical Analyses'. Bureau of Mineral Resources, Canberra, Report 246.

Milsom, J.S., 1970. 'Woodlark Basin, a Minor Center of Sea-Floor Spreading in Melanesia', *Journal of Geophysical Research*, 75, pp. 7335–39.

Moore, J.G., 1967. 'Base Surge in Recent Volcanic Eruptions', *Bulletin Volcanologique*, 30, pp. 337–63.

Nelson, H.E., 1971. 'Disease, Demography, and the Evolution of Social Structure in Highland New Guinea', *Journal of the Polynesian Society*, 80, pp. 204–16.

——, 1976. *Black, White and Gold: Goldmining in Papua New Guinea 1878–1930*. The Australian National University Press, Canberra.

Newstead, G., 1969. 'Keeping Watch on Volcanoes', *Hemisphere*, 13, no. 1, pp. 32–37.

Pigram, C.J., R.W. Johnson & G.A.M. Taylor, 1977. 'Investigation of Hot Gas Emissions from Koranga Volcano, Papua New Guinea, in 1967', *BMR Journal of Geology and Geophysics*, 2, pp. 59–62.

Powell, M., 1969. 'Volcanic Disturbances — Esa'ala'. Department of Public Health, Samarai, Milne Bay District. Photocopy of unpublished memorandum supplied by A. Speer (personal communication, 2011).

Reynolds, M.A., 1956. 'The Gomwa Bay (D'Entrecasteaux Islands) Earthquakes: July–September, 1955', Bureau of Mineral Resources, Canberra, Record 1956/7.

——, 1957. 'Volcano—Seismic Phenomena in Eastern Papua since 1939', Bureau of Mineral Resources, Canberra, Record 1957/14.

Sapper, K., 1931. 'Volcanoes, their Activity and their Causes'. Physics of the Earth — I: Volcanology, *Bulletin of the National Research Council*, National Academy of Sciences, 77, pp. 1–33.

Sillitoe, R.H., E.M. Baker & W.A. Brook, 1984. 'Gold Deposits and Hydrothermal Eruption Breccias Associated with a Maar Volcano at Wau, Papua New Guinea', *Economic Geology*, 79, pp. 638–55.

Sinclair, J., 1981. *Kiap: Australia's Patrol Officers in Papua New Guinea*. Pacific Publications, Sydney.

Skinner, R.I., 1967. Wau Evacuation Plan — Advice to Residents. Civil Defence and Emergency Services, Territory of Papua and New Guinea, 17 June.

Smith, I.E.M., 1976. 'Peralkaline Rhyolites from the D'Entrecasteaux Islands, Papua New Guinea', in R.W. Johnson (ed.), *Volcanism in Australasia*. Elsevier, Amsterdam, pp. 275–85.

———, 1981. 'Young Volcanoes in Eastern Papua', in R.W. Johnson (ed.), *Cooke-Ravian Volume of Volcanological Papers*. Geological Survey of Papua New Guinea Memoir, 10, pp. 257–65.

Sydney Morning Herald, 1969a. 'N.G. Volcanoes Active: Relief Ship on Way as Precaution', Wednesday 19 April, p. 3.

———, 1969b. 'Canoes Ferry Islands from Volcano Threat', Friday 11 April, p. 5.

Taylor, G.A., 1955. 'Notes on Volcanic Activity and Thermal Areas in the D'Entrecasteaux Islands', Bureau of Mineral Resources, Canberra, Record 1955/75.

———, 1966. 'The Surveillance of Volcanoes in the Territory of Papua and New Guinea', *South Pacific Bulletin*, 16, no. 2, pp. 15–20.

———, 1967. 'Notes on an Evacuation Plan for Wau in the Events of Volcanic Emergency'. Unpublished typescript, R.W. Johnson collection, Canberra, 13 June.

———, 1971. 'An Investigation of Volcanic Activity at Doma Peaks'. Bureau of Mineral Resources, Canberra, Record 1971/137.

———, 1972. 'Karkar Island', in Johnson et al., *Geology and Petrology of Quaternary Volcanic Islands off the North Coast of New Guinea*. Bureau of Mineral Resources, Canberra, Record 1972/21, pp. 63–70.

Watson, J.B., 1963. 'Krakatoa's Echo?', *Journal of the Polynesian Society*, 72, pp. 152–55.

11. Cooke-Ravian and a Volcanic Resurgence: 1971–1979

Once upon a time, in olden days, men saw that to the south the whole land was covered with dark clouds … it was raining ashes in those parts, so that the people could not go out to dig up their food crops … . The ash-storm reached them and they had to stay inside their houses for four or five nights. By this time they were either terribly hungry or else they actually did starve to death … . People today do not know that the ash-storm once took place.

Ko, a villager from near Mount Hagen in the highlands of New Guinea Island (Vicedom & Tischner, 1943, p. 91)

New Eruption Time Cluster

Bureau of Mineral Resources (BMR) geophysicist R.J.S. 'Rob' Cooke, an Australian, joined Rabaul Volcanological Observatory (RVO) in mid-1971. He had for many years been involved with routine geophysical surveys measuring the gravity fields of parts of Australia and Antarctica, and the shift into practical volcanology was an exciting and dramatic career change. Cooke arrived at Rabaul in time to experience the two major, north Solomon Sea earthquakes in July 1971, and to witness the damage from the resulting tsunamis in the harbour at Rabaul. Cooke developed a passion during his career for volcanology and particularly for unearthing historical documents on volcanic eruptions and volcano observations in Near Oceania. He was a persistent researcher, tenacious in argument, and disciplined in the writing of reports on his research results.

Cooke took over leadership of the observatory in 1972, and was on hand to manage a prominent time cluster of volcanic eruptions in the Bismarck Volcanic Arc in 1972–1975. The RVO team of volcanologists at this time included Cooke's cousin, C.O. 'Chris' McKee, who arrived in 1973 at the start of a career in geophysics and volcanology in present-day Papua New Guinea that still continues there after 40 years. Leslie Topue was still working at RVO in the 1970s, but the scientific staff now represented a new generation of men, building on the legacy left by G.A.M. 'Tony' Taylor and his team, who were starting afresh in gaining personal volcanological experience. Furthermore, a new cadre of younger Melanesians was now being employed as volcanological assistants, including Elias Ravian, a Tolai from Tavui No. 1 village, Rabaul. Outstanding amongst these new recruits was Benjamin P. Talai from the Duke of York Islands. Talai in 1977 became the first Melanesian employee at RVO to obtain a university degree and, in 1978, to become a professional volcanologist at the observatory.

Eight volcanoes in Near Oceania were in eruption between 1972 and 1975 in a remarkable time clustering of eruptive activity.[1] Bagana and Kavachi in the extreme east appear to have been active frequently over a longer period — perhaps *constantly* active in the case of Bagana — and conceivably they are not truly part of the eruption time cluster. Eruptions at the remaining six, however, represent an apparently strong volcanic 'pulse' along the Bismarck Volcanic Arc in 1972–1975. Five of these six — Manam, Karkar, Long, Ritter and Langila — are all in the western sector of the arc and all were in eruption within an eight-month period in 1974. Ritter and Karkar had not been reported to be in eruption since the late nineteenth century, and the brief eruption at Ritter in 1972 represents the only one for that year belonging to the 1972–1975 cluster in the western Bismarck Volcanic Arc. The sixth volcano of the cluster is Ulawun in the eastern part of the arc, but it started a pronounced period of major but intermittent eruptions in early 1970 — that is, well before the beginning of the 1972–1975 cluster and before the four major tectonic earthquakes of late 1970 and 1971.

Two other volcanoes became 'restless' during the 1970s. Each of these generated concerns about the likelihood of new eruptions taking place and the impact these might have on nearby communities. This history, too, raises the general question of whether volcanoes that develop precursory, and worrying, signs of impending eruptive activity, but which do not go on to produce eruptions, should be used to define volcanic pulses. One of the two volcanoes was Kadovar Island, about 25 kilometres west of Bam volcano, where a new area of hot ground appeared on the south-eastern slope of the volcano in September 1976 — that is, later than the defined 1972–1975 eruption 'pulse'. Appearance of the hot ground closely followed a period of increased, local, tectonic earthquake activity, but the thermal area died away after another phase of tectonic earthquakes, possibly meaning that a trend towards an eruption at Kadovar had been arrested.[2]

The other period of geophysical 'unrest' was at Rabaul. Earthquake 'swarms' started to be recorded by RVO in Rabaul Harbour in November 1971, only four months after the major north Solomon Sea earthquakes of July that year.[3] The swarms lasted from ten minutes to several hours each and included hundreds of events. Furthermore, the swarms throughout the 1970s tended to include increasing numbers of earthquakes and, even more striking, their epicentres formed a distinctive pattern — an elliptical annulus.

---

1 Cooke (1975) and Cooke et al. (1976).
2 Wallace et al. (1981).
3 Mori et al. (1989).

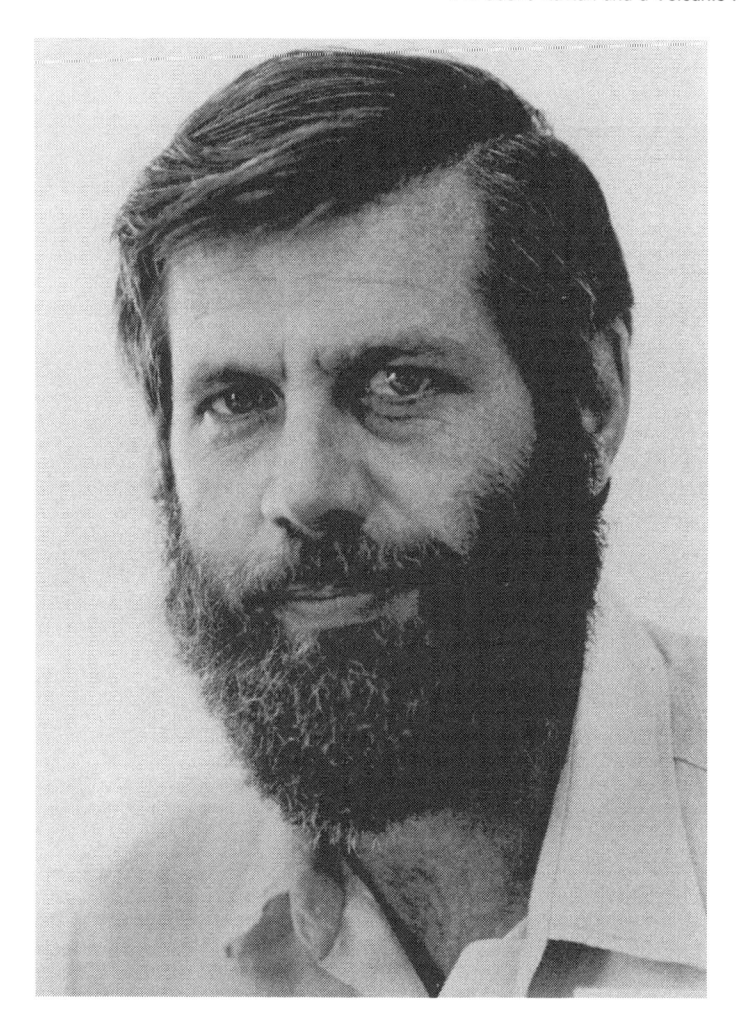

11. Cooke-Ravian and a Volcanic Resurgence: 1971–1979

Figure 88. R.J.S. Cooke in the 1970s.

Source: Photograph supplied by H. Cooke.

Deciding which eruptions belong to so-called 'time clusters' is not as straightforward a process as it might seem. There are, firstly, difficulties in establishing the precise start date of a cluster. For example, should the 1950s cluster be restricted to just 1953–1957 by eliminating the Lamington eruption of 1951, and should the severe eruption at Ulawun in 1970 be added to the 1972–1975 cluster bringing the start time forward by two years? The finish date can be problematic too. Thus, should the 'restlessness' at Kadovar in 1976 and a prominent eruption at Ulawun in 1978 be added to the 1970s cluster? A graph can be drawn of the number of eruptions against each year, but there are commonly problems in clearly separating the clusters or 'peaks' from normal background 'noise' — that is, from the normal, ongoing, and perhaps low-level eruptions at frequently active volcanoes such as Bagana, Manam and Kavachi. There is, in addition, the question of whether the clusters are simply a statistical anomaly, rather than representing a geophysically significant increase in the number of volcanic eruptions. Nevertheless, the volcanologists whose workloads and periods away from home increased significantly during 1953–1957 and 1972–1975, proposed that the clusters had geophysical causes rather than being aberrations of statistics.

Table 4. Volcanoes in Eruption in Near Oceania from 1972 to 1975[4]

Manam 1974–1975. Eruptive activity in June–August 1974 produced block-and-ash flows and lava flows down the South East Valley followed by lesser activity that died away in early January 1975.
Karkar 1974–1975. Eruptions from Bagiai cone on the floor of the inner summit caldera produced mainly lava flows that eventually covered about 75 per cent of the uninhabited caldera floor.
Long 1973–1974. Five separate phases of explosive activity at Motmot Island in Lake Wisdom included early surtseyan explosions as well as eruption of some lava flows.
Ritter 1972 and 1974. Brief submarine eruptions, lasting only a few hours each, were noted on 9 October 1972 and 17 October 1974 from just west of the island.
Langila 1973–1975. Explosive eruptions between May 1973 and May 1974 were accompanied by the eruption of lava flows, and were followed by further activity — including lava flows — from December 1974 to February 1975.
Ulawun 1973. An eruption in October was similar to a severe one from Ulawun in 1970, but it lasted only 14 days compared to 28 days in 1970. Ulawun was again in severe eruption in 1978.
Bagana 1972–1975. Lava continued to flow from Bagana during the cluster period and there were notable explosive eruptions in 1975.[a]
Kavachi 1972–1975. Shallow-water explosive eruptions were reported at times.[b]

a. Bultitude (1979).
b. Johnson & Tuni (1987).

Cooke and colleagues addressed the question of whether tectonic earthquakes were the triggers for the subsequent volcanic pulses of the 1950s and 1970s.[5] They concluded that there was indeed a marked increase in regional earthquake activity in 1951–1953 — that is, prior to the dominant eruptive period of 1953–1957 — but noted that other periods of earthquake increase were not followed

---

4   Cooke & others (1976).
5   Cooke (1975) and Cooke et al. (1976).

by eruption pulses. Furthermore, the correlation was less notable for 1970–1972, although Cooke was convinced there was indeed a connection between the Madang 1970 and Josephstaal 1972 earthquakes series and the outbreak of eruptions at Karkar and Manam in the western Bismarck Volcanic Arc in 1974, as Taylor had forecasted. Cooke noted also, however, that strong pulses of eruptive activity at both Manam and Karkar in early June 1974 took place just before the solstice, hinting that earth tides may have influenced at least the intensity of the eruptions.

A general conclusion does seem to be that nearby tectonic earthquake activity in the Bismarck Volcanic Arc in particular may, from time to time, 'trigger' volcanic unrest and subsequent eruptions, such as in the 1880s, 1950s and 1970s. Such a broad generalisation, however, does not preclude the idea that both tectonic earthquakes and volcanic eruptions are simply two different expressions of a common period of general 'geodynamic' unrest or stress build-up of whatever origin, that one of them — earthquakes or eruptions — does not necessarily always have to precede the other and that, indeed, one may take place without the other.

Ulawun and the Threat of Cone Collapse

New eruptions broke out from the summit crater of the imposing stratovolcano of Ulawun on the north coast of New Britain in January 1970, and again in 1973, 1978 and 1980.[6] Each of these four eruptive periods was notable for the spectacular expulsion of block-and-ash pyroclastic flows and lava flows, the first time that these phenomena had been identified and recorded at the volcano. Indeed, Ulawun up to 1970 had not been the focus of much volcanological interest at all because most of its few reported eruptions had been weak and, apparently, infrequent. Other volcanoes and their crises had required much greater attention. Nevertheless, Ulawun in the 1960s had been producing some small-scale explosive eruptions, and the volcano was known to have produced larger eruptions in the distant past, particularly in 1915. Father J. Stamm in 1915 was based at Toriu on the New Britain coast, about 50 kilometres northeast of Ulawun, and in 1961 he recalled the following:

> suddenly came the eruption about April 1915. At the beginning we could not see anything of it, there was a heavy Northwest storm, but we felt the ashes falling down, covering the ground three to four inches[at Toriu] … . Shortly after this … . A column of fire [from Ulawun] rose high up in the air, as if a giant gun was fired straight up. Such a blast came every few minutes. At daytime fire could not be seen, only a black cloud that became shining white when the sun shun [sic] upon it.[7]

---

6   Johnson et al. (1972), Cooke et al. (1976), McKee et al. (1981b), and McKee (1983, 1989).
7   Stamm (1961), p. 2. See also Cooke (1981).

Villagers from the coastal strip on the north-western slopes of Ulawun later related to Stamm how most of their houses had collapsed under the weight of ash, but that there had been no casualties. A Roman Catholic mission and sawmill were established in this same coastal area in 1928–1929 — at Ulamona, which today is the main community centre along this stretch of vulnerable coast.

Volcanologists N.H. Fisher and C.E. Stehn in August 1937 climbed the then 2,220-metre-high volcano,[8] evidently the first Europeans to do so, and I undertook ten days of geological work on the volcano in 1969 — that is, the year before the start of a long period of intermittent eruptive activity which has led to Ulawun becoming recognised today as one of the more frequently active and higher-risk volcanoes in Near Oceania.

Figure 89. This is how Ulawun appeared from the north-west in 1967. Ulamona Mission and Sawmill are in the foreground. The deep north-western valley can be seen on the volcano itself, as well as the east-west escarpment low on the southern flank to the far right.

Source: R.F. Heming. Geoscience Australia (GA-8687-1).

None of the recent pyroclastic flows and lava flows from Ulawun have reached the coast and impacted on the communities there, although a pyroclastic flow in 1980 came menacingly close to Sule village and airstrip. Furthermore, ash falls have not caused as much damage as they appear to have done in 1915. High-rising ash clouds from Ulawun are, however, a threat to aviation, and there are concerns about the structural stability of Ulawun.

---

8   Fisher (1937).

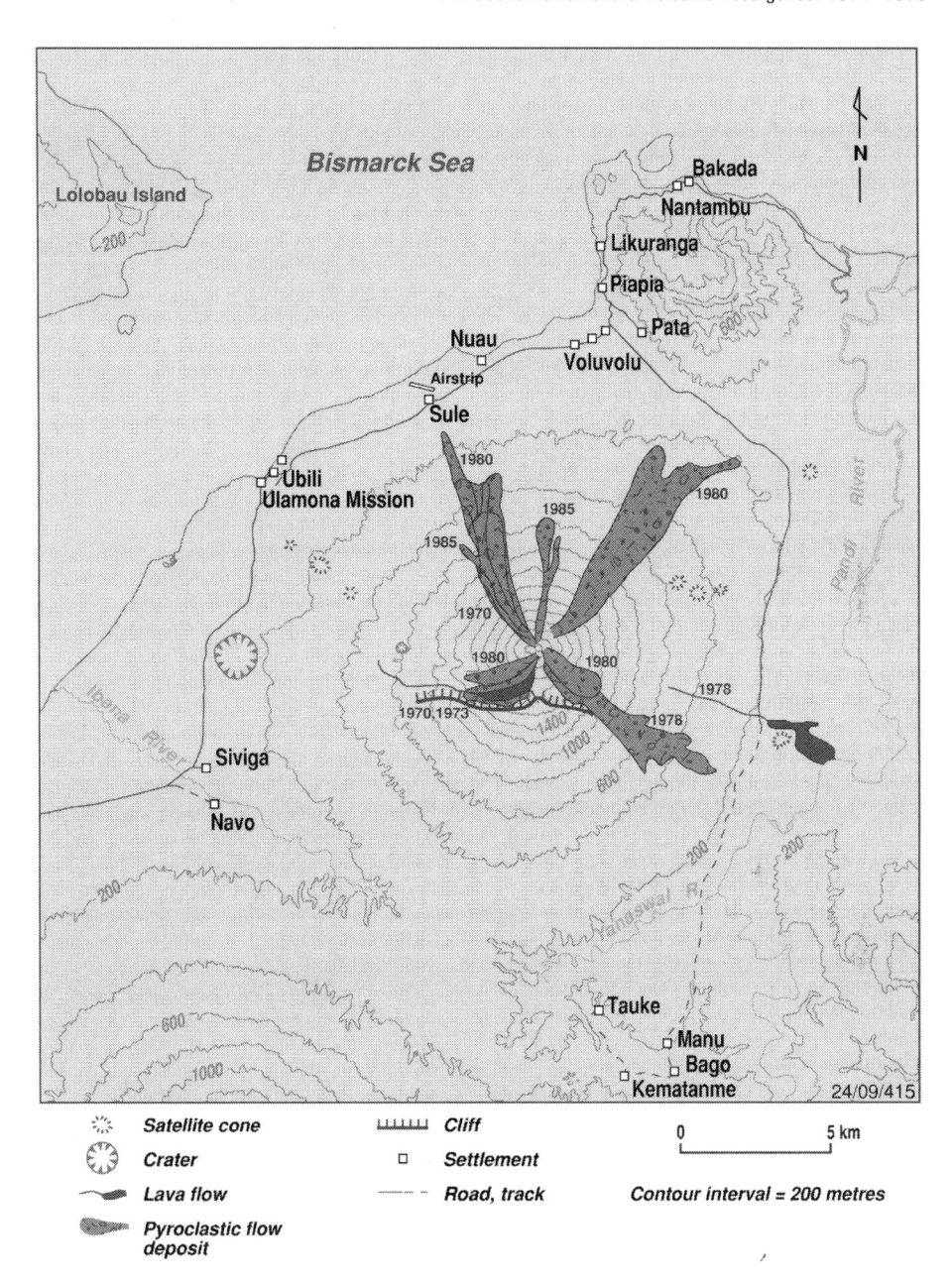

Figure 90. Ulawun volcano showing areas covered by 1970–1985 pyroclastic flows and lava flows. The volcano to the north-east of Ulawun is Likuruanga or North Son.

Source: Ulawun Workshop Report (1989, figure on p. 16).

Fire Mountains of the Islands

Figure 91. Incandescent lava issues from several active vents along the fissure on the eastern flank of Ulawun volcano on 13 May 1978. The lava is seen from the west coalescing into a single, eastward-moving, and non-incandescent flow.

Source: R.J.S. Cooke. McKee et al. (1981b, Figure 9). Geoscience Australia (no registered number).

The structurally unstable condition of Ulawun began to be suspected during the 1970s and particularly during and after the 1978 eruption.[9] An 'apparent splitting' of the volcano was observed high on the steep south-eastern side of the young cone on 7 May, and lava flowing from the top of the fissure and,

---

9 McKee et al. (1981b). See also Johnson et al. (1983).

reportedly, disappearing back into the volcano near the bottom of it. An even more unusual but significant phenomenon took place on 10 and 11 May when lava began flowing from a large number of vents on a linear fissure low down on the eastern flank of the volcano. This represents the first time that such a 'flank' eruption had been observed and recorded at any active volcano in Near Oceania. The flank fissure trended towards the summit of Ulawun, implying that the volcano is capable of producing magma along radial fractures deep within the volcano. A line of four, old, satellite cones that runs south-eastwards from near Ulamona on the western flank of Ulawun is evidence for another radial fracture, and there seems to be another radial alignment of small cones on the north-eastern flank too.

The danger at Ulawun, therefore, is that gravitational failure of the apparently over-steepened and over-heightened cone might take place, especially at times when magma injected into radial and other deep-seated fractures causes the volcano to swell and then fail. The danger is not just to the nearby coastal communities but also to settlements elsewhere in the Bismarck Sea, if a large debris avalanche entered the sea and caused a Ritter-like tsunami. This issue was addressed, 20 years after this period of activity, when a major international workshop on Ulawun volcano and volcanic-cone collapses was held at Walindi in West New Britain Province.[10]

Long Island Disaster and Tibito Tephra

Volcanic eruptions from sources within Lake Wisdom on Long Island are known to have started in 1953, 1955, 1968, 1973, 1976 and 1993, and they all seem to have been similar in character. An active volcano on the floor of the lake builds up to lake level, and produces what in most cases probably should have been strombolian eruptions, but which — because of the near-surface interaction of the magma with the lake water — are actually surtseyan, at least until an island is well formed. These eruptions are spectacular to witness and photograph from aircraft, particularly against the stunning background and natural beauty of Lake Wisdom and its surrounds, but so far they have not been disastrous for the people who live on Long Island, particularly as there are no villages on the lake shore. Nevertheless, some volcanic eruptions at Long Island in the not-so-distant past were not so benign.

Visitors to the seashore of Long Island cannot help but be struck by the youthful nature and extensive amounts of unconsolidated pumice and ash in the eroding coastal cliffs. These deposits include carbonised logs that are so well preserved

---

10 Ulawun Workshop Report (1989).

they could be mistaken for the remnants of a recent forest fire, were they not buried by the pumice. Stories collected from coastal people on both Long Island and the mainland also refer to the youth of a major volcanic eruption that devastated the entire island. Taylor, for example, wrote of being informed by an Administration officer

> that the eruption was of comparatively recent origin as stories of the escape from Arop [Long Island] are still evident among natives of the surrounding islands. It seems evident that some very alarming warning phenomena preceded this eruption as a considerable number of natives appear to have escaped from the island before the catastrophic eruption took place. [He] believes that the Siassi island people originally come from Long Island, and has found, on the harsher parts of the neighbouring New Guinea coast, settlements of natives who are also evacuees. One group, he believes, settled near the Lutheran Anchorage on northern Umboi but were subsequently wiped out by the 1888 eruption [sic] of Ritter Island.[11]

Early warning, evacuation, and recovery are also contained in a traditional story in which a long-haired man came to Long where the islanders wanted to kill him. A *bikman* of Long intervened, saved him, and then helped him build a canoe so that he could leave the island. The long-haired man in response warned the *bikman* of an impending disaster and advised him to flee with his family, which he did — to Crown Island. The eruption at Long wiped out those who remained on the island, but the *bikman* and his family eventually returned to establish a new community after which the island became bountiful in pigs, dogs and wild fowl.[12]

People came back to Long Island, apparently mainly in the latter half of the nineteenth century, presumably well after vegetation had reappeared on the island.[13] The implication is that the eruption was so recent that it must have taken place since the first Europeans entered the region in the sixteenth century. There are, however, no records of this devastating eruption having been witnessed by early European voyagers, and information about the eruption comes from geological studies of the eruption deposits and from Melanesian oral history.

The young pumice and ash deposits that are exposed on the coast of Long Island have been called the Matapun Beds.[14] Much of the material is ignimbrite — that is, pyroclastic material laid down by large pumice flows similar to those produced during the Krakatau eruption of 1883. The eruption, like Krakatau's,

---

11 Taylor (1953), pp. 4–5.
12 Harding (1967), note, p. 133.
13 Ball & Hughes (1982).
14 Pain et al. (1981).

was clearly a large one. Its total volume exceeded ten cubic kilometres and almost certainly the eruption was related to the formation of the large summit caldera that contains Lake Wisdom. Furthermore, the pumice and ash from the eruption would have been distributed well beyond the shores of Long Island.

Two academic geomorphologists, Colin F. Pain and Russell J. Blong, both New Zealanders, in 1970 began developing their research interest in the thick volcanic-ash beds found in the central highlands on the New Guinea mainland. They, and others, demonstrated through systematic mapping that many of the old tephras in the highlands originated, as expected, from the volcanoes of the mainland Fly-Highlands province.[15] Australian archaeologists in 1970, however, were studying prehistoric drainage ditches exposed in the Kuk Tea Plantation east of Mount Hagen, and were curious about the origin of pale, thin, inorganic layers that were exposed by their excavations. They invited the opinion of Pain and Blong, who quickly deduced that the layers were volcanic ash. Did these ash layers also originate from the highlands volcanoes? This question triggered a great deal of scientific effort in mapping the ash layers. The youngest of the thin tephras intrigued Blong, in particular, and this set him off on a path of major research that resulted, finally, in publication in 1982 of *The Time of Darkness*, a book devoted to the subject of the single ash layer.[16] This painstaking and time-consuming research involved further field mapping, ash-sample collection, analysis of the physical and chemical characteristics of the ash, and collection of numerous stories from Melanesian oral history,[17] all aimed at elucidating the source and age of the ash and what effect it may have had on highlands communities. The ash was Tibito Tephra.

Tibito Tephra is not well preserved. The ash fell mainly on mountainous, forest-covered terrain characterised by heavy tropical rainfall, erosion, washouts, and reworking by people in gardens, and therefore identifying and tracking the tephra across country was much more difficult than had it fallen more conveniently in, say, arid central Australia. But Blong demonstrated that the ash thickened towards the east, and that the size of the ash particles increased in the same direction. He tracked Tibito Tephra to Long Island, and concluded that it was the distal equivalent of the Matapun Beds. This tephra was only one of many ash layers in the highlands that are thought to have originated from the volcanoes off the north coast of New Guinea in recent geological times. Discriminating between these ashes during fieldwork in the highlands is not easy, especially in the absence of subsequent chemical analyses, as stressed by more recent studies in the highlands by palaeoecologists.[18]

---

15 Pain & Blong (1976).
16 Blong (1982).
17 Blong (1979). See also, for example, Vicedom & Tischner (1943).
18 Haberle (1998).

Figure 92. The pumice and ash of the Matapun Beds are here exposed in a five-metre-high cliff on the north-eastern coast of Long Island in 1970. The lower two-thirds of the cliff consist mainly of ignimbrite, which is overlain by layers of airfall ash.

Source: R.W. Johnson. Geoscience Australia (M-1114-8.tif).

The many Melanesian stories about the Long Island eruption and Tibito Tephra fallout are different in detail, but are valuable in making some general conclusions, particularly where matched with the scientific information about the pyroclastic materials preserved today on Long Island and in the highlands, and about the known effects of eruptions elsewhere in the world. The most disastrous effects almost certainly would have been experienced by any people who did not evacuate from Long Island. Human bones have been reported from the Matapun Beds,[19] perhaps the remains of victims. Severe effects would also have been experienced by people in the mainland coastal areas to the west and south-west of the island where ash accumulations of five–ten centimetres or more probably caused house collapses in villages. Tsunamis too may have inundated settlements on long sections of the mainland coastline. Ash thicknesses on the mainland decrease to the west yet deposits there may have caused some deaths through ash accumulation and roof collapses — similar to the coastal

---

19   Egloff & Specht (1982).

communities — and by mudflows, but more importantly may have destroyed gardens and triggered food shortages and probably even famine over wider parts of the affected highlands.

The Matapun-Tibito eruption from Long is by far the largest known volcanic eruption to have taken place in Near Oceania in the last 500 years and, indeed, is one of the largest known anywhere in the world over the last 1,000 years or so. But when exactly did it take place? This question was debated intensely during the 1970s when biologists interested in post-eruption biological colonisation, archaeologists and anthropologists interested in past peoples, and geoscientists interested in the volcanic tephras, all combined in interdisciplinary field research on Long Island. Genealogical evidence pointed consistently to a time after 1700, the year when William Dampier passed by the island, noting that it was covered by lush vegetation.[20] But such deductions from oral history are plagued with chronological uncertainty, especially where generations 'drop out' of genealogical sequences and where the lengths of generations are uncertain. Nevertheless, the first 'radiocarbon ages' to be obtained on the Matapun Beds — including some obtained by Taylor in the 1950s and 1960s — also pointed to a 'modern' post-1700 date, and the apparently late reoccupation of Long Island was consistent with a recent eruption too.

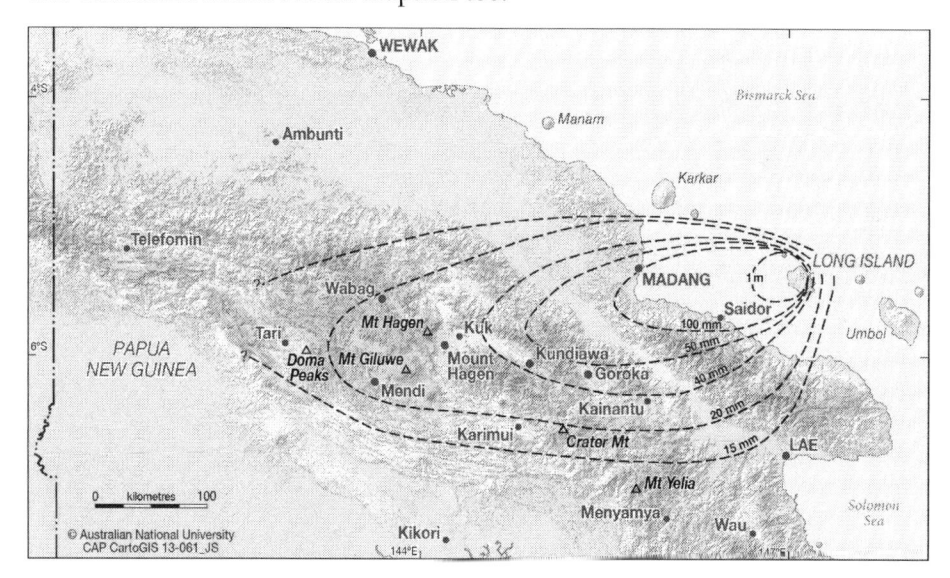

Figure 93. The inferred original thicknesses of Tibito Tephra, which was mapped by Russell Blong, are shown by the dashed lines extending from Long Island westwards across the present-day towns of the highlands region. Only a few of the Fly-Highlands volcanoes are shown.

Source: Adapted from a redesign by J. Golson and R.J. Blong of the original map published by Blong (1982, Figure 29).

---

20   Ball & Johnson (1976) and Blong (1982).

Dampier published a sketch of how Long Island looked in profile in 1700 — low and flat-topped, more or less the way it is today — but this cannot be taken necessarily as evidence that the Matapun Beds had been laid down and a single Long Island caldera had already formed, by 1700. This is because the young Matapun–Tibito eruption was not the only one from which pumice and ash are preserved on Long Island. Older, voluminous, fragmental deposits are exposed at and near the coast, but also — and most noticeably — in the unscaleable walls of the Lake Wisdom caldera itself. In other words, major explosive eruptions have characterised the later geological history of Long Island and the caldera, therefore, may have formed by a *series* of collapses or subsidence events rather than by the single engulfment of an 'ancestral mountain'. Also, was Dampier in fact correct about the shape of the island being 'long' in 1700, and did a post-1700 eruption change the outline of the island in plan into its present-day more equant form? An eruption sometime in the eighteenth or early nineteenth century — before Dumont D'Urville sailed passed the vegetation-covered island in 1827 — at first seemed most likely.[21]

Additional radiocarbon ages were obtained during the 1970s on the Matapun Beds and Tibito Tephra, by which time techniques for collecting carbon samples had improved significantly. Furthermore, the radiocarbon-dating method itself had become more sophisticated both in analytical technique and in interpretation of results. Blong stressed the inadequacies for dating purposes of oral history and genealogies, and argued for an eruption date somewhere between 1640 and 1670 on the basis of the radiocarbon results.[22] His argument depended largely on unravelling the statistical uncertainties that had to be considered before converting the radiocarbon ages to the true dates of the modern calendar, but Blong's general conclusion was supported strongly and independently by a radiocarbon specialist, the late Henry Polach, whom I had asked to investigate.[23] Blong himself, however, remained cautious, stressing the uncertainties still involved and concluding that a post-1700 date for the eruption could not be ruled out completely.[24] Such interpretations are still a common problem with young carbon samples, particularly in recognising that there will be a natural spread in radiocarbon dates of samples when an old forest, including trees perhaps more than a century old, are overwhelmed by the products of a severe explosive eruption. The Long–Tiboto data represent a good example of all the uncertainties involved.[25]

---

21 Ball & Johnson (1976).
22 Blong (1982).
23 Polach (1981).
24 Blong (1982). See also, Haberle (1998).
25 A more modern radiocarbon-dating technique that should be considered for Long Island carbon samples is the so-called 'wiggle-match' method where 14C dates are obtained for multiple samples taken from core

Sulphuric-acid aerosols injected into the stratosphere by the Matupun–Tibito eruption at Long Island may well have caused the same kind of brilliance in worldwide sunsets seen after the Krakatau eruption in 1883. Volcanic aerosols from large equatorial eruptions are slow to settle out of the atmosphere. They drift to the polar regions of the Earth over a period of many months, where eventually they are caught up in snow falls and become incorporated in the sequence of annual layers that can be counted in ice cores drilled in glaciers. A prominent acid-rich layer has been identified in ice cores from Greenland, dated at 1645 AD, and attributed to Long Island.[26] This is close to the year of 1643 when Abel Tasman sailed past Long Island, but he made no reference to the state of the island.

Yomba and Cook: Two 'Mystery' Volcanoes

The Long Island story has an addendum. Villagers on the coastlines near Madang on the mainland facing Long Island, have stories — collected in the 1970s — of an island called Yomba that sank beneath the sea eight–ten generations previously.[27] Some tellers of the stories said that their ancestors lived on the island and escaped to the mainland and other islands where their descendants live today. Some versions of the story refer to a large devastating tsunami, some to a volcanic eruption and accompanying time of darkness. All of the informants were in agreement that Yomba was not Long Island and that it had disappeared before the latest major eruption at Long.

Some of the informants said that Yomba Island was where Hankow Reef is today, about 50 kilometres north-west of Long and more or less in line with the other volcanoes of the western Bismarck Volcanic Arc. The coral reef at Hankow almost certainly has a volcanic basement so the oral history relating to the existence of a volcanic Yomba Island is generally consistent with this inferred geology. The Yomba stories carry elements of major geophysical catastrophe, but the scientific evidence for this is yet to be found. A major volcanic collapse at Yomba within, say, the last 1,000 years — either caldera formation and an accompanying explosive eruption, or major gravitational collapse such as at Ritter in 1888 — is considered unlikely on the basis of the results of a major side-scan sonar survey of the sea floor around Hankow Reef in 2004.[28] Furthermore, no tephras have been found that might be traced back specifically to Yomba. The reputed

---

to rim on individual logs, and then finding the best match. for the sequence of dates against a calendar-year versus 14C-date calibration curve. C.O. McKee (personal communication, 2012) recently applied this technique to the dating of a carbonised log from Rabaul.
26   Zielinski et al. (1994).
27   Mennis (1981).
28   Silver et al. (2009).

volcanic island could have been low-lying and therefore, perhaps, vulnerable to the effects of a major earthquake or tsunami, but again, the supporting evidence has yet to be found. Nevertheless, the strength of the oral history from the Madang area cannot easily be swept aside and the questions about the existence and disappearance of Yomba Island remain unanswered.

Claims about the origin of, and recent volcanic activity at, another 'shoal' in Near Oceania have also been scrutinised. Cook submarine volcano in the New Georgia Group of the western Solomon Islands was marked on a British Admiralty map as a 36-metre-deep shoal after a British naval vessel, HMS *Cook*, reported an uncharted submarine disturbance, including a sulphuretted smell, on 14 December 1963. The disturbance took place just a few kilometres off Munda Point on southern New Georgia Island. The shoal itself, however, was detected at night using only primitive, backup, depth-sounding instruments. A strong earthquake took place in the same general area 20 days afterwards and, in May 1965, two young villagers in a canoe saw a large mass of water rise from the same area, then fall back again into the sea. A submarine volcano was postulated and called 'Cook' by John Grover, head of the British Solomon Islands Geological Survey in Honiara.[29] Detailed marine geophysical surveys in later years, however, failed to detect any sea floor volcano large enough to have a summit as shallow as 36 metres below sea level.[30] The cause of the sea-surface disturbances may have been a hydrothermal blowout from the sea floor, perhaps related to seismic activity and sea floor faulting. Cook submarine volcano, therefore, appears to be modern myth.

Fatal Eruption on Karkar in 1979

The RVO observation camp near the southern rim of the inner caldera of Karkar had a spectacular vantage point from which to view the 1974–1975 eruptions at Bagiai Cone, about 260 metres high and two kilometres away, down on the floor of the caldera.[31] Bagiai, or Bagia, is also the traditional name for the 3.2-kilometre-wide inner caldera. The panorama from the southern rim of the caldera and its lava-field covered floor was also enhanced by the escarpment of an older and outer caldera forming a forested backdrop in the distance.

---

29   Grover (1968).
30   Exon & Johnson (1986).
31   McKee et al. (1976) and McKee & Wallace (1981).

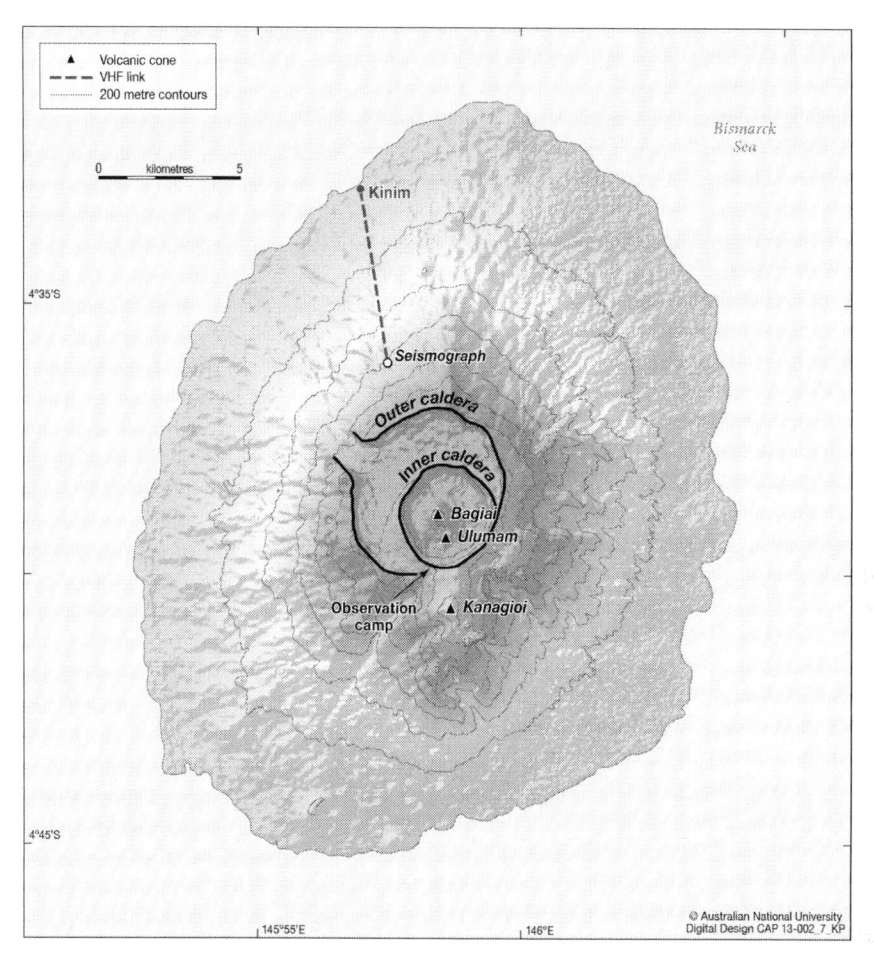

11. Cooke-Ravian and a Volcanic Resurgence: 1971–1979

Figure 94. The summit of Karkar Island is dominated by the peak of Kanagioi and by the two calderas. Kanagioi is either a satellite volcanic cone like many others found on the northern and southern slopes of the island — but not identified in this map — or else is the last remaining one of a cluster of summit volcanoes that were largely engulfed when the calderas formed.

Source: Adapted from McKee et al. (1976, Figure 1).

Vulcanian–strombolian eruptions in February–August 1974 and December 1974 to February 1975 were impressive enough during the day, but at night the pyrotechnic sprays of glowing lava fragments from Bagiai could be truly impressive. Great volumes of new lava and fallouts of ash in 1974–1975 also spread out over most of caldera floor, transforming its former vegetation-covered aspect.

The summit area of the island, including the highest peak Kanagioi, is unpopulated and traditionally was regarded as being inhabited by deities and the spirits of dead ancestors as well as fiends who exist alongside humans in the physical universe. An Australian anthropologist working on Karkar in the mid-1960s found that her informants had traditional belief systems which incorporated elements of Christianity, aspirations for acquiring European knowledge and material wealth, efforts to understand the origin of Europeans, and elements of cargo cultism. This syncretic approach to acquiring a new world view also involved the volcanic features at, and inferred volcanic sounds from, the summit of Karkar Island:

> Spirits of the dead went to live first in the volcanic crater, Bagia. Here, Kulbob [a mythical creation deity], assisted by the ancestor, Karkar [the first human being], supervised factories and workshops, whose sounds of production had long been a part of local legend … . Here, too Jesus imparted 'real' knowledge, and Misken, the minor mountain deity, acted as guardian and companion of the spirits as she had done traditionally. Heaven was Kanagioi itself …[32]

A permanent seismograph had been installed by RVO in June 1975 on the north-western flank of the island and was connected by VHF telemetry to a manned receiving post at Kinim to the north. Evidence for volcano reactivation was detected between November 1977 and July 1978 when there was an increase in volcanic earthquakes, and then water vapour and gas emissions became more noticeable from the caldera floor near the south-east foot of Bagiai and the surrounding area in September 1978. Explosive activity eventually broke out there on 12 or 13 January 1979. RVO staff were exchanging periods of duty at the observation camp on the caldera rim, and Cooke and volcanological assistant Elias Ravian were there on the evening of 7 March, equipped with a voice radio, and about to be relieved on 11 or 12 March. The details of what happened during the night are not clear, but three large volcanic earthquakes were felt at the coast between 0100 and 0132 hours on the 8 March, accompanied by a dense black cloud rising up from the summit area.

Cooke and Ravian did not make radio contact the next morning and a ground party climbed to the caldera rim to investigate.[33] They found evidence that a powerful volcanic explosion had taken place from near the foot of Bagiai. The explosion had blasted laterally and south-eastwards, and had stripped vegetation from the caldera wall and from a crescent-shaped area of rain forest on the caldera rim itself. The RVO observation camp was just within the western tip of the crescent. The bodies of Cooke and Ravian were found at the camp, buried

---

32 McSwain (1977), p. 171.
33 Geological Survey Staff (1979).

under about 15 centimetres of ash, the area clearly having been bombarded by hot ash and by rocks up to a metre in diameter. Large forest trees showing signs of charring were shredded, their limbs broken off and reduced to bare trunks. McKee and Talai flew from Rabaul to Karkar on 9 March and joined a group who arranged for the bodies of Cooke and Ravian to be removed from the summit by helicopter and returned by air to Rabaul. Cooke had been working on several volcanological manuscripts and had them with him at the campsite. These were recoverable, although many pages had been severely charred by the hot lateral blast.

Figure 95. Devastation at the RVO observation campsite is seen in this photograph, taken on 9 March 1979, in what previously was lush tropical rain forest. Bagiai and the inner caldera are down to the right.

Source: C.O. McKee, Rabaul Volcanological Observatory.

The fatal eruption of 8 March created anxiety on Karkar Island because of fear that the intensity of the eruptions might escalate and affect coastal communities. Discussions were held about the possibility of an evacuation of the island being necessary, but the volcanic threat died away in the following weeks and months. Furthermore, the March eruption was assessed to be of localised hydrovolcanic type, strongly influenced by the access of rainwater to a body of magma beneath Bagiai. A new crater at the foot of Bagiai had been greatly widened and deepened, and it appeared to be a maar, much like the ones formed by hydroexplosive activity at Goropu in 1943–1944. McKee and colleagues were later able to correlate periods of heavy rainfall at Karkar with smaller explosive phases that continued from the maar after the deadly explosion of 8 March.[34] The Karkar crisis of 1978–1979 was the closest that RVO came during the 1970s to considering the possibility of an evacuation, in contrast to the 1950s when four volcanic crises involved the displacement of people.

Figure 96. Chris McKee is seen here at the observation camp on the caldera rim at Karkar in 1979 after the fatal eruption of 8 March. Bagiai is in the foreground, together with the new maar-like crater emitting water vapour. The walls of both the inner and outer caldera are visible in the background.

Source: Photograph supplied by C.O. McKee.

The sudden deaths of Cooke and Ravian devastated not only their immediate families, friends and colleagues, but also shocked the wider community in both Papua New Guinea and Australia, as well as volcanologists internationally who had known Rob personally. Cooke was buried in the European cemetery in Rabaul, and Ravian at his village, but the two names are linked in the memorial that was erected for them at RVO headquarters and in a volume of scientific papers, including several by Cooke, that was later published in their honour.

---

34 McKee & others (1981a).

References

Ball, E.E. & I.M. Hughes, 1982. 'Long Island, Papua New Guinea — People, Resources and Culture', *Records of the Australian Museum*, 10, pp. 463–25.

Ball, E.E. & R.W. Johnson, 1976. 'Volcanic History of Long Island, Papua New Guinea', in R.W. Johnson (ed.), *Volcanism in Australasia*. Elsevier, Amsterdam, pp. 133–47.

Blong, R.J., 1979. 'Time of Darkness Legends from Papua New Guinea', *Oral History*, 7, no. 10, pp. 1–135.

——, 1982. *The Time of Darkness*. The Australian National University Press, Canberra.

Bultitude, R.J., 1979. *Bagana Volcano, Bougainville Island: Geology, Petrology, and Summary of Eruptive History between 1875 and 1975*, Geological Survey of Papua New Guinea Memoir 6.

Cooke, R.J.S., 1975. 'Time Variations in Recent Volcanism and Seismicity along Convergent Boundaries of the South Bismarck Sea, Papua New Guinea', *Australian Society of Exploration Geophysicists Bulletin*, 6, pp. 77–78.

——, 1981. 'Notes on the Activity of Ulawun Volcano, 1700–1958: Results of a Literature Search', in R.W. Johnson (ed.), *Cooke-Ravian Volume of Volcanological Papers*. Geological Survey of Papua New Guinea Memoir, 10, pp. 147–51.

Cooke, R.J.S., C.O. McKee, V.F. Dent & D.A. Wallace, 1976. 'Striking Sequence of Volcanic Eruptions in the Bismarck Volcanic Arc, Papua New Guinea, in 1972–75', in R.W. Johnson (ed.), *Volcanism in Australasia*. Elsevier, Amsterdam, pp. 149–72.

Egloff, B.J. & J. Specht, 1982. 'Long Island, Papua New Guinea — Aspects of the Prehistory', *Records of the Australian Museum*, 10, pp. 427–46.

Exon, N.F. & R.W. Johnson, 1986. 'The Elusive Cook Volcano and other Submarine Forearc Volcanoes in the Solomon Islands', *BMR Journal of Australian Geology and Geophysics*, 10, pp. 77–83.

Fisher, N.H., 1937. Geological Report on an Ascent of the Father Volcano, New Britain. Territory of New Guinea. Australian Department of External Territories, Canberra. National Australian Archives, Canberra, Series A518, Item AB836/4 (New Guinea: Volcanic Eruption — Visit of Stehn, Cilento and Woolnough), three unnumbered folios.

Geological Survey Staff, 1979. 'Fatal Eruption of Karkar Volcano 8 March 1979', Geological Survey of Papua New Guinea Report 79/10.

Grover, J.C., 1968. 'Submarine Volcanoes and Oceanographic Observations in the New Georgia Group 1963–1964', *British Solomon Islands Geological Record*, 3, pp. 116–25.

Haberle, S.G., 1998. 'Dating the Evidence for Agricultural Change in the Highlands of New Guinea: the Last 2000 Years'. *Australian Archaeology*, 47, pp. 1–19.

Harding, T.G., 1967. *Voyagers of the Vitiaz Strait: a Study of a New Guinea Trade System*. University of Washington Press, Seattle.

Johnson, R.W., R.A. Davies & A.J.R. White, 1972. *Ulawun Volcano, New Britain*. Bureau of Mineral Resources, Canberra, Bulletin 142.

Johnson, R.W., R.P. Macnab, R.J. Arculus, R.J. Ryburn, R.J.S. Cooke & B.W. Chappell, 1983. 'Bamus Volcano, Papua New Guinea: Dormant Neighbour of Ulawun, and Magnesian-Andesite Locality', *Geologische Rundschau*, 72, pp. 207–37.

Johnson, R.W. & D. Tuni, 1987. 'Kavachi, an Active Forearc Volcano in the Western Solomon Islands: Reported Eruptions between 1950 and 1982', in B. Taylor & N.F. Exon (eds), *Marine Geology, Geophysics, and Geochemistry of the Woodlark Basin–Solomon Islands*. Circum-Pacific Council for Energy and Mineral Resources Earth Science Series, 7, pp. 89–112.

McKee, C.O., 1983. *Volcanic Hazards at Ulawun Volcano*. Geological Survey of Papua New Guinea Report, 83/13.

——, 1989. 'Ulawun', in L. McClelland, T. Simkin, M. Summers, E. Nielsen & T.C. Stein (eds), *Global Volcanism 1975–1985*. Prentice Hall, Englewood Cliffs, New Jersey, pp. 175–76.

McKee, C.O., R.J.S. Cooke & D.A. Wallace, 1976. '1974–75 Eruptions of Karkar Volcano, Papua New Guinea', in R.W. Johnson (ed.), *Volcanism in Australasia*. Elsevier, Amsterdam, pp. 173–90.

McKee, C.O. & D.A. Wallace, 1981. 'Lava Fields in the Inner Caldera of Karkar Volcano', in R.W. Johnson (ed.), *Cooke-Ravian Volume of Volcanological Papers*. Geological Survey of Papua New Guinea Memoir, 10, pp. 49–62.

McKee, C.O., D.A. Wallace, R.A. Almond & B. Talai, 1981a. 'Fatal Hydroeruption of Karkar Volcano in 1979: Development of a Maar-like Crater', in R.W. Johnson (ed.), *Cooke-Ravian Volume of Volcanological Papers*. Geological Survey of Papua New Guinea Memoir, 10, pp. 63–84.

McKee, C.O., R.A. Almond, R.J.S. Cooke & B. Talai, 1981b. 'Basaltic Pyroclastic Avalanches and Flank Effusion from Ulawun Volcano in 1978', in R.W. Johnson (ed.), *Cooke-Ravian Volume of Volcanological Papers*. Geological Survey of Papua New Guinea Memoir, 10, pp. 153–65.

McSwain, R., 1977. *The Past and Future People: Tradition and Change on a New Guinea Island*. Oxford University Press, Melbourne.

Mennis, M.R., 1981. 'Yomba Island: A Real or Mythical Volcano?', in R.W. Johnson (ed.), *Cooke-Ravian Volume of Volcanological Papers*. Geological Survey of Papua New Guinea Memoir, 10, pp. 95–99.

Mori, J., C. McKee, I. Itikarai, P. Lowenstein, P. de Saint Ours & B. Talai, 1989. 'Earthquakes of the Rabaul Seismo-deformational Crisis September 1983 to July 1985: Seismicity on a Caldera Ring Fault', in J.H. Latter (ed.), *Volcanic Hazards*. IAVCEI Proceedings in Volcanology, 1, pp 429–62.

Pain, C.F. & R.J. Blong, 1976. 'Late Quaternary Tephras around Mount Hagen and Mount Giluwe', in R.W. Johnson (ed.), *Volcanism in Australasia*. Elsevier, Amsterdam, pp. 239–51.

Pain, C.F., R.J. Blong & C.O. McKee, 1981. 'Pyroclastic Deposits and Eruptive Sequences of Long Island. Part 1: Lithology, Stratigraphy, and Volcanology', in R.W. Johnson (ed.), *Cooke-Ravian Volume of Volcanological Papers*. Geological Survey of Papua New Guinea Memoir, 10, pp. 101–07.

Polach, H., 1981. 'Pyroclastic Deposits and Eruptive Sequences of Long Island. Part 2: Radiocarbon Dating of Long Island and Tibito Tephras', in R.W. Johnson (ed.), *Cooke-Ravian Volume of Volcanological Papers*. Geological Survey of Papua New Guinea Memoir, 10, pp. 108–13.

Silver, E., S. Day, S. Ward, G. Hoffmann, P. Llanes, N. Driscoll, B. Applegate & S. Saunders, 2009. 'Volcanic Collapse and Tsunami Generation in the Bismarck Volcanic Arc, Papua New Guinea', *Journal of Volcanology and Geothermal Research*, 186, pp. 210–22.

Stamm, J., 1961. Unpublished and untitled letter to G.A.M. Taylor, 4 May. RVO File V.F/9A (R.O.1/1/4), Folios 48 & 49.

Taylor, G.A.M., 1953. 'Notes on Ritter, Sakar, Umboi and Long Island Volcanoes'. Bureau of Mineral Resources, Canberra, Record 1953/43.

Ulawun Workshop Report, 1989. *Volcanic Cone Collapses and Tsunamis: Issues for Emergency Management in the Southwest Pacific Region*. IAVCEI Workshop on Ulawun Decade Volcano, Papua New Guinea. International Decade for Natural Disaster Reduction, Emergency Management Australia, Canberra.

Vicedom, G.F. & H. Tischner, 1977 (1943). *Myths and Legends from Mount Hagen*, Andrew Strathern (trans.). Institute of Papua New Guinea Studies, Port Moresby.

Wallace, D.A., R.J.S. Cooke, V.F. Dent, D.J. Norris & R.W. Johnson, 1981. 'Kadovar Volcano and Investigations of an Outbreak of Thermal Activity in 1976', in R.W. Johnson (ed.), *Cooke-Ravian Volume of Volcanological Papers*. Geological Survey of Papua New Guinea Memoir, 10, pp. 1–11.

Zielinski, G.A., P.A. Mayewski, L.D. Meeker, S. Whitlow, M.S. Twickler, M. Morrison, D.A. Meese, A.J. Gow & R.B. Alley, 1994. 'Record of Volcanism since 7000 B.C. from the GISP2 Greenland Ice Core and Implications for the Volcano-Climate System', *Science*, 264, pp. 948–52.

12. Eruption Alert at Rabaul Caldera: 1971–1994

In a statement issued in Papua New Guinea on Monday [23 January 1984], the principal volcanologist, Dr P. Lowenstein, said that 'evidence is accumulating to suggest that the volcano has embarked on an irreversible course towards the next eruption and that it is only a matter of time before this occurs ... the eruption that was previously only a possibility is now much more likely to occur within the next few months'.

Peter Hastings (1984)

Crisis Build-up and Stage-2 Alert

Villagers living near the south-eastern end of Matupit Island, Rabaul, were by 1970–1971 aware that nearby coastal cliffs had encroached perilously close towards their homes as a result of sea-wave erosion. Their concerns were alleviated after 1971, however, when a new beach began to form at the foot of the pumice cliffs which gradually became stranded inland. The south-eastern end of the island was rising episodically, following the two, major, Solomon Sea earthquakes that had shaken Rabaul in July 1971. Rabaul Volcanological Observatory (RVO) staff led by Rob Cooke began measuring the amount of ground uplift in 1973 using a survey line that ran southwards from Rabaul town to the end of the island. About 60 centimetres of uplift had been detected by 1979, the year of Cooke's death, and to more than a metre by 1983.[1]

Matupit is a low-lying island that in 1971 was already well known for its vertical oscillations, most noticeably at times of major earthquakes or tsunamis when a causeway linking the island with the shore might disappear and then reform.[2] The island is made up of flat-lying pumice beds, but there is no persuasive geological evidence that it is, or ever was, an eruptive centre. The island rather has risen out of the waters of Rabaul Harbour, perhaps very recently in geological terms, as early European observers of Matupit Island had commented on the island's youthful vegetation. Furthermore, emergence and occupation by people of the island only a few centuries previously at most, is supported

---

1 McKee et al. (1985).
2 Fisher (1939, p. 18), for example, reported the causeway forming after an earthquake in 1919, and continuing to rise slightly up to the time of the 1937 eruption. The causeway and its road were destroyed in 1971 as a result of tsunamis generated by the 1971 regional earthquakes, but were subsequently reconstructed by civil works.

by genealogical evidence.[3] The growth of Matupit's size may indeed have encouraged increases in population. Matupit, by the 1970s–1980s, supported more than 2,000 people who, in common with other Tolai communities — and Melanesian communities in general — had a strong link with their land. They were, however, volcanically vulnerable, being only about two kilometres from the western slopes of Tavurvur volcano across the entrance of Greet Harbour. Collection of eggs from the buried, volcanically heated, nests of megapode birds near Tavurvur was a source of some cash income for the Matupit people.

Uplift of the south-eastern end of Matupit Island in 1971–1983 was accompanied by increased numbers of local earthquakes, some of them felt in Rabaul town, and many taking place in 'swarms' that became more pronounced as this 12-year period progressed. Cooke was the first to illustrate the unusual, possibly unique, pattern that these earthquakes formed where plotted as a map of epicentres.[4] The pattern he concluded had a broad D-shape, rather like a doughnut flattened straight along its north-western side, and enclosing a mainly submarine area in the harbour that was more-or-less earthquake-free. The straight segment covered the south-eastern part of Matupit and trended directly — and rather menacingly — towards Rabalanakaia volcano. Cooke thought that this linear segment was a possible fault zone, and he identified the complementary curved segment of the 'D' as part of the seismically active margin of Rabaul Caldera. The D-pattern, however, was based on relatively few earthquake data obtained from the inferior number and distribution of recording stations that existed in the early 1970s.

The uplift and seismic swarms began to raise concerns about future volcanic activity at Rabaul, and geophysical monitoring was enhanced by RVO. Cooke, a specialist in gravity measurement, introduced surveys of the gravity field to complement the surveying lines, and new recording stations were established to improve monitoring of the earthquake activity. The likelihood of a volcanic eruption similar to those in 1878 and 1937 was discussed, but larger-scale eruptions were possible too. This point was highlighted by timely publication in 1974 of a study of the volcanic geology of Rabaul, which demonstrated for the first time that the volcano was built up in part by major ignimbrites formed by huge, caldera-forming, explosive eruptions, the most recent one only about 1,400 years ago.[5] Such a major Krakatau-type eruption taking place in, say, 1980 — the year of a National Census — would impact severely on most of the 100,000 people living in the north-eastern part of the Gazelle Peninsula, and particularly the 15,000 people living in Rabaul town itself.

---

3   Sack (1987).
4   Cooke (1977).
5   Heming (1974).

12. Eruption Alert at Rabaul Caldera: 1971–1994

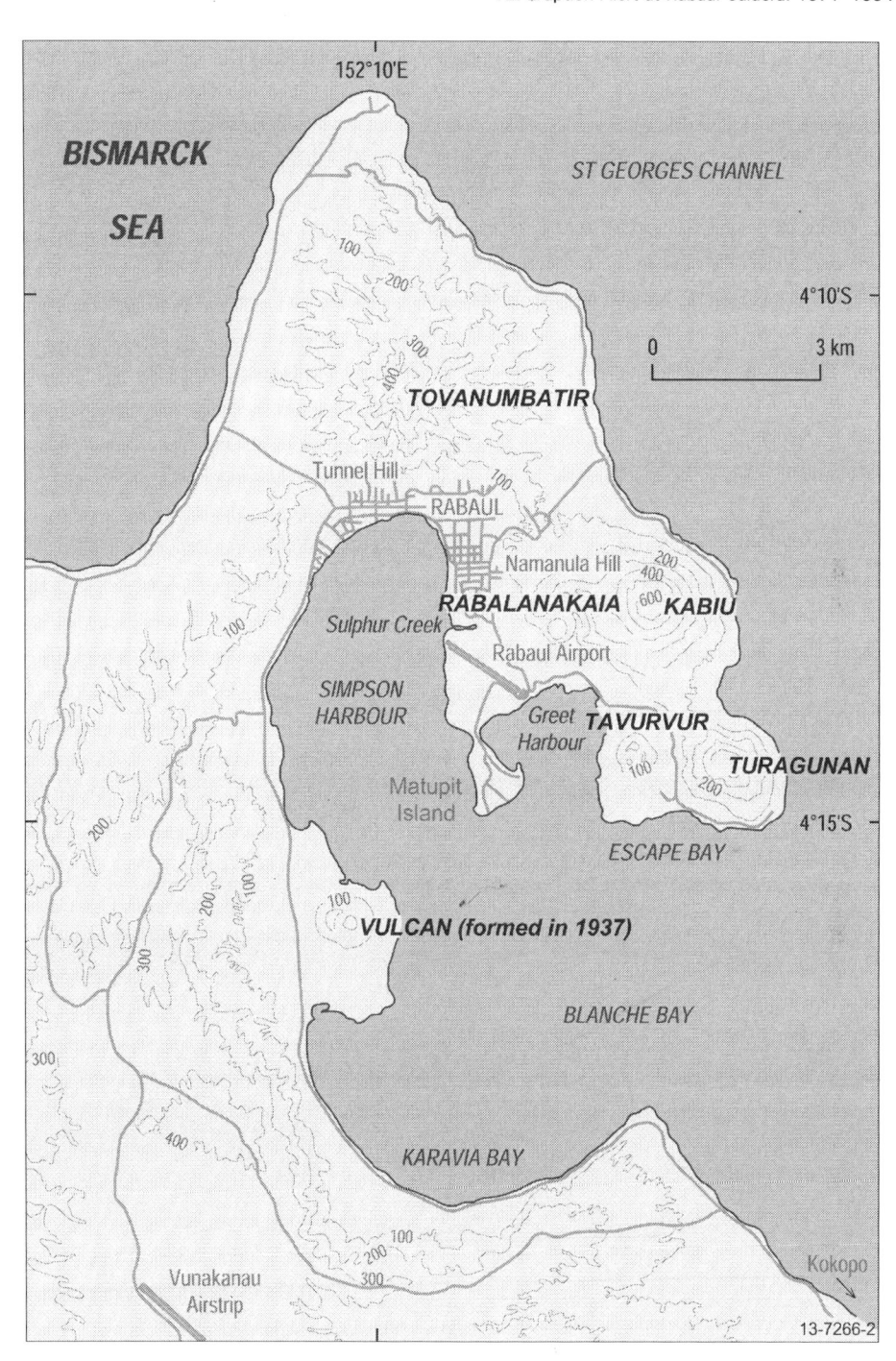

Figure 97. Villages are not shown in this simplified map of the main features of Blanche Bay and Rabaul town in the 1970s.

Source: Adapted from Johnson & Threlfall (1985, figure on p. 4).

Peter L. Lowenstein, a geochemist and Englishman, took over leadership of RVO after Cooke's death in 1979. Lowenstein, like his predecessors in this position, had had no training in applied volcanology, but volcanologists Chris McKee and Ben Talai by this time had gained practical experience in monitoring many eruptions in Papua New Guinea. Lowenstein came to Rabaul with natural organisational and management strengths that would prove invaluable in the years ahead as a major volcanic crisis developed in the Rabaul area. He could also use his impeccable English accent and sense of humour to great effect in clearly articulating his views, especially opinions that he held strongly. A new volcanological team was established under Lowenstein's leadership after French volcanologist Patrice de Saint Ours, Papua New Guinean seismologist Ima Itikarai, American–Japanese seismologist Jim Mori, and surveyor Malcolm Archibald, an Australian, joined the staff at RVO.

A dramatic and exponential increase in volcano unrest at Rabaul Caldera began in late August 1983 and became particularly noticeable on 19 September 1983 when an intense earthquake swarm was felt in Rabaul, accompanied by a sharp increase in the uplift rate at Matupit Island. These events marked the start of what came to be referred as the 'seismo-deformational crisis' at Rabaul, and which would demand virtually the sole attention of RVO staff.[6] A disaster plan had been prepared for East New Britain Province a few months earlier by a United Nations disaster-management specialist, in consultation with national and provincial government authorities, including RVO, and together with the private sector. The plan assumed eruptions only of the scale of those in 1878 and 1937, and it included a scheme of four stages of volcanic alert — Stage 1 where a volcanic eruption was expected within years to months, up to Stage 4 where one was expected within just days to hours.[7]

Table 5. Stages of Volcanic Alert at Rabaul

	Possible eruptive activity	Summary of meaning
1	Within years to months	Risk exists. No immediate cause for alarm.
2	Months to weeks	Increased risk. Still no public action necessary.
3	Weeks to days	Risk is serious. Precautionary actions required.
4	Days to hours	Situation is critical. Event imminent. Public Red Alert.

The East New Britain Provincial Disaster Committee (PDC), which was chaired by the Secretary of the Department of East New Britain, Nason Paulius, held its inaugural meeting on 13 April 1983, and government authorities practised the first of several evacuation exercises in late May and early June. A range of other public awareness-raising activities were initiated, including — in the

---

6   Lowenstein (1988).
7   East New Britain Provincial Disaster Plan (1983).

background in Canberra — preparation of my own book, with Rev. Neville Threlfall, on the 1937–1943 eruptions at Rabaul. This publication was sponsored by the Insurance Underwriters' Association of Papua New Guinea, and was aimed at illustrating how the previous eruptions had affected the Rabaul area.[8]

Figure 98. Principal features of the north-eastern Blanche Bay area are seen clearly in this computer-enhanced aerial photograph mosaic compiled in the early 1980s, before the uplift of the south-eastern end of Matupit Island became strongly noticeable.

Source: Adapted from Johnson & Threlfall (1985, dust cover).

8   Johnson & Threlfall (1985).

A Stage Two volcanic alert indicative of a possible volcanic eruption within only weeks to months was declared on 29 October 1983 following further intense earthquake swarms. The crisis period escalated up to April 1984, and by 15 May major seismic swarms had taken place since the previous September.[9] The number of recorded earthquakes in April reached a monthly maximum of more than 13,000, and the southern end of Matupit Island had risen a total of about 1.6 metres above its 1973 height. These events caused considerable apprehension amongst the communities of Rabaul town and the surrounding region. There was a partial and voluntary evacuation of the town itself, including businesses.[10] Many villagers moved to land outside the caldera area, including perhaps as many as 40 per cent of those people living in the highest risk areas south of the main business district. Furthermore, blocks of government-owned land south of the Warangoi River were made available for settlement, including at Sikut. Other people left the province altogether — for West New Britain, for example — and some expatriates departed for Australia. An inevitable topic of debate in the community and media, as well as at different levels of government, concerned the suitability of Rabaul as a place for a town and provincial capital — a repetition of the debates that had taken place after the 1937 eruption and after the Second World War. Prime Minister Michael Somare entered the debate by announcing in mid-February 1984 that, should an eruption take place, Kokopo would be developed as a new administrative centre and that Rabaul had insufficient land for expansion anyway.[11]

Disaster-preparedness activities intensified in early 1984. Maps of danger areas in Rabaul were posted on public notice boards together with information about points of assembly and refuge. Roads were cleared of overhanging tree branches, and old airstrips and wharfs outside the caldera were upgraded. Drafting of legislation led to the passing in March of eight Acts of Parliament to cater for disaster preparations, including the *Natural Disaster Act*, which resulted in the formal establishment of both the National Disaster Centre and Provincial Disaster Committees. This national legislation was indicative of the prominence of the Rabaul area in both the national economy and psyche of Papua New Guinea. RVO and the PDC established a close working relationship, and scores of situation reports and information bulletins on the condition of the volcano were provided by RVO to the PDC as a basis for official decision-making.[12] Evacuation rehearsals continued at Rabaul and the crisis began to receive international attention.

---

9   McKee et al. (1984).
10 Lowenstein (1988), Blong & Aislabie (1988) and Neumann (1996).
11 Darius (1994).
12 Lowenstein (1988).

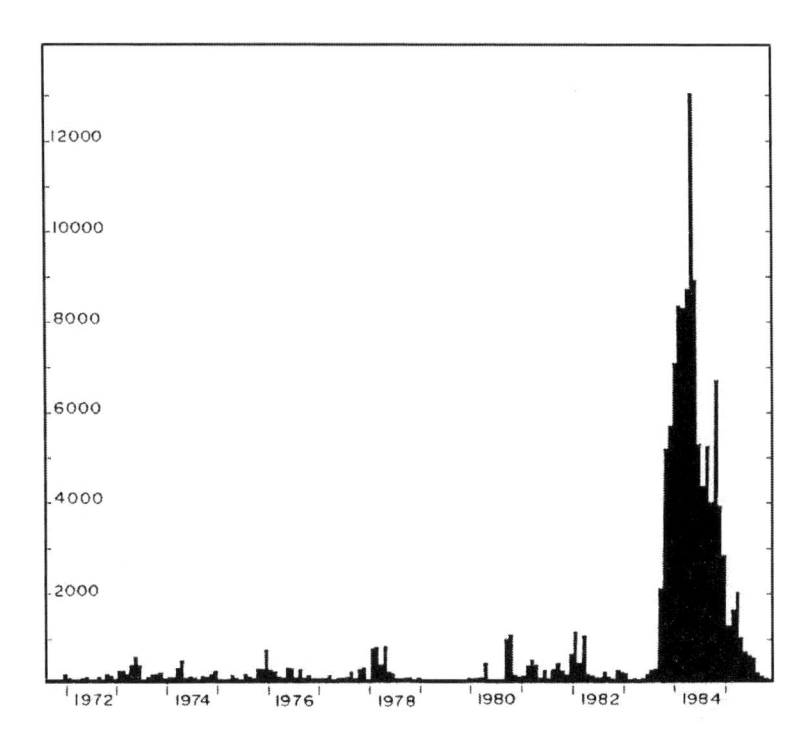

Figure 99. The number of harbour earthquakes recorded monthly at Rabaul increased dramatically in 1983–1985.

Source: Mori et al. (1989, Figure 2). Reproduced with the permission of Springer-Verlag.

There was high expectation by early 1984 that an eruption would take place — that, indeed, the volcano had embarked on an 'irreversible course' towards a likely eruption.[13] Yet the crisis for RVO in practice represented the challenge of management of scientific uncertainty, of there being no scientifically based and precise way of presenting accurate predictions or even definitive forecasts about the expected eruption, much as the public and media would have liked or expected it. Information about the likelihood and scale of the expected eruption was at first restricted, but people in the town and in both the national and international media, increased the demand for the latest information, particularly situation statements from RVO directly. RVO staff, however, were already overworked trying to keep up-to-date with the increased schedules of instrumental monitoring, data collection and interpretation. Reports had to be written for government authorities, particularly the PDC, which was responsible for official information releases rather than RVO itself. Having to deal directly with the visits of many individual media personnel and anxious citizens concerned about Rabaul volcano, was a time-consuming burden.

---

13   Hastings (1984).

A group of three, including myself, from the Bureau of Mineral Resources (BMR) in Canberra, and funded by the Australian International Development Assistance Bureau (AIDAB), was sent to Rabaul in January 1984 to assist RVO. I made visits to Manam and Langila volcanoes, and to the Esa'ala area in Milne Bay Province, and New Zealander volcanologist Brad Scott visited Manam, to investigate reports of increased activity that could not be checked by RVO staff because of work commitments in Rabaul.[14] My visit to Esa'ala in February 1984 was in response to local reports of increased activity in the thermal areas bordering Dawson Strait area, including coral-reef die-off at Dobu Island, and therefore of eruption fears which were reported in national newspapers. No evidence was found for any impending eruptions, but volcano fears in Milne Bay had clearly been heightened by the prominent reporting of the happenings in Rabaul.

Some news media began to spread misinformation, or sensationalised reports, as well as criticisms of the government and agencies — including RVO — about the management of the crisis. One widely distributed article was written by a journalist for the influential Australian weekly magazine the *Bulletin*, in which the authorities grappling with the Rabaul situation were portrayed as secretive, indecisive, bungling and stressed, and involved in political infighting.[15] The front page of the magazine blazed the words 'Rabaul's killer volcano' and showed a weather cloud above old Kabiu volcano, backlit by the dawn light, which gave the false impression of an actual volcanic eruption. The situation was made even more difficult in RVO by the natural reluctance or anxiety of some scientists — as seen in scientific agencies in other countries — of being misinterpreted or over-interpreted by aggressive, story-seeking media, or publically stepping beyond their areas of expertise and authority, or wanting to avoid perceptions amongst scientific peers of 'grandstanding'. Science communication today is still a challenge for many scientists in other parts of the world who are employed in observatories that have mandates for early warnings of hazard events.

The public relations situation had become unmanageable, and it was not mitigated until the PDC established a Public Information Unit, and until an Australian geologist, Dr Hugh L. Davies, who had had many years of experience working in Papua New Guinea, was appointed for some months as a volcanological liaison officer. An information newsletter funded by the private sector and called 'Rabaul Gourier' — a combination of *Post-Courier*, a national newspaper, and *guria* meaning earthquake — was only one of several initiatives developed to keep the citizenry and media informed.

---

14 Johnson (1984).
15 Stannard (1984).

Scientific Responses to the Caldera Unrest

Instrumental monitoring of Rabaul Caldera by RVO had increased dramatically by 1984. Additional funds for the purchase of new equipment were made available by the national government and by the development-assistance agencies of foreign countries, particularly Japan, Australia, New Zealand and the United States. The number of earthquake recorders increased and several, different, ground-survey methods for measuring the type and extent of ground deformation — uplift, tilt and horizontal extension —were deployed.[16] These included electronic-distance measurements using laser beams shot to numerous reflectors positioned around the shores of Rabaul Harbour. Tide gauges were used to measure seawater depths, which decreased as parts of the coastline rose. Never before, or since, has such an impressive array of geophysical monitoring instruments been deployed in Rabaul Harbour. Even the heights of emerged and stranded marine barnacles were mapped in an attempt to determine the extent of any longer term, harbour-wide uplift.[17]

The large amount of volcano-monitoring data collected by RVO before and during the seismo-deformational crisis was used not only for attempts at eruption forecasting, but also for interpretations of what the internal structure of Rabaul Caldera might look like. The major scientific analysis to emerge from this work was later published by Jim Mori and McKee, in association with several co-workers, in numerous peer reviewed papers in the geoscientific literature.[18] Their important interpretation was based on the following five observations:

1. Most earthquake epicentres plotted on a map defined a 'seismic annulus', an elongate doughnut shape about ten kilometres long from north-south, rather than the D-shape noted by Cooke. This annulus was thought to mark the ring-like boundary of a block of coherent rock that was presumed to have subsided during the latest caldera-forming event, about 1,400 years ago.

2. The earthquakes were mostly less than about four kilometres deep, and they defined the top of a postulated, large, underlying magma reservoir.

3. The sides of the subsided block seemed to dip steeply outwards as if the block were an inverted keystone.

4. Much of the detected uplift, tilt and horizontal extension could be explained by a smaller magma body that had intruded upwards from the larger magma reservoir into the block to within about two kilometres of the surface. This

---
16 Lowenstein (1988).
17 De Saint Ours et al. (1991).
18 See, in particular, Archbold et al. (1988), McKee et al. (1989) and Mori et al. (1989).

smaller magma body or *intrusion* was causing the bulging of the sea floor centred south-east of Matupit Island.

5. Some of the ground-deformation data hinted at the existence of a second, shallow source of upward intrusion beneath the sea floor east of Vulcan and located over the seismic annulus.

The conclusion that a 'bulge' existed close to Matupit Island led to some speculation that the doming there represented the site of possible future submarine eruptions which might build a new volcano above the sea, rather than the anticipated eruptions taking place at either Tavurvur or Vulcan, as in 1878 and 1937. The final hypothesis favoured by the RVO volcanologists, however, involved lateral, underground migration of magma from the proposed shallow magma body to beneath both of these well established volcanoes. This interpretation has some characteristics similar to those for Glen Coe-type calderas, particularly existence of a coherent, near-cylindrical block rather than the disintegrated and highly fractured rocks envisaged for Krakataua-type calderas. Speculation also centred on an especially threatening aspect of the caldera block. Could it subside again, perhaps by a reduction of magmatic pressure, and the 'inverted keystone' drop catastrophically into the large, underlying magma reservoir, generating another major eruption the size of the one 1,400 years ago?

Attention was paid not only to the monitoring data but to improving fundamental knowledge about the structure and evolution of Rabaul Caldera and how its volcanoes behave, both now and in the past. A marine geophysical survey of the Rabaul Harbour floor was undertaken by the United States Geological Survey (USGS) in 1982, including the area of the 'bulge' south-east of Matupit Island. Up-domed and folded sediments, slumps, and steep faults were discovered on seismic-reflection profiles, adding support to the RVO theory.[19] Questions that arose, however, concerned both the age and full extent of these structures and the age of the bulge itself. Did they all start forming in 1971 or were they much older in origin, including the bulge, which might represent only the latest episode of the long, oscillating, uplift history of Matupit Island? Another question was why there was no cluster of earthquake epicentres marking the area of the bulging and fracturing? Such a cluster might *not* be expected if the rocks above the inferred magma body were soft and weak, but in this case upward streaming of heat, and even magma, might have been expected to reach the surface more easily, particularly if the bulge was a long-lived feature. One highly respected volcanologist known for his bold thinking even suggested to me in the early 1980s that small volcanic eruptions may have taken place already on the deeper parts of the sea floor.

---

19 Greene et al. (1986).

12. Eruption Alert at Rabaul Caldera: 1971–1994

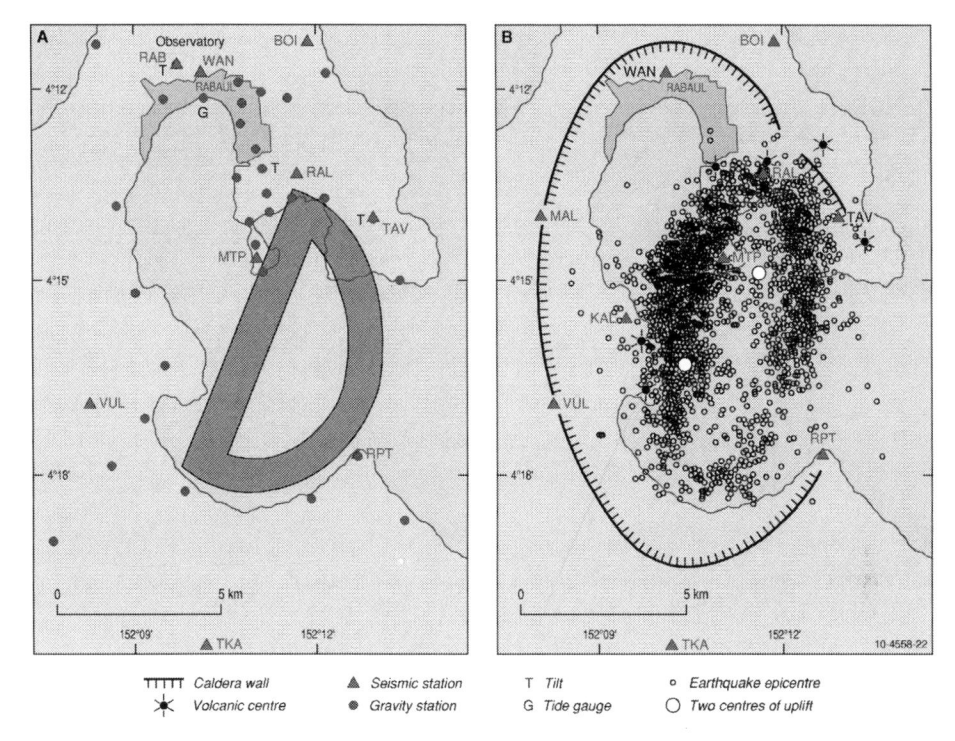

Figure 100. Two different patterns for the 'seismic annulus' were obtained by mapping the epicentres of earthquakes in the Blanche Bay area. The D-shape on the left was produced by Cooke (1977, adapted from Figure 2) and encompasses the epicentres of earthquakes recorded in the early 1970s. The pattern on the right is for the epicentres of more than a thousand earthquakes recorded from 1971 to 1983 (Mori et al. 1989, Figure 3).

Source: Both figures were redrawn from the originals and presented together by Johnson et al. (2010, Figure 22).

I arranged for a Remotely Operated Vehicle equipped with a video camera and temperature sensor to undertake dives at 11 different sites over the bulge in 1985, but we found no evidence of elevated water temperatures, hot springs, upward streaming of gas bubbles, or submarine craters or eruptions.[20] This inspection of the sea floor, however, in turn raised the need for systematic surveys of the discharge or flow of heat from the sea floor of Rabaul Harbour, as a means of understanding how volcanic heat is distributed in relation to the submarine centres of uplift and to the much deeper, larger, magma body. A series of AIDAB-funded heat-flow surveys — the main one in 1992 — provided no evidence that the main 'bulge' south-east of Matupit Island coincided with an area of elevated heat flow.[21]

---

20   Johnson (1986).
21   Graham et al. (1993).

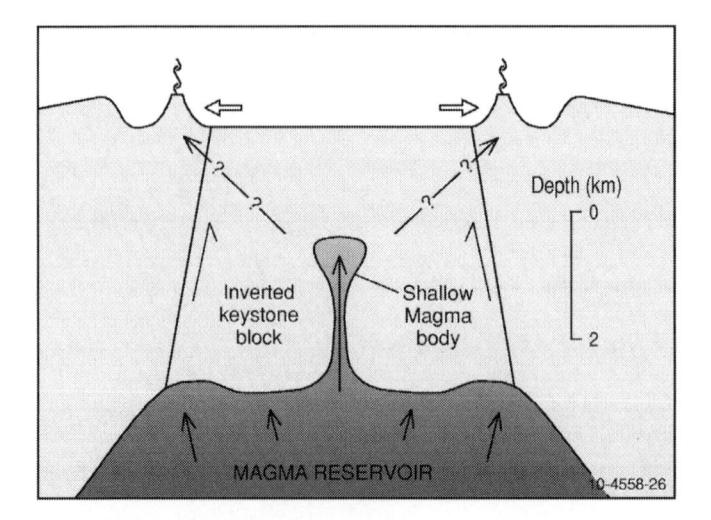

Figure 101. The four-kilometre-deep magma reservoir beneath Rabaul is shown in this cartoon of a vertical 'slice' through Rabaul Caldera, as envisaged by RVO volcanologists. The magma reservoir presses up on the inverted-keystone block, helping to keep it in place. Magma from the smaller, shallower body moves obliquely through the block and its boundary — as shown by the lines of question marks — in order to feed Tavurvur and Vulcan volcanoes on the surface.

Source: Mori et al. (1989, adapted from Figure 22 by Johnson et al, 2010, Figure 26).

Major advances in understanding the land geology of Rabaul Caldera were also made as a result of geological mapping in 1979 and 1984.[22] One important conclusion of these studies was that as many of five, but possibly nine, major ignimbrite eruptions of Krakatau-type may have taken place at Rabaul during the previous 20,000 years or so, meaning that such catastrophic eruptions take place once every 2,000 to 3,300 years, more or less. The latest of these major eruptions was the one about 1,400 years ago, corresponding to an eighth century AD age. It deposited the so-called Rabaul Pyroclastics, the pumiceous pyroclastic flows of which were particularly energetic. These extended out from Rabaul to at least 50 kilometres, flowing over the sea and reaching Watom Island. There were erroneous speculations that the eruption was responsible for major atmospheric effects in the northern hemisphere and for dateable acidity layers in polar ice cores.[23] The Rabaul Pyroclastics are of *dacite* composition, and evidently are a sample of the large magma reservoir beneath Rabaul Harbour. Furthermore, the large number of ignimbrites at Rabaul can also be taken as evidence that the

---

22   Walker et al. (1981) and Nairn et al. (1995).
23   Stothers (1984). The erroneous correlation was based on uncalibrated radiocarbon dates — a calibrated age of AD 720–750 ± 20 is regarded as the best current estimate for the '1400 BP' eruption (C.O. McKee personal communication, 2012).

caldera, like the one at Long Island, probably originated by a *series* of collapses, and therefore that a high 'ancestral mountain' may never have existed, at least above sea level.

Worldwide Volcanic Crises and Developments in Risk Awareness

The Rabaul crisis achieved considerable prominence internationally, even at this time — in the early to mid-1980s — when disastrous eruptions that might otherwise have diverted attention were taking place at other volcanoes worldwide, and when there were threatening signs of unrest at calderas in other countries. This was a period when important contributions by volcanologists in the United States achieved worldwide prominence and impact, triggered particularly by the highly publicised volcanic eruption at Mount St Helens, Washington State, on 18 May 1980. Government volcanologists in the USGS had accumulated many years of experience monitoring the relatively 'passive' volcanoes of Hawaii, but the explosive activity at Mount St Helens — a subduction-zone type volcano on the continental mainland — was a new experience for them. Managing the 1980 disaster raised similar challenges to those facing RVO staff at Rabaul during the Stage Two period. These included handling the critical but sensitive relationship between volcanologists, who were dealing with scientific uncertainty and ambiguity in their monitoring of volcanic unrest, and the authorities, public and media, who had their own need for clear and decisive information.[24] There were also conflicts with some US academic volcanologists who wanted to study the unrest and eruption themselves and who believed that any restrictions imposed of them were a violation of their right to unfettered scientific freedom.

A further issue for USGS volcanologists was being able to continue gaining experience in monitoring explosive-type volcanoes in the United States during those long periods when those volcanoes are dormant. The USGS therefore created the Volcanic Crisis Assistance Team (VCAT), which was to be partly funded by the US Agency for International Development (USAID), and would provide a rapid-response capability for those foreign countries who invited its assistance during a volcanic crisis.[25] The response included donation of volcano-monitoring equipment to the country concerned, as well as temporary secondment of USGS staff for in-country training, help in instrument installation, and advice on volcanic hazards and eruption forecasting, if requested. USGS volcanologist Norman G. Banks — founder of VCAT — came to Rabaul in 1984

---

24 Peterson (1988).
25 Tilling & Punongbayan (1989), Ewert et al. (1998) and Thompson (2000).

accompanied by technicians as part of this USAID-funded arrangement. They helped install the laser-beam, electronic-distance measurement equipment for use by RVO, and provided training.

The 1980 eruption at Mount St Helens was of moderate size and was assigned a VEI of five. VEI stands for 'Volcanic Explosivity Index', a concept proposed by two US-based volcanologists in 1982.[26] VEI is a semi-quantitative measure of eruption 'size' by which a score, or index, is assigned largely on the basis of the volume of magma erupted by a particular explosive volcanic eruption. Non-explosive eruptions, even if large in volume, are arbitrarily assigned a VEI of zero. Use of the VEI system increased greatly during the 1980s, especially when scientists of the Smithsonian Institution, Washington D.C., adopted it in a global database of volcanic eruptions that they had been developing.[27] The Mount St Helens eruption of 1980 has a higher VEI than does, for example, the Mount Lamington activity of 1951 — as shown in the following short list of selected historical eruptions.[28] Its VEI, however, is less than those for the caldera-forming eruptions at Rabaul in the eighth century and at Long Island in the seventeenth century. Furthermore, all of these examples have smaller VEI than does the great eruption at Tambora in 1815. It, impressively, has a VEI of seven, the only historical eruption rated so highly.

Table 6. VEI Values for Major Eruptions

Tambora, 1812–1815, Indonesia	VEI 7
Rabaul, eighth century, Papau New Guinea	VEI 6
Long Island, seventeenth century, Papau New Guinea	VEI 6
Krakatau, 1883, Indonesia	VEI 6
Tarawera, 1886, New Zealand	VEI 5
Mount St Helens, 1980–1985, United States	VEI 5
El Chichón, 1982, Mexico	VEI 5
Pelée, 1902–1905, Caribbean	VEI 4
Rabaul, 1937, Papau New Guinea	VEI 4
Lamington, 1951–1956, Papau New Guinea	VEI 4
Taal, 1965, Philippines	VEI 4
Galunggung, 1982–1983, Indonesia	VEI 4
Rabaul, 1878, Papau New Guinea	VEI 3
Manam, 1956–1958, Papau New Guinea	VEI 3
Nevado del Ruiz, 1985–1991, Colombia	VEI 3

---

26 Newhall & Self (1982).
27 Siebert et al. (2010).
28 Uncertainties in the VEI assignments made by the Smithsonian Institution for some of these eruptions are not shown in this list.

The above list includes two other eruptions of relevance from 1980–1985. The first of these was at Galunggung, Indonesia, in 1982 when Boeing 747 aircraft — belonging to two different international airlines — flew, at separate times, into drifting ash from the volcano, causing all four engines to stall on each aircraft.[29] These ash/aircraft encounters mark the start of considerable international effort to mitigate the threat of high-rising ash clouds to both international and domestic aviation. Many such ash clouds could be seen on images taken from satellites, but not where regions were covered by extensive weather clouds, such as those in monsoonal tropical areas like Indonesia and Papua New Guinea. There had been, however, considerable advances by the mid-1980s in the ways that eruption clouds could be observed from polar-orbiting and geostationary satellites. These included development of multispectral scanners that allowed for the satellite detection and tracking of the sulphur-dioxide gas in volcanic clouds. Furthermore, CSIRO scientists in Australia in the early 1980s began developing a multispectral-scanner technique for discriminating volcanic-ash clouds — initially those from Galunggung in 1982 — from clouds of meteorological origin, by using the contrasting infra-red characteristics of each type of cloud.[30] Volcanic aerosols injected into the stratosphere could also be identified and tracked using 'lidar'— meaning Light Detection and Ranging — instruments on board satellites and at ground-based stations. Aerosols from the 1980 eruption at Ulawun, for example, were detected in this way.[31]

The second important eruption in the above list was that at Nevado del Ruiz, Colombia, in 1985–1991 when a small VEI 3 eruption caused melting of the snow and ice cap on the summit of the Andean volcano and produced lahars. Highly destructive lahars swept down river valleys late at night on 13 November 1985, killing more than 22,000 people, and resulting in the second largest volcanic disaster of the twentieth century after Mont Pelée in 1902, in terms of numbers of human fatalities.[32] A major disaster-management question to emerge from the aftermath of the Ruiz disaster was why people had not evacuated before the lahars overwhelmed them. A volcanic-hazard map for Ruiz had been made available to local authorities a month earlier, and an early warning had been issued by scientists when the summit eruption began. The USGS Volcanic Disaster Assistance Program (VDAP), the successor to VCAT, came into formal existence in 1986 prompted by the 1985 Ruiz eruption and its disastrous outcome. VDAP continues to operate internationally, including in Papua New Guinea.

Pioneering techniques for mapping volcanic hazards had been developed by USGS volcanologists at Mount St Helens long before the 1980 eruption[33]

---

29 Johnson & Casadevall (1991).
30 Honey (1991).
31 McCormick (1985).
32 See, for example, Voight (1990).
33 Crandell & Mullineaux (1978).

and applied elsewhere, but the importance of such maps seems to have been underestimated by the Colombian authorities at Ruiz volcano. Volcanic-hazard maps are neither 'geological' nor 'evacuation' maps. Rather, they portray those areas most likely to be affected by particular kinds of volcanic processes — pyroclastic flows, lahars, ash falls, and so on — and, ideally, they are developed also for eruptions having different VEI. Geological maps of volcanoes, in contrast, show areas where volcanic rocks of different types and ages are exposed. Evacuation maps simply provide designation of escape routes, such as roads, tracks, airstrips and wharfs for shipping, on or near active volcanoes, without explicit — though commonly inferred — reference to volcanic hazards.

Alarming earthquake activity, ground deformation, and new gas emissions in one case, were also being reported in the early 1980s from the large, active calderas at Long Valley, eastern California, United States, and at Phlegrean Fields, west of Naples in Italy. The concurrent restlessness of these two calderas, together with Rabaul Caldera, in three different parts of the world, was coincidental, but were compared with considerable interest by volcanologists internationally. The three calderas also featured prominently in a major scientific literature review in the 1980s that was undertaken by two USGS volcanologists. These authors summarised almost 1,300 episodes of historical 'unrest' at 138 large calderas worldwide. Their conclusions included the following:

> The remarkable unrest at Rabaul Caldera that began in 1971 is perhaps the most threatening example in this compilation.[34]

A fundamentally important point that also emerged internationally in the 1980s was the distinction that must be made in volcanic-disaster management between 'risk' and 'hazard'.[35] Volcanic *hazard* refers simply to the size and frequency of a volcanic event and carries no implications about its impact on people and their settlements. *Risk*, on the other hand, is a more complex concept because it involves definition of *what* is being threatened by a particular volcanic hazard — people, schools, businesses, critical infrastructure, lifelines, and agricultural land. All of these different 'at-risk' categories, or 'elements', may in total be called 'exposure' or 'value', and defining risk depends on knowing and ideally measuring the *vulnerability* of each one of them. Thus, the small volcanic eruption at Ruiz in 1985 had high risk because of the large number of people and villages who would be, and were, affected by it, rather than because of the size of the eruption. Conversely, a large volcanic eruption in an unpopulated area may have low risk, but might have higher risk to aircraft flying overhead. Risk can be summarised by means of the following disarmingly simple formula:

> Risk = The hazard (their magnitudes and frequencies) x Elements exposed to risk x Their separate vulnerabilities

---

34 Newhall & Dzurisin (1988), p. 227.
35 Fournier d'Albe (1979).

An obvious corollary of this formula is that large populations of vulnerable people, together with agricultural lands and built environments that are also at-risk, have the potential to be affected by the greatest disasters. Reducing such disasters became a worldwide theme for the United Nations in the late 1980s, and the 1990s were declared an International Decade for Natural Disaster Reduction (IDNDR), a worthy venture that was successful particularly in those countries where funding could be identified for significant, national, IDNDR activities. The late 1980s and 1990s were, for me, a time of career transition when I greatly reduced my scientific interests in volcanic petrology, geochemistry and the tectonic settling of volcanoes, and became more involved in work on disaster-risk reduction in both Papua New Guinea and Australia.

'Natural disaster', even in the IDNDR title, carries an implication that disasters are caused in the first place by natural hazards, just as technological disasters are 'caused' by hazardous events such as industrial fires, oil-depot explosions, or chemical leaks. The affected communities are portrayed as unfortunate victims unable to defend themselves adequately against ferocious and uncontrollable external forces. 'Acts of God', the phraseology still enshrined in legalese and insurance policy, is a similar quasi-fatalistic expression. Both terms belie the fact that disasters follow only where the communities are ill-prepared for the impact of, say, a volcanic eruption, have not organised themselves to be appropriately resilient, and are vulnerable because of basic sociological characteristics such as weak economies, poverty, ineffective political leadership, and poor governance. The hazards themselves are not to blame, being without harmful intentions towards societies.

This international shift towards comprehensive risk and hazard assessments was reflected in an array of disaster-management studies in Rabaul. RVO volcanologists and others produced the first volcanic hazard maps for Rabaul,[36] and Lowenstein and Talai determined a risk rating for the numbers of at-risk people on or near the young volcanoes of Papua New Guinea, noting that the top five volcanoes — in order of decreasing risk — were Rabaul, Lamington, Manam, Karkar, and Ulawun.[37] Two geoscientists from the New Zealand Department of Scientific and Industrial Research (DSIR), undertook an assessment of volcanic, earthquake and tsunami hazards at Rabaul and included the first consideration of the numbers of people at-risk from the different volcanic hazards in the Rabaul area.[38] Furthermore, geographer Russell Blong and economist Colin Aislabie collaborated in assessing what would be lost in economic terms by considering the impacts of three different kinds of eruption

---

36 McKee et al. (1985).
37 Lowenstein & Talai (1985).
38 Latter & Hurst (1987). J.H. Latter, a volcanologist at RVO in the late 1960s, also published on the numbers of at-risk people on active volcanoes in the member countries of the Economic and Social Commission for Asia and the Pacific, or ESCAP (Latter, 1988).

Fire Mountains of the Islands

on the north-eastern Gazelle Peninsula.[39] Finally, an Australian geographer, Ken Granger, demonstrated how digital geo-referenced information — where loaded onto computer-based geographic information systems (GIS) — could be used for practical crisis-management purposes in the Rabaul area — and elsewhere.[40] GIS was evolving at this time as a valuable tool for risk-assessment purposes.

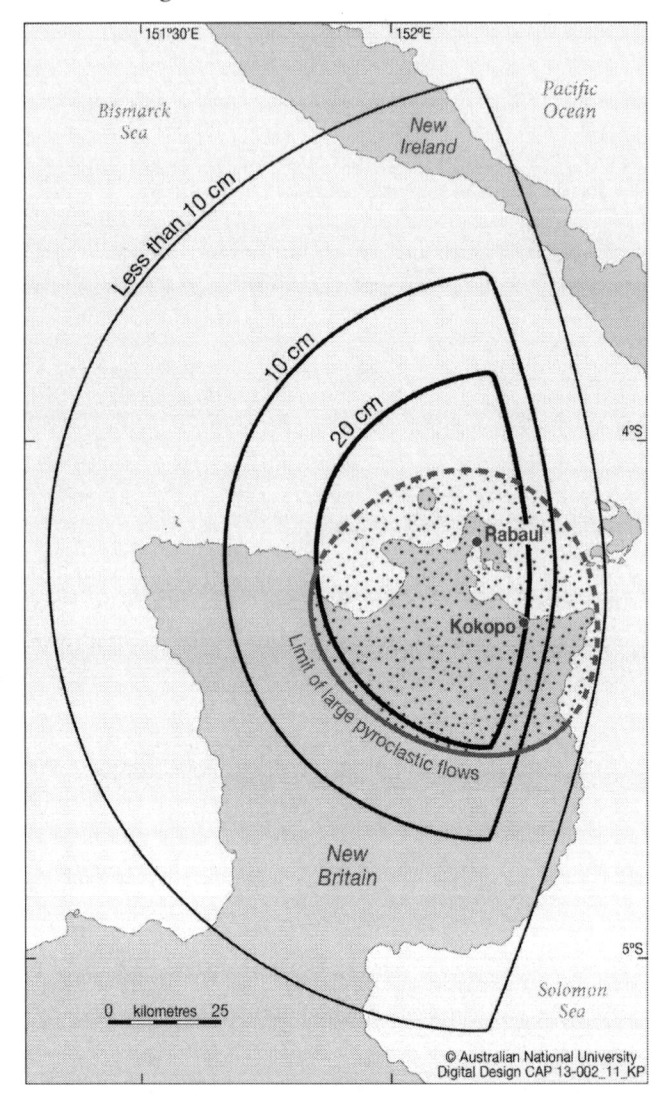

Figure 102. The north-eastern Gazelle Peninsula would be affected by airfall-ash deposits thicker than 20 centimetres and by far-reaching pyroclastic flows in the case of a large eruption of about VEI 6, as shown in this simple volcanic hazard map for the area.

Source: Adapted from McKee et al. (1985, Figure 13).

---

39   Blong & Aislabie (1988).
40   Granger (1988, 1990).

Costs, Benefits and Crisis Decline: 1985–1994

Rabaul in April 1984 was prepared for an imminent volcanic eruption. Then, unexpectedly, the number of earthquake swarms and their intensities began to decline. Next, in November 1984, the Stage Two alert was called off, reverting to Stage One, and the seismicity by mid-1985 returned to its pre-1983 levels. Perceptions grew amongst parts of the Rabaul community that the crisis had been a false alarm or 'failed eruption' and that RVO had not delivered an accurate prognosis. There were concerns at RVO itself that community complacency might set in, leading to a more blasé approach towards any further crisis periods.[41] The two USGS volcanologists who had undertaken their review of geophysical unrest at active calderas worldwide, noted that residents and scientists in such circumstances are faced with a difficult problem:

> Unrest [at calderas] is abundantly clear but hard to interpret and more likely than not to subside rather than lead directly to an eruption. On the other hand, the potential always exists for a truly devastating eruption to occur with little additional warning. Worrisome changes that do not lead to an eruption are likely, and the public must be willing to accept false alarms if they wish to receive timely warning of an eruption.[42]

The total cost of the emergency preparations during the Rabaul crisis and incurred by the national government and private enterprise, may well have exceeded 20 million PNG kina.[43] There were substantial losses of business revenue. Insurance premiums during and after the crisis increased greatly for volcano, earthquake and tsunami cover, or cover was not offered at all. This meant that banks would not approve loans, and so a development stand-off gripped the high-risk town. School closures had disrupted the education of children. Some shortages of imported foodstuffs were caused by stocks being purchased for emergency supplies, which, in the end, were not needed. There were fire sales. Agricultural productivity also declined in the province during the crisis, although this was offset, fortuitously, by higher world prices for copra and cocoa. Many people thought, overall, that the emergency was a waste of money, given that no eruption had eventuated.[44]

There was some criticism, too, about the 1983 provincial disaster plan and its effectiveness. Granger, for example, in promoting the future use of GIS-based data for crisis management, criticised the way in which information had been collected and managed at Rabaul:

---

41 Lowenstein (1988) and De Saint Ours (1993).
42 Newhall & Dzurisin (1988), p. 26.
43 Blong & Aislabie (1988) and Lowenstein (1988).
44 K. Neumann (personal communication, 2008).

> In some instances key information was unavailable and had to be collected. In other instances information was available but in a form that required considerable manipulation to make it useable, or was so out-dated that it was of dubious value. [Furthermore] ... there was no guarantee that the information being used by one planner was the most current or most accurate available. Nor was it clear that it was the same information as that being used by another planner working on the same, or an associated, aspect in another agency. Under such circumstances it is not surprising that the 1983 Plan contains gaps in essential information and errors or inconsistencies in the data provided.[45]

Interest in future volcanic activity at Rabaul retained some momentum after the end of the Stage Two alert. The 50th anniversary of the 1937 eruption was commemorated at a well-attended ceremony,[46] and some new instrumental monitoring was introduced by RVO. In particular, the University of Queensland, funded by AIDAB, in 1987 installed new tide gauges in Rabaul Harbour. Gauges were deployed on metal frames constructed on each of the two centres of sea floor uplift — the main one south-east of Matupit Island and the possibly secondary one east of Vulcan. These instruments, importantly, were to be linked by satellite communications to RVO and to the university in Brisbane, so providing the potential for near 'real time' monitoring of ground-height changes. The momentum of such volcano-focused activities, however, fell away after about 1987, including ongoing funding for the real-time tide-gauge system. In addition, there was reduced attention from the media which, understandably, was motivated more by the excitement and uncertainty of the Stage 2 crisis early in 1984, than by the 'let down' of its aftermath. Attention from international volcanologists was also diverted by other events, including particularly the major, VEI 6, eruption at Pinatubo volcano, Philippines, in June 1991 and the persistent threat there from lahars in the following years.

The amount of financial support that could be provided to RVO by the national government declined following the 1983–1985 crisis. Lowenstein resigned from RVO in January 1989, largely because of this reduced support. McKee took over as head of RVO, assisted by Talai as deputy. Leslie Topue retired in 1989, after 38 years of service, and died in 1992. The observatory, by the late 1980s, experienced difficulties in keeping all of the monitoring instruments up-to-date and operational. Some initiatives could not be started, including the high-priority construction of an emergency observatory in case RVO headquarters became inoperable during an eruption. Similarly, proposals to install telemetry links for the automatic measurement of temperature at key points in the caldera could not be funded. The routine, manual measurement of temperatures on the

---

45 Granger (1988), p. 6.
46 See, for example, Neumann (1996).

volcanoes was labour intensive, and it declined sharply during and after the crisis period, mainly because of other work demands. Neither was monitoring of changes in volcanic-gas emissions at Tavurvur possible. Six villagers were killed by gas asphyxiation in a crater on the southern side of the volcano on 24 June 1990 when three villagers, who had been collecting megapode eggs in the crater, were overcome by carbon dioxide, and another three villagers were killed the day after when they tried to rescue them.[47]

Figure 103. Dignitaries lay wreaths during a *matamatanai* or commemorative ceremony near Tavana village, west of Vulcan, on the morning of 29 May 1987, the 50th anniversary of the 1937 eruption. Villagers also staged a *balabalaguan*, a mortuary ceremony, in honour of the victims who years before had been memorialised in the old monument shown here.

Source: R.W. Johnson. Geoscience Australia (no registered number).

Some benefits, however, emerged from the 1983–1985 crisis.[48] The extensive emergency planning in particular was a positive result — not only the eight Acts of Parliament concerned with national and provincial disaster-management arrangements, but also the general raising of community awareness of the volcanic hazards that threaten Rabaul, and what should be done in responding to the different stages of alert. The community in effect was made more risk-aware and therefore more resilient. Nevertheless, some parts of the community

---

47   Rabaul Volcanological Observatory (1990).
48   Lowenstein (1988).

— for example, the European expatriates — may have been much better informed of what was happening than, say, Sepiks and other non-Tolai people in the settlements. Disaster preparations in the province that were funded by the national and foreign governments also resulted in improvements to infrastructure, including airstrips, wharves, roads, bridges, and water and power supplies. For example, an old airstrip at Tokua, 12 kilometres east of Kokopo, was developed in view of the considerable vulnerability of the Lakunai Airport within the caldera just to the north-west of Tavurvur volcano.

Benefits also accrued directly to RVO. The observatory had for a time received considerable and increased attention and resources from foreign governments, and from the national government, including increases to its annual recurrent budget. RVO also benefited from the practical experiences of volcanic-crisis management and of enhanced instrumental monitoring, especially the use of new geophysical instruments and techniques. Furthermore, its management of the crisis in hindsight can be judged as successful — despite the criticisms levelled at it at the time — in that it kept the community and authorities informed about the state of the volcano and what its expectations were for a possible eruption, despite the uncertainty under which it was operating. The pressure to raise the alert to Stage Three, or even Stage Four, had been considerable, but Lowenstein and his team resisted this.

Authorities at Rabaul, to their credit, continued to rehearse their emergency-management responsibilities. The rehearsals by April 1994, however, did not involve the communities themselves, which, in retrospect, can be regarded as an unfortunate disengagement from the population at large. Memories of the 1983–1985 crisis and its intense earthquake activity faded. Instrumental monitoring continued at Rabaul, although the volcano seemed now to be relatively 'quiet'. McKee and some other RVO staff spent several periods away from Rabaul, monitoring eruptions at Manam in 1992 and Long Island in 1993, and establishing monitoring networks at Ulawun and Karkar. McKee and others were also occupied in mapping the geology of Lolobau and Hargy volcanoes in West New Britain, work that brought out McKee's particular strengths as a careful scientific observer in the field. RVO in 1993 also hosted a group of overseas volcanologists who had been attending a major international volcanological conference in Canberra and who visited Rabaul on a field excursion. RVO life almost seemed back to 'normal'.

Yet, the level of earthquake activity in the seismic annulus at Rabaul in 1992–1994 had not declined to pre-1971 levels. Indeed, there were indications of slight increases in the number of earthquakes, although not nearly to the levels seen in 1983–1985. Furthermore, some earthquakes in 1992 were detected beneath Namanula Hill and, later, further out into St Georges Channel — that is, well outside of the seismic annulus. These formed a zone that came to be referred to as the 'northeast earthquakes'. In addition, the south-eastern end of Matupit Island continued to rise slightly, and increasingly so after 1991. In other words, there had still been no deflation of the 'bulge'. All of these slight changes were insufficient to justify RVO raising the alert level at Rabaul, but they would later, in hindsight, become more significant for what would follow in 1994.

Volcanic Alert on Simbo Island

Interest in the volcanic crisis at Rabaul, Papua New Guinea, may have declined by 1993, but the opposite was true for people living on remote Simbo Island, across the international border in the western Solomon Islands, about 400 kilometres west-north-west of the capital Honiara. Simbo is a geologically youthful volcanic island on which, in the south, there are active hot springs, fumaroles, and solfataras, including at Pakeru or Ove Crater and Lake at the south-western end of the island.[49] There are no confirmed reports of historical eruptive activity from any of the three main eruptive centres on Simbo, but radio reports from the island on 11 February 1993 were sufficiently dramatic to suggest to expatriate government geologists, R. Addison and M.G. Petterson, from the Department of Natural Resources in Honiara, that an eruption may have started at Ove Crater.

The radio reports were of 'volumes of black smoke, a strong smell of sulphur, explosions and fire' and, later on the same day, of 'villagers complaining … of increased smoke and fumes causing nausea, chest pains and stringing eyes. The ground was reported to have been shaken by seismic shocks' and there was 'a white glow around the vent'.[50] Advice from officials in Honiara was provided to the villagers on how to cope with volcanic eruptions and there were discussions on evacuation procedures that may need to be adopted. Honiara officials, including Addison and Petterson, flew to the island on both 12 and 13 February. Observations were made from helicopter on 12 February:

> The circuits of the crater revealed … a number of active fumarolic vents (fissures between rocks) from which steam and smoke were issuing to be carried in low billows downwind in a northeasterly direction. The area from which most of smoke or steam appeared to be issuing was black, as if discoloured by volcanogenic sublimates … . A moderately strong smell of sulphur dioxide was detected from within the helicopter … a second series of circuits of the crater were [sic] made, during which the

---

49   Guppy (1887, Chapter 4), Grover (1955, Chapter 13) and Dunkley (1986).
50   Petterson et al. (2008), pp. 149–51.

smoke was perceived to have thickened in consistency and to be flowing over the crater rim, down the slope of the cone, to be carried away as a ground-hugging cloud across the ridge NE of Ove Lake.[51]

A ground inspection revealed that the discolouration seen from the air appeared to be the result of burning caused by heat or acid attack. Trees and leaves were discoloured within the crater and trees there had shed large amounts of dead leaves. A key observation on the northern side of the crater wall was of sulphur that seemed 'to have been burnt and melted. Patches and pools of black and yellow sulphur appear to have flowed and solidified in small pools and lobate flows in hollows of the wall.'[52]

The conclusion reached was that combustible sulphur had been ignited as a result of a bush or grass fire lit, evidently by tourists, on the top of the crater rim, and the burning had produced the clouds of 'smoke', sulphur dioxide and vapour. The 'white glow' may refer to the flames of the sulphur burning. Seismologist A.K. Papabatu had installed a seismometer near Ove on 14 February and had maintained recording there for four weeks, but no local, volcano-related earthquakes were recorded on the seismometer. The emissions had in fact died away after rainfalls on 12 and 13 February, which evidently extinguished any further burning. There had been no volcanic eruption, and the Simbo event was yet another, although unusual, example of a volcanic false alarm.

References

Archbold, M.J., C.O. McKee, B. Talai, J. Mori & P. de Saint Ours, 1988. 'Electronic Distance Measuring Network Monitoring during the Rabaul Seismicity/Deformational Crisis of 1983–1985', *Journal of Geophysical Research*, 93, no. B10, pp. 12, 123–136.

Blong, R. & C. Aislabie, 1988. 'The Impact of Volcanic Hazards at Rabaul, Papua New Guinea', Institute of National Affairs Discussion Paper 33, Port Moresby.

Cooke, R.J.S., 1977. 'Rabaul Volcanological Observatory and Geophysical Surveillance of the Rabaul Volcano', *Australian Physicist*, February, pp. 27–30.

Crandell, D.R. & D.R. Mullineaux, 1978. 'Potential Hazards from Future Eruptions of Mount St. Helens Volcano', Washington, United States Geological Survey Bulletin 1383-C.

---

51 Petterson et al. (2008), p. 151.
52 Petterson et al. (2008), p. 153.

Darius, W., 1984. 'Move to Kokopo is On', *Papua New Guinea Post-Courier*, 16 February, p. 1.

De Saint Ours, P., 1993. 'Lessons Learned by Rabaul Volcano Observatory in Volcano Crisis Management', *World Organisation of Volcano Observatories Workshop*, Guadaloupe, French West Indies, Program and Abstracts (unpaginated).

De Saint Ours, P., B. Talai, J. Mori, C. McKee & I. Itikarai, 1991. 'Coastal and Seafloor Changes at an Active Volcano: Example of Rabaul Caldera, Papua New Guinea', Workshop on Coastal Processes in the South Pacific Islands Nations, Lae, Papua New Guinea. *SOPAC Technical Bulletin*, 7, pp. 1–13.

Dunkley, P.N., 1986. 'The Geology of the New Georgia Group, Western Solomon Islands'. British Geological Survey Overseas Directorate, British Technical Cooperation Report, Western Solomon Islands Geological Mapping Project, 21, Report MP/86/6.

East New Britain Provincial Disaster Plan, 1983. East New Britain Provincial Government, Rabaul.

Ewert, J., C.D. Miller, J.W. Hendley II & P.H. Stauffer, 1998. 'Mobile Response Team Saves Lives in Volcano Crises', United States Geological Survey Fact Sheet 064-97.

Fisher, N.H., 1939. *Geology and Vulcanology of Blanche Bay, and the Surrounding Area, New Britain*. Territory of New Guinea Geological Bulletin 1.

Fournier d'Albe, E.M., 1979. 'Objectives of Volcanic Monitoring and Prediction', *Journal of the Geological Society of London*, 136, pp. 321–26.

Graham, T.L., M.G. Swift, R.W. Johnson, J. Pittar, P. Musunamasi & I. Kari, 1993. *Rabaul Harbour Heat Flow Project 1993 Papua New Guinea: Final Report*. Australian International Development Assistance Bureau.

Granger, K.J., 1988. 'The Rabaul Volcanoes: An Application of Geographical Information Systems to Crisis Management'. Master of Arts thesis, The Australian National University, Canberra.

——, 1990. 'Process Modelling and Geographic Information Systems: Breathing Life into Spatial Analysis', *Mathematics and Computers in Simulation*, 32, pp. 243–47.

Greene, H.G., D.L. Tiffin & C.O. McKee, 1986. 'Structural Deformation and Sedimentation in an Active Caldera, Rabaul, Papua New Guinea', *Journal of Volcanology and Geothermal Research*, 30, pp. 327–56.

Grover, J.C., 1955. 'Simbo Volcano', in *Geology, Mineral Deposits and Prospects of Mining Development in the British Solomon Islands Protectorate*. Interim Geological Survey of the British Solomon Islands Memoir, 1, pp. 46–48.

Guppy, H.B., 1887. *The Solomon Islands: Their Geology, General Features, and Suitability for Colonisation*. Swan Sonnenschein, Lowrey, London.

Hastings, P., 1984. 'Volcano Set to Blow: 20,000 Plan Island Escape', *Sydney Morning Herald*, 26 January, p. 1.

Heming, R.F., 1974. 'Geology and Petrology of Rabaul Caldera, Papua New Guinea', *Geological Society of America Bulletin*, 85, pp. 1253–64.

Honey, F.R., 1991. 'Passive, Two-Channel, Thermal-Infrared Imaging Systems for Discrimination of Volcanic Ash Clouds', in T.J. Casadevall (ed.), *Volcanic Ash and Aviation Safety: Proceedings of the First International Symposium on Volcanic Ash and Aviation Safety*. United States Geological Survey Bulletin, 2047, pp. 347–50.

Johnson, R.W., 1984. 'Volcanological Investigations in Papua New Guinea, February 1984'. Geological Survey of Papua New Guinea Report 84/4.

——, 1986. 'Underwater Video Survey of the Volcanic Bulge on the Floor of Rabaul Harbour, Papua New Guinea, December 1985', *CCOP/SOPAC Proceedings of 15th Session, Rarotonga, Cook Islands*, pp. 138–39.

Johnson, R.W. & T.J. Casadevall, 1991. 'Aviation Safety and Volcanic Ash Clouds in the Indonesia–Australia Region, in T.J. Casadevall (ed.), *Volcanic Ash and Aviation Safety: Proceedings of the First International Symposium on Volcanic Ash and Aviation Safety*. United States Geological Survey Bulletin, 2047, pp. 191–97.

Johnson, R.W., I. Itikarai, H. Patia & C.O. McKee, 2010. 'Volcanic Systems of the Northeastern Gazelle Peninsula, Papua New Guinea: Synopsis, Evaluation, and a Model for Rabaul Volcano', Rabaul Volcano Workshop Report, Papua New Guinea Department of Mineral Policy and Geohazards Management, and Australian Agency for International Development, Port Moresby.

Johnson, R.W., & N.A. Threlfall, 1985. *Volcano Town: The 1937–43 Rabaul Eruptions*. Robert Brown and Associates, Bathurst.

Latter, J.H., 1988. 'Quantitative Volcanic Risk in Asia and the Pacific', in *Urban Geology in Asia and the Pacific*. United Nations Economic and Social Commission for Asia and the Pacific Atlas of Urban Geology, 2, pp. 16–23.

Latter, J.H. & A.W. Hurst, 1987. 'An Assessment of Volcanic, Seismic, and Tsunami Hazard at Rabaul and Neighbouring Areas of New Britain, Papua New Guinea', New Zealand Department of Scientific and Industrial Research, Wellington, Contract Report 25.

Lowenstein, P.L., 1988. 'Rabaul Seismo-Deformational Crisis of 1983–85: Monitoring, Emergency Planning and Interaction with the Authorities, the Media and the Public'. Geological Survey of Papua New Guinea Report 88/32.

Lowenstein, P.L. & B. Talai, 1985. 'Volcanoes and Volcanic Hazards in Papua New Guinea', Geological Survey of Japan Report, 263, pp. 315–31.

McCormick, M.P., 1985. 'Aerosol Observations for Climate Studies', *Advances in Space Research*, 5, no. 6, pp. 67–73.

McKee, C.O., P.L. Lowenstein, P. de Saint Ours, B. Talai, I. Itikarai & J. Mori, 1984. 'Seismic and Ground Deformation Crises at Rabaul Caldera: Prelude to an Eruption?', *Bulletin of Volcanology*, 47, pp. 397–11.

McKee, C.O., R.W. Johnson, P.L. Lowenstein, S.J. Riley, R.J. Blong, P. de Saint Ours & B. Talai, 1985. 'Rabaul Caldera, Papua New Guinea: Volcanic Hazards, Surveillance, and Eruption Contingency Planning', *Journal of Volcanology and Geothermal Research*, 23, pp. 195–37.

McKee, C., J. Mori & B. Talai, 1989. 'Microgravity Changes and Ground Deformation at Rabaul Caldera, 1973–1985', in J.H. Latter (ed.), *Volcanic Hazards: Assessment and Monitoring*. IAVCEI Proceedings in Volcanology, Springer-Verlag, Berlin, 1, pp. 399–428.

Mori, J., C. McKee, I. Itikarai, P. Lowenstein, P. de Saint Ours & B. Talai, 1989. 'Earthquakes of the Rabaul Seismo-Deformational Crisis September 1983 to July 1985: Seismicity on a Ring Fault', in J.H. Latter (ed.), *Volcanic Hazards: Assessment and Monitoring*. IAVCEI Proceedings in Volcanology, Springer-Verlag, Berlin, 1, pp. 429–62.

Nairn, I.A., C.O. McKee, B. Talai & C.P. Wood, 1995. 'Geology and Eruptive History of the Rabaul Caldera Area, Papua New Guinea', *Journal of Volcanology and Geothermal Research*, 69, pp. 255–84.

Neumann, K., 1996. *Rabaul Yu Swit Moa Yet: Surviving the 1994 Volcanic Eruption*. Oxford University Press.

Newhall, C.G., & D. Dzurisin, 1988. *Historical Unrest at Large Calderas of the World*, United States Geological Survey Bulletin 1855.

Newhall, C.G. & S. Self, 1982. 'The Volcanic Explosivity Index (VEI): An Estimate of Explosive Magnitude for Historical Volcanism', *Journal of Geophysical Research*, 87, no. C2, 1231–238.

Peterson, D.W., 1988. 'Volcanic Hazards and Public Response', *Journal of Geophysical Research*, 93, no. B5, pp. 4161–170.

Petterson, M.G., D. Tolia, S.J. Cronin & R. Addison, 2008. 'Communicating Geosciences to Indigenous People: Examples from the Solomon Islands', in D.G.E. Liverman, C.P.G. Pereira & B. Marker (eds), *Communicating Environmental Geoscience*. Geological Society, London, Special Publications, 305, pp. 141–61.

Rabaul Volcanological Observatory, 1990. 'Rabaul Caldera', *Global Volcanism Network Bulletin*, 15, no. 6, pp. 8–9.

Sack, P., 1987. 'The Emergence and Settlement of Matupit Island', *Bikmaus, Journal of Papua New Guineas Affairs, Ideas and the Arts*, 7, pp. 1–14.

Siebert, L., T. Simkin, & P. Kimberley, 2010. *Volcanoes of the World*. 3rd edn. Smithsonian Institution, Washington D.C., University of California, Berkeley.

Stannard, B., 1984. 'Rabaul Trembles as Fears of Big Bang Grow', *Bulletin*, 104, no. 5405, 28 February, pp. 44–49.

Stothers, R.B., 1984. 'Mystery Cloud of AD 536', *Nature*, 307, pp. 344–45.

Thompson, D., 2000. *Volcano Cowboys: The Rocky Evolution of a Dangerous Science*. St. Martin's Griffin, New York.

Tilling, R.I. & R.S. Punongbayan, 1989. 'Scientific and Public Response', in R.I. Tilling (ed.), *Volcanic Hazards, Short Course in Geology, 1*, American Geophysical Union, Washington D.C., pp. 103–06.

Voight, B., 1990. 'The 1985 Nevado del Ruiz Catastrophe: Anatomy and Retrospection', *Journal of Volcanology and Geothermal Research*, 44, pp. 349–86.

Walker, G.P.L., R.F. Heming, T.J. Sprod & H.R. Walker, 1981. 'Latest Major Eruptions of Rabaul Volcano', in R.W. Johnson (ed.), *Cooke-Ravian Volume of Volcanological Papers*. Geological Survey of Papua New Guinea Memoir, 10, pp. 181–93.

13. Eruptions at Rabaul: 1994–1999

Many of our old folk (our *patuana*) knew that an eruption was imminent. The strength of the earthquakes told them that an eruption was only a matter of days or hours away … . But the government authorities had not said anything about an eruption … . The Volcanological Observatory, as we have always been told, has some of the most modern and sophisticated monitoring equipment which can predict an eruption to the minute. How come these machines have not said anything about an imminent eruption?

Derol Ereman, a Boisen High School student from Matupit Island (quoted by Neumann, 1995, pp. 2–3).

First Three Weeks

Independence Day celebrations for the 19th national birthday of Papua New Guinea were interrupted by earthquake activity over the weekend in Rabaul beginning at 2.50–2.51 am on Sunday 18 September 1994. Two earthquakes about 40 seconds apart — one near Tavurvur the other near Vulcan — were felt strongly throughout the harbour area. Aftershocks and ground shaking continued, particularly in the Vulcan area. Rabaul Volcanological Observatory (RVO) volcanologists suspected, for about 12 hours, that the earthquake activity represented another 'seismic swarm', similar to many of those experienced in Rabaul during the 1970s and 1980s. The ground shaking continued and, by Sunday afternoon, villagers near Tavurvur on Matupit Island had begun a spontaneous evacuation into Rabaul town, encouraged by older people who recalled the 1937 volcanic eruption. Hundreds of other Matupits, however, remained on their threatened island.[1]

The number of people moving along the road grew, fed by other nearby communities, and by evening thousands of evacuees had gathered at floodlit Queen Elizabeth Park, an evacuation assembly point prescribed in the Rabaul Disaster Plan. Vehicles organised by the Provincial Disaster Committee (PDC), plus many that were provided by local businesses or hot-wired and taken

---

[1] Principal sources for both written and photographic information produced during the first three years following the Rabaul volcanic disaster of September 1994 include the following: AIDAB (1994), Durieux & de Wildenberg (1994), Smithsonian Institution (1994), Lokinap (1994, 1995), McKee et al. (1994), World Publishing (1994), Blong & McKee (1995), Davies (1995a, 1995b), Dent et al. (1995), Finnimore et al. (1995), Lauer (1995), Lindley (1995), Nairn & Scott (1995), Neumann (1995), Rabaul Petrology Group (1995), Rose et al. (1995), Tomblin & Chung (1995), Neumann (1996a, 1996b, 1997), Roggensack et al. (1996) and Williams (1996).

from, for example, the closed transport pool of the Department of Works, began transporting evacuees out of Rabaul. Traffic streamed out of the caldera particularly over Tunnel Hill. The PDC, on RVO's recommendation, declared a Stage 2 alert at 6.15 pm, which was announced to the general community on the provincial radio station. Looting, mainly by men and youths, began that night in Rabaul as vehicles coming into town were loaded up with alcohol, television sets, VCRs, food, and whatever else could be taken from the abandoned homes and businesses. Vandalism followed, as well as later scavenging of building materials.

RVO volcanologists during the night of 18–19 September became convinced that a volcanic eruption would take place. They recommended to the PDC that a Stage 3 alert be announced, signifying that an eruption was likely to take place within days to weeks, but a decision was made to postpone any such declaration until daybreak, given that the spontaneous evacuation was already underway and proceeding, if not in perfect orderly fashion, then at least without any noticeable mass panic. The National Disaster Emergency Services (NDES) in Port Moresby, however, was informed from Rabaul that an eruption was thought to be imminent, and its director, Leith Anderson, that night passed on this information to other parts of the government — including Prime Minister Sir Julius Chan — as well as to aviation authorities and to the media, which began broadcasting the news nationwide including in Rabaul. Meanwhile, spontaneous evacuation of villages near Vulcan, such as Tavana and Valaur, was taking place, especially after about 1.00 am on Monday 19 September, when escaping aircraft could be heard taking off from the Lakunai Airport in the clear moonlight. People knew that such aircraft movement was unusual because the airstrip had no runway lights for approved night-time departures or landings.

The provincial radio station in Rabaul had closed during the night and did not resume service until early next morning, after which the same Stage 2 announcement continued to be broadcast, including the advice that people need not evacuate. RVO volcanologists by this time — at dawn — were convinced, belatedly, that an eruption could be expected immediately because they could see from Observatory Ridge that sea floor on the eastern side of Vulcan had risen several metres out of the water, as it had in the hours before the eruption in 1937. The south-eastern end of Matupit Island had risen too. Then, at 6.06 am, only 27 hours after the two felt earthquakes, Tavurvur volcano began issuing vapour and ash, its first eruption in 51 years. Voluminous damp ash in dense dark-grey clouds began to be driven by the strong, and strongly directional, south-east trade winds over the eastern part of Rabaul town, enveloping it in darkness. Rain turned the ash to mud.

13. Eruptions at Rabaul: 1994–1999

Figure 104. Ash clouds from Tavurvur volcano spread across the eastern part of Rabaul town on 19 September 1994, as seen here from Observatory Ridge at 7.30 am. The lower of the two layers of cloud represents an earlier phase of the eruption.

Source: Lauer. (1995, p. 15). CPD Resources, Queensland (Img0051). Digitally enhanced copy provided by N. Lauer, Brisbane.

Vulcan broke out in eruption at 7.17 am, first as surtseyan activity at the waters' edge but soon also from further vents upslope on the volcano, from which pyroclastic flows spilled across part of the harbour. Strong jetting and plinian activity next developed at Vulcan, the column entraining sea water and reaching heights of at least 20 kilometres, well into the stratosphere. The clouds from Tavurvur, in contrast, mostly rose no more than about six kilometres. Vulcan continued to issue pyroclastic flows and surges, and tsunamis from the Vulcan activity washed up along the harbour shoreline causing damage — to houses on Matupit Island for example. Rabaul Harbour became clogged with floating pumice which was trapped there by the south-east winds for many weeks. The simultaneous eruptive activity from the two volcanoes in the south-east season of 1994 was therefore similar to the 1937 Rabaul eruption, except that Vulcan was the first to become active in 1937. Furthermore, the damage inflicted on Rabaul Town in 1994 was much greater than in 1937.

Figure 105. Vulcan is seen from the south jetting pumice, ash, and water vapour out from an inclined vent at shortly after 7.30 am on 19 September 1994. Pyroclastic flows are moving to the north-east across the waters of Rabaul Harbour. The water surface in the foreground is calm but would soon become more disturbed after the pyroclastic flows began crashing down onto the water, generating volcanic tsunamis.

Source: M. Phillips, B. Alexander and S. McGrade, Rabaul.

Early evacuation of people and communities to safer places meant that few were caught out in the areas of serious volcanic fallout — particularly in abandoned Rabaul town — and none were overwhelmed by the Vulcan pyroclastic flows. Numerous places of assembly became hastily erected camps throughout the north-eastern Gazelle Peninsula in patterns not matching those set out in the formal disaster plan. The camps became 'care centres' ranging in size from large formal ones — as in the case of the Kokopo Showground, Vunapope Mission, and at Kerevat — to small spontaneous encampments at or near the unaffected homes of friends and *wantoks* in hamlets, villages, or in Kokopo itself, for example. Others stayed at schools, churches, missions, and community centres. Many families and villages were split between different care centres in the chaos of the evacuations. More than 100,000 people are thought to have been affected by the eruption and many had to be accommodated in the care centres and their needs attended to by the authorities.

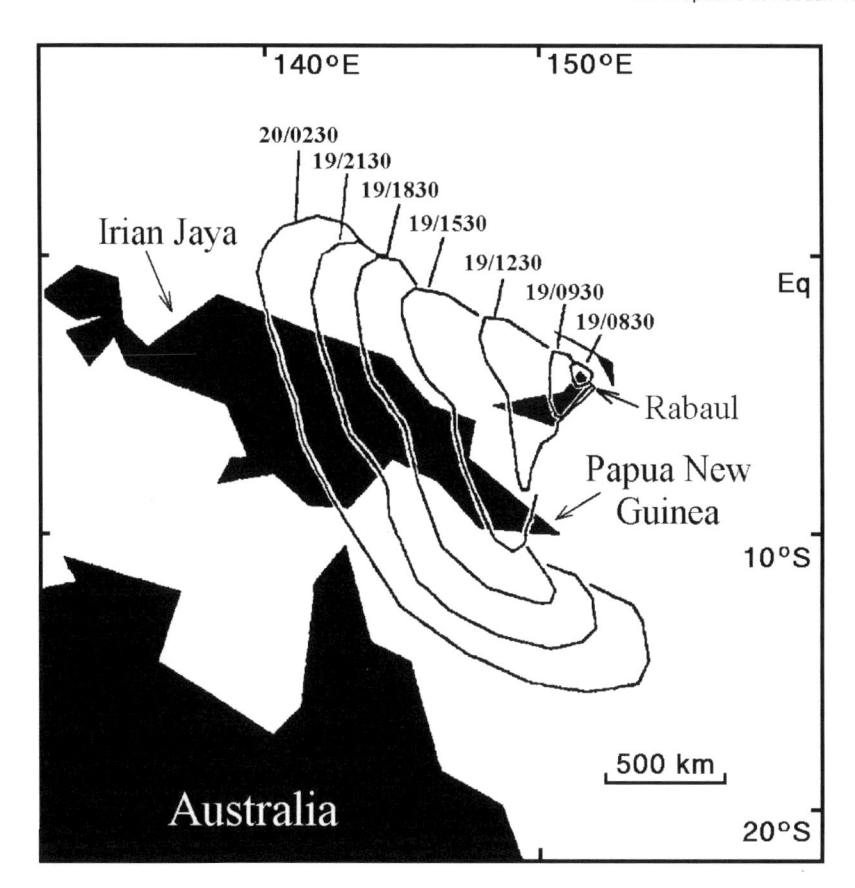

Figure 106. The westward progress of the Vulcan eruption cloud was tracked by the US National Ocean and Atmosphere Administration using satellite imagery. '19/0830' refers to 8.30 am on 19 September 1994, and so on. Tracking clouds in this way assists aviation authorities. There were no encounters between international jets and the drifting ash clouds from Rabaul in 1994, but airline companies sustained additional fuel costs because of aircraft diversions on longer routes.

Source: McKee et al. (1994, Figure 4). The map was supplied by J. Lynch, National Ocean and Atmosphere Administration.

Stage 3 and 4 alerts were declared officially, but redundantly, after the start of the Tavurvur eruption, and a massive national response was initiated, supported by resources from many international agencies, including churches and non-governmental organisations. The Australian military began using a C130 Hercules aircraft to fly in relief supplies to Tokua Airport and to take out evacuees to Port Moresby, as did other flights. Many other evacuees left for other parts of Papua New Guinea by sea. Anderson, of NDES, had been appointed Controller of the relief effort almost immediately, but the PNG military took charge a few days later after the national executive council rescinded Anderson's appointment.

'Operation Unity' was initiated, its name evidently carrying a signal from the national government in relation to recent discussions concerning secession by the Islands Region from the rest of Papua New Guinea. Brigadier General Rochus I. Lokinap was the new Controller. His command structure of four senior men included deputy controller Ellison Kaivovo, Secretary of the Department of East New Britain and chairman of the PDC. A disaster control centre was established at the Ralum Golf Club on the western outskirts of Kokopo, near the decaying stone steps of Queen Emma's old mansion overlooking St Georges Channel. Access to destroyed Rabaul town was restricted by a checkpoint manned by police and military at the foot of Tunnel Hill. The general public, including former town residents, had to obtain passes before entry was permitted.

The Rabaul eruption received widespread attention from both national and international media, and accuracy and appropriateness in the newspaper coverage were mixed. International television coverage was immediate. I was in the United Kingdom visiting my father in the third week of September when we saw the spectacular images from Rabaul on BBC television, the first indication for me that the long-expected eruption had finally taken place.

Figure 107. Eruption clouds from both Vulcan and Tavurvur were photographed on 19 September 1994 by astronauts at the International Space Station. The larger, higher, Vulcan cloud is seen spreading fan-like westwards across the Bismarck Sea. The lower and less pronounced cloud from Tavurvur — indicated by the two arrows — extends north-westwards.

Source: Smithsonian GVN Bulletin (August 1994, Figure 4). US National Aeronautics and Space Administration (STS064-116-64).

Physical Damage

Several of us were invited to Rabaul to assist with the ongoing assessment of the eruption and disaster. I arrived there on Friday 7 October, nearly three weeks after the catastrophe. Hugh Davies, now Professor of Geology at the University of Papua New Guinea, was already in Rabaul as the assigned volcanological liaison officer, a position he had also held during the 1983–1985 seismo-deformational crisis. Davies met me at Tokua Airport and we travelled by helicopter to the busy disaster control centre at Raluan, and then over devastated Rabaul town to RVO headquarters on Observatory Ridge. Tavurvur was still in mild eruption, but the volcanic activity at Vulcan had ceased on 2 October and had greatly diminished even by 24 September. Russell Blong was in Rabaul town, toiling in the dusty and deserted streets collecting building damage data for his PNG insurance-industry sponsors.[2] Furthermore, a Volcanic Disaster Assistance Program (VDAP) team of three from the United States Geological Survey (USGS) was based at RVO headquarters. VDAP-USGS had accepted a formal request from the Government of Papua New Guinea for assistance in re-establishing the instrumental monitoring network at Rabaul, which had been largely lost as a result of ash damage, looting and tsunamis. Other arrivals had been self-motivated and curious university researchers, including most notably, the well-known American volcanologist Stanley N. Williams and one of his PhD students.

Chris McKee and I drove down Tunnel Hill Road from RVO that Friday afternoon, passing through the army checkpoint at the foot of the hill, and then east along Malaguna Road where the ash-fall damage to buildings and the wharf was not severe. Ash thickness increased substantially, however, beyond St Francis Xavier Cathedral and then, turning right into Mango Avenue, the full extent of the town's devastation could be appreciated. McKee was still as deeply affected after several visits to Mango Avenue as I was shocked by my first close-up sight of the ash-mantled buildings and roads where, formerly, there had been a thriving business district and town life. Many roofs had collapsed in the days after the initial eruption at Tavurvur on 19 September, but not those, for example, of the Hamamas Hotel whose owners, the McGrade family, had promptly removed ash from the hotel's roofs and gutters. The roads were eerily deserted and silent, the ash deadening the sounds of tyres on what had been humming tarmac. Cross-cutting streets had been bulldozed down to the harbour's edge in order to make channels for floods and mudflows from the walls of the caldera that were now stripped of vegetation. Powerlines were down. Rainfall gullying had destroyed the road up Namanula Hill, which previously had linked the town with the villages on the St Georges Channel coast to the east. The north-coast road at Vuvu and parts of the Kokopo Road had also been cut by torrential flooding following the Vulcan eruption.

---

2 Blong (2003).

Figure 108. The extent of damage to buildings in the northern part of Rabaul town is seen in this aerial view in October 1994. Roofs of many ash-covered buildings in the foreground have collapsed or buckled. Trees have been stripped of foliage, including those that once shaded the central strip of Malaguna Road. By this time graders had cleared roadways and streets, leaving ash piled in ridges on either side of their tracks.

Source: R.W. Johnson. Geoscience Australia (no registered number).

Houses in Malaytown and Rapindik, south of the main Rabaul business district, were the worst affected, partially buried by about 1.5 metres of ash. Prime Minister Chan's nearby home was badly damaged too, as were the buildings, airstrip and the few remaining fixed-wing aircraft and helicopters at Lakunai Airport. McKee and I, that same evening, used flashlights to visit his looted and deserted home at the top of Tunnel Hill Road. Personal possessions, including those of his two young daughters, had been trashed underfoot repeatedly, by what seems to have been dozens of looters who had trampled indiscriminately through the clothes and mementos of a family's life. I saw in later days that damage in the ash-desert area around and to the north-west of Vulcan, was also severe. The villages of Valaur and Tavana, for example, were buried, including the Tavana memorial to those killed by the 1937 eruption — as well as the memorial's now ash-interred words, 'The earth buried them'. Several metres of ash deposition here had buried several kilometres of the sealed Kokopo–Rabaul trunk road, and vehicles had to negotiate an unsealed and dusty track that had been hurriedly re-engineered to reconnect the two towns.

13. Eruptions at Rabaul: 1994–1999

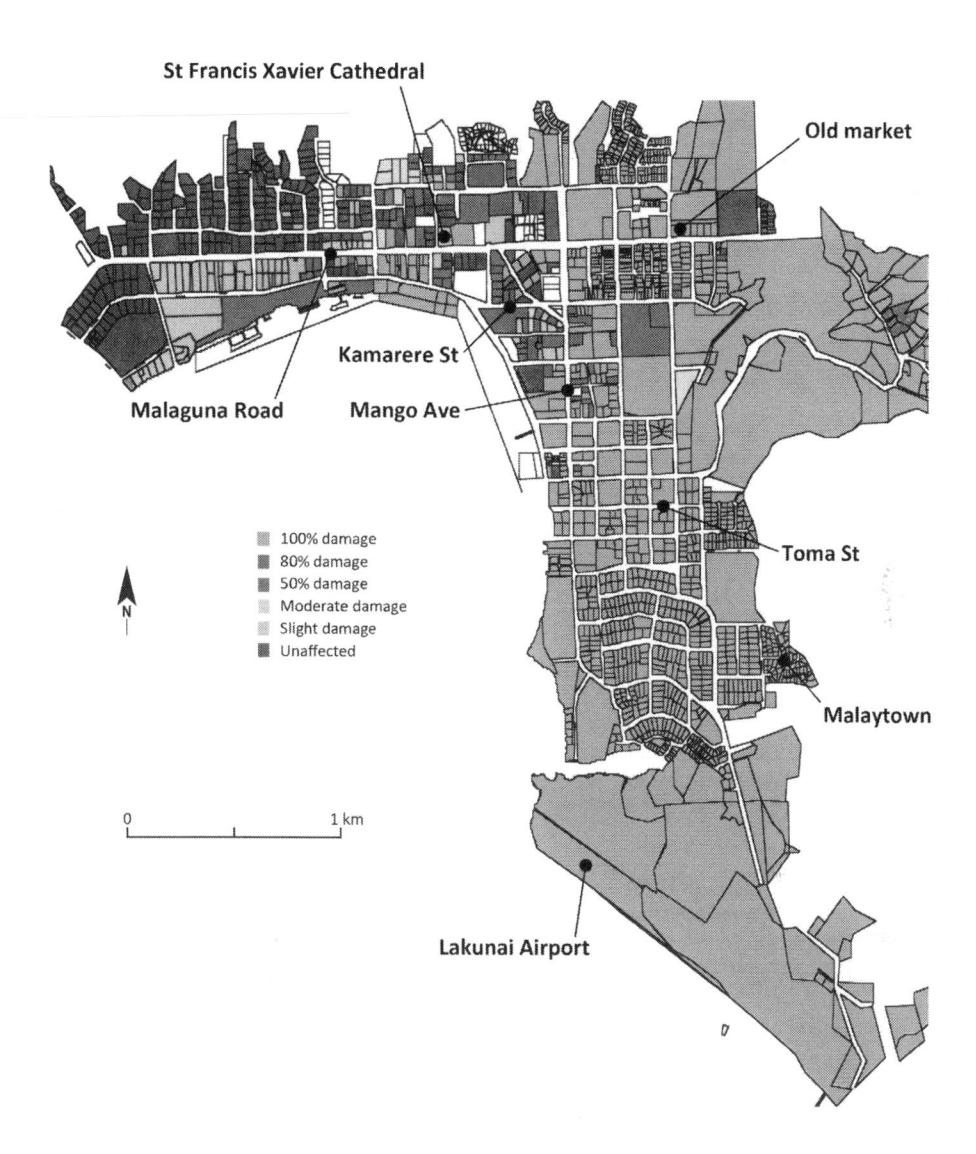

Figure 109. Officers of the East New Britain Provincial Administration produced this map of relative ash damage in Rabaul town. Blong & McKee (1995, Figure 13) published a version of it, but questioned the definition of six different categories used for the damage assessment. The main features of the damage are, however, clear: maximum damage in the east as far south as the airport, and very little damage on either side of Malaguna Road in the west, including the wharfs area.

Source: Adapted from Lokinap (1995, front-cover illustration).

How much direct volcanic damage had been inflicted on the physical investment in the north-eastern Gazelle region? An Australian International Development

Assistance Agency (AIDAB) mission estimated infrastructure losses to government-owned assets to be about 100 million PNG kina, but together with private-sector losses — which were more difficult to estimate — the total may have been of the order of 300 million.[3] The damage from the ash falls was therefore significant. Nevertheless, Blong and McKee reported that

> there can be no doubt that looters trashed the town. One insurer estimated that two-thirds of the eruption-related claims under the company's policies were for looting ... . Many buildings were looted many times. Contents that were not stolen were often vandalised, strewn around, trampled in the ash and mud, or abandoned in the streets as the loot became too heavy ... . Many who lost possessions believe that the looting was well organised ... . [It] continued for weeks.[4]

Only four people lost their lives from the ash falls in Rabaul, plus one other killed by a lightning strike south of Vulcan. This is a remarkably low death toll, bearing in mind the potential for a much higher number had an evacuation not taken place in time. The toll is low, too, compared to the almost 3,000 people who were killed by the Lamington eruption of 1951 and to the 500 deaths in the Vulcan area in 1937. The Rabaul eruption of 1994, however, destroyed much more property and infrastructure and therefore cost more in financial terms.

The ravaged Mango Avenue and Malaytown sectors of Rabaul town were virtually unrepairable and therefore untenable. Decisions on the future now had to be made at all levels of society: individuals, families, villages, communities, businesses, churches, non-governmental organisations and the national and provincial governments. The Provincial Government's headquarters in Rabaul had been destroyed, and the Administration moved temporarily to Vunadadir, west of Kokopo. Indeed, a general population shift to the Kokopo area as the main regional centre, and to settlements farther away, was virtually unstoppable, even as some politicians and business people argued for the rebuilding of Rabaul. Grief and nostalgia for 'old Rabaul' was widespread and palpable.[5] Much of western Malaguna Road, however, had received only light ash falls, the nearby wharfs were still operable, and three hotels were still operating on Mango Avenue. Parts of the town would therefore survive. The disaster relief phase ended during the first months of 1995 and some relative normality returned when Operation Unity ceased, care centres were closed, schools reopened, people moved to resettlement areas or returned to those villages that could rebuilt, and local agricultural products began to be exported again from the wharfs.

---

3   AIDAB (1994) and Blong & McKee (1995).
4   Blong & McKee (1995), p. 30.
5   Neumann (1997).

13. Eruptions at Rabaul: 1994–1999

Post Mortem and New Directions for RVO

Many expatriate or foreign professionals — as individuals or in partnerships — assessed the 1994 disaster at Rabaul and derived lessons learnt for disaster management, social and environmental recovery, regional development, and business and private insurance in the few years following the disaster. In addition, as provincial Secretary, Kaivovo conceived the idea of a collection and assessment of perceptions, experiences and opinions from the Gazelle communities who had been so seriously affected by the eruptions. Dr Klaus Neumann, a Kuanua-speaking Pacific historian of German origin, followed up on this proposal, visiting the Rabaul area, interviewing hundreds of people, and drawing on more than 2,000 written contributions by community and high-school students in East New Britain.[6]

Two of the several recurrent themes in these written reviews are, firstly, the remarkably effective and 'spontaneous' evacuation of the communities before the eruptions began, and second, an antithesis — the intensity and longevity of the looting. One view is that the successful self-evacuation represents a volcano-respecting community of people who had learnt lessons from the government-sponsored hazard-awareness raising activities during the 1980s. An alternative view, however, is that the communities were so resilient, strong and intrinsically intelligent that the evacuation would have taken place anyway, without any organised approach to awareness-raising from authorities who, in the end, had failed to provide adequate warnings of the impending eruption. There was indeed a sense that significant parts of the Tolai communities did not have great respect, *variru*, for the authorities who, rather, were perceived as being ineffective and secretive, if not weak and condescending. The European, Chinese and other Papua New Guinean parts of Rabaul society responded differently to the dominant Tolai sector, and indeed received different degrees of attention from the authorities. The most vulnerable and least resilient were probably Papuan New Guineans from other parts of the country — called *vaira* by the Tolai. Some of them were squatters. Some were the 'children of Rabaul' — those born in Rabaul of non-Tolai, or only one Tolai, parents.[7]

Many people were surprised and shocked by the extent and intensity of the looting. Others, of a more pragmatic and realistic persuasion, were not. Even undamaged government buildings, including Nonga Hospital and community schools, were targeted, but not, significantly, the churches. One view is that looting is a normal aspect of any disaster anywhere in the world and that nothing special should be interpreted from it in the case of Rabaul. Indeed,

---

6   Neumann (1995, 1996a, 1996b, 1997).
7   See, especially, Neumann (1996a), chapter 31.

many communities worldwide are, in different degrees, potentially fractious, in part reflecting systemic issues of cohesion and discontent. These issues may, therefore, at times of disaster, lead to a sudden breakdown of law and order and therefore to opportunities for looting, unless well-resourced police and military interventions are implemented quickly and effectively and vigilance is maintained. This assumes of course that police and military personel themselves do not take part in the looting. Breakdown in communication infrastructure, such as takes place in any society impacted by a major disaster, is also a contributing factor.

A full analysis of the motivation for the looting has not been made in the case of the Rabaul volcanic disaster of 1994. Could the looting have been mitigated by allowing townspeople to return immediately to their properties and so protect their assets from looters? Can the repeated trashing of the town and other facilities be taken as a sign of community frustration at the government agencies? There were also, however, signs of increased crime in the Rabaul area even before the eruption. Did, therefore, unemployment, limited access to material wealth and disaffection in the villages represent an opportunity for strong criminal elements to manipulate a situation for their own ends as the disaster unfolded? Both community resilience and a decision-making independence from government authority were therefore demonstrated in 1994. The impressive and life-saving spontaneous evacuation of a resilient people, and the criminal opportunism if not anarchistic overtones of those involved in the long-lasting and vehement looting may, in fact, have been grounded in a similar lack of regard and trust for government authorities, including advisers such as the RVO.

The feeling amongst the RVO volcanologists was that they had let the community down, a conclusion that was confirmed by the bluntness of some of Neumann's words:

> The RVO and PDC were both accused of not having alerted people in time. It had been widely believed that the RVO would be able to forecast an eruption several days, if not weeks, in advance. The belief had not been publicly discredited by the RVO ... . [Furthermore, the] widely known incident of people from Valaur visiting RVO on Sunday morning [18 September 1994] only to be told not to worry has often been cited to me by Tolai villagers as a particularly telling example of the ignorance of the scientists at the RVO ... . [The] PDC is also accused of being responsible for broadcasting the message that people should not panic and stay at home on Sunday night and early Monday morning ... at a time when Tavurvur was already visibly emitting ash is often quoted as evidence for the alleged incompetence of members of the PDC.[8]

---

8 Neumann (1996b), p. 7.

Some in the community were even more cynical. The crisis in 1983–1985 had been a waste of money; the RVO was wrong in its forecasts then, so why should it be expected to get it right in 1994?[9] Warnings had not, indeed, been issued sufficiently early by RVO and the PDC in order to permit businesses and householders to remove belongings and goods, store them safely, secure abandoned buildings, remove perishables from refrigerators, and so on — and then to evacuate safely. A period of less than 27 hours was far too short. Even an earlier announcement of a Stage 4 alert — say, at dawn on the Sunday morning, 24 hours before the start of the eruption — could not have provided the much longer preparation period required. RVO had been in a hopeless situation.

RVO staff were interviewed by many visitors after the disaster and were disarmingly open and honest about their role during the nightmare events of 18–19 September that had enveloped them so rapidly. They had, as scientists, been caught during the critical 27-hour period in part by the desire to base a final decision on Stage 4 on the best possible factual evidence. This understandable caution was in striking contrast to the attitudes of the communities themselves who were quite willing to sacrifice the need for caution, precision, and certainty for a timely decision to self-evacuate.

RVO acknowledged that there was a need for change to their *modus operandi*. Six main strategic issues can be identified:[10]

1. The first, and perhaps most obvious, issue was that the four-stage alert scheme was inadequate. Its main weaknesses had in fact been noted during the 1983–1985 seismic crisis — that RVO may well have to raise the level of alert very quickly. This meant, for example, that pronouncement of a Stage 2 alert, signifying an eruption within weeks to months, might change to Stage 4 in a matter of hours — as happened in September 1994. But this implied, in turn, that RVO must have had inadequate or wrong information and understanding to announce the Stage 2 alert in the first place. A new alert scheme was needed, one that did not signify a definite timeframe of eruption expectancy.

2. PNG's volcanological service had deteriorated over a period of several years after the 1983–1985 crisis. Even the long-established program of methodical temperature measurement at key locations in the harbour had ceased. Furthermore, staff levels had declined and instruments on high-risk volcanoes nationwide were in need of replacement. The Rabaul monitoring network in 1994 had been adequate for the general 'research' kind of approach that involves sufficient time for the slow, careful, and systematic

---

9 K. Neumann (personal communication, 2008).
10 Lessons learnt from the 1994 eruption are discussed by McKee (1999) in an unpublished manuscript. Fourteen needs and six volcanological lessons were identified.

analysis of data, but not for the immediate and stressful demands of short lead-up periods when *real-time* monitoring was essential.

3. Much more needed to be discovered about the caldera system at Rabaul in order to progress a fuller scientific understanding of how it actually operates. Questions included: What does the underground magmatic 'plumbing' at Rabaul look like? Can it be mapped? Would any forthcoming answers help in understanding where the magma resides, how much there is of it and, more particularly, what signals does the magma provide when it moves towards the surface?

4. RVO needed to have a better organised strategy for community engagement. Many individual scientists and technicians at RVO, such as Leslie Topue and Ben Talai, had for years been prepared as individuals to answer questions, and give talks, about volcanoes in general and RVO's work in particular, to local groups and communities. There was now, however, the need for a whole-of-agency commitment to the broader area of volcanic-risk assessment and reduction, in partnership with at-risk communities themselves, rather than the work being focused mainly on instrumental monitoring, hazard mapping and research.

5. The RVO building was erected in 1940 on Observatory Ridge, largely because of the panoramic view that could be obtained of Rabaul Harbour and its active volcanoes, but that visibility was lost during the 1994 eruption by the enveloping ash clouds from Tavurvur and Vulcan. The question of the suitability of the Observatory Ridge site needed to be addressed, in an era when any loss of visual observation capacity can be compensated by more modern instrumental monitoring — including video cameras — based at a central observatory in a less vulnerable location.

6. The final issue involved the management of well-meaning, but uninvited, scientists — who arrive during volcanic crises for their own research purposes and who do not always contribute constructively to the priority needs of crisis management and effective liaison with the public, officials and media. This was dealt with internationally after the Rabaul 1994 eruption and after volcanic crises elsewhere.[11]

The Secretary of the Department of Mining, RVO's host department in Port Moresby, announced in the second week of October 1994 that Talai would be taking over as head of RVO, the first Melanesian to do so. This represented a reversal of management roles between Papua New Guinean Talai and expatriate McKee, who later moved to Port Moresby and became head of the Geophysical Observatory there.

---

11   IAVCEI Subcommittee for Crisis Protocols (1999).

Figure 110. Ben Talai at the Rabaul Volcanological Observatory in about 1999.

Source: S.J. Saunders.

AIDAB completed its 'Needs Assessment' mission and, supported by the Papua New Guinea Government, followed up immediately on the first recommendation — to help strengthen Papua New Guinea's national volcanological service centred on RVO as a matter of urgency. AIDAB, soon to become known as AusAID — the Australian Agency for International Development — then accepted a full project proposal that was prepared by several of us in RVO, AusAID, and the Australian Geological Survey Organisation — formerly BMR and now called Geoscience Australia. Thus began the 'Papua New Guinea – Australia Volcanological Service Support Project' (VSS Project), which ran under the joint guidance of Talai and myself from 1995 to 1999. The AusAID-funded VSS Project purchased new volcano-monitoring equipment for RVO, provided funds for community-awareness campaigns to be run by RVO throughout Papua New Guinea, and supported a major deep-crustal geophysical survey of the Blanche Bay caldera region.

Talai's appointment as leader of RVO in October 1994 was a landmark event, but in 1999 at the age of only 51 — and having accumulated more than 25 years of volcanological experience — he left the organisation. Seismologist Ima Itikarai, from Central Province, who had joined RVO in 1984, took over as the new head of RVO.

Ongoing Eruptions and New Insights

Explosive eruptions continued at Tavurvur after the main eruptions of 19 September 1994. These explosions were of different strengths and were anything from minutes to several months apart, but their cumulative effect on the town of Rabaul was unrelenting, particularly during the dry season when the south-east trades blew yet more ash over the beleaguered town. Gas, vapour, and aerosol from Tavurvur, together with accompanying rain caused acid attack on metal surfaces, vegetation was prevented from rejuvenating, and some trees died altogether. The town at times seemed to regenerate and become greener, especially during the north-west monsoon when the ash from the volcano drifted south-eastwards, but the additional rainfall caused floods of mud and water from the caldera walls, which made roads and streets in the town and on the road to Kokopo impassable to traffic. Wharfs at Rabaul were still operating but siltation of shallow waters began to be a problem both there and in other parts of Blanche Bay. The ash drifted in a south-easterly direction during the north-west monsoons, which produced respite for Rabaul, but as a result the ash instead fell on the developing town of Kokopo and the airport at Tokua, which on occasions had to be closed to air traffic. Dried-out ash and dust in either season could be re-suspended on windy days, or lofted by fast traffic on unsealed roads, causing discomfort if not distress to those ingesting the contaminated air, particularly people already experiencing bronchial ailments.[12] There was no indication, by the end of 1999, that the explosive eruptions from Tavurvur would stop.

Residents became familiar with the sudden explosive rise of ash clouds high above Tavurvur. RVO volcanologists recognised these, scientifically, as being mainly vulcanian, but there were times when the eruptions had a definite strombolian character. These were identifiable clearly after dark when incandescence in the crater was impressively vivid and the explosive eruptions included sprays of glowing blocks and lumps of lava that crashed onto the outer slopes causing broken blocks to cascade down towards the foot of the volcano. In particular, eight brief strombolian events were identified, interspersed with 'normal' vulcanian activity between May 1996 and 17 August 1997. Lava flows were produced on four of these occasions. This distinctive period of strombolian activity came to be known informally as 'phase 2' of the ongoing eruption.[13]

Staff at RVO, under Itikarai's leadership, and others, began compiling data on the geophysical precursors to the 1994 eruption at Rabaul. These were used in conjunction with results of ongoing research and from the different surveys

---

12 Dent et al. (1995). Longer term health aspects of the ongoing eruption at Rabaul are still being assessed. See for example, Le Blond et al. (2008).
13 Patia (2004).

of Rabaul Caldera, eventually to reach an improved understanding of how eruptions take place at Rabaul. Itikarai illustrated the geophysical build-up from 1968 to the eruption in 1994 by plotting both the number of earthquakes recorded monthly at Rabaul and the amount of uplift at the southern end of Matupit Island, and by dividing 1968–1994 into five successive periods.[14] The 1983–1985 seismic crisis, or Period 3, at Rabaul stands out prominently, providing the basic rationale for the view that an eruption was imminent at the time. No eruption took place, but note that uplift continued at Matupit Island after Period 3, together with smaller earthquake swarms, before the outbreak of the 1994 eruption. In retrospect, therefore, the caldera at Rabaul can be judged to have continued being active, and to be still inflating, after 1985.

Figure 111. Strombolian eruption at Tavurvur volcano as seen from Kaputin Point, Matupit Island, on 14 March 1997. A new lava flow is moving down the southern slopes of the volcano.

Source: S.J. Saunders.

Another feature of the pre-eruption earthquakes that was noted, especially by Itikarai, was that not all of the earthquakes were in the 'seismic annulus' which had so characterised the earthquake pattern before the 1994 eruption. Some earthquakes between 1992 and 1994 — and probably before that, too — were in a zone trending roughly north-eastwards out of the caldera area. Furthermore, profound changes in the earthquake pattern took place in the years following the 1994 eruption. The seismic annulus virtually disappeared and more 'north-east earthquakes' began to be recorded, defining a zone trending into St Georges Channel during a time when eruptions were continuing at Tavurvur.

---

14 Itikarai (2008). Note that these pre-1994 'periods' are defined differently from 'phase 2' of the eruption itself, as identified by Patia (2004).

Fire Mountains of the Islands

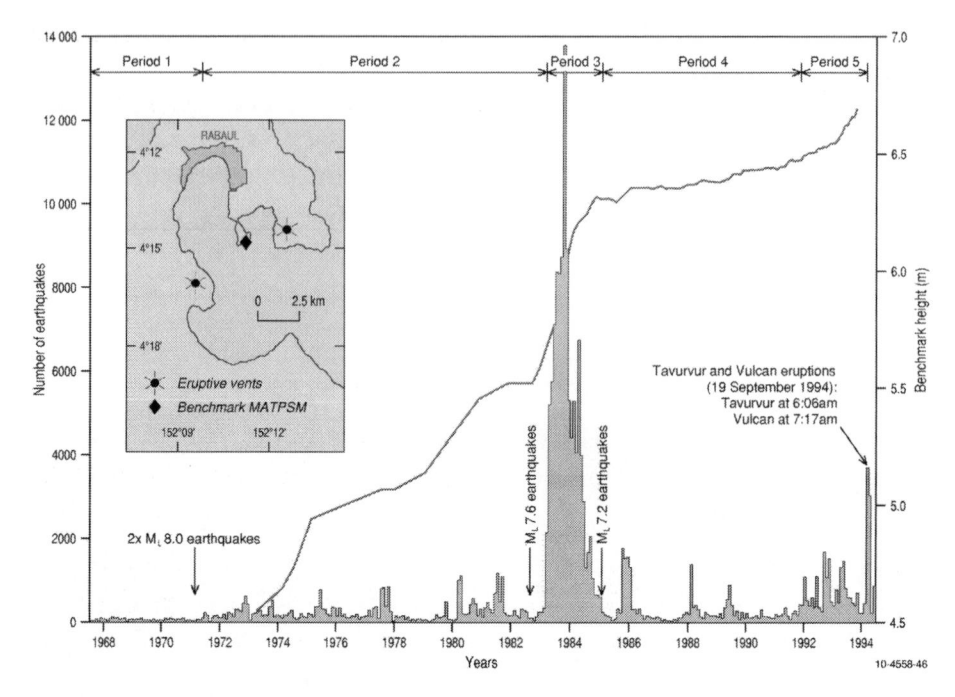

Figure 112. The numbers of earthquakes recorded each month at Rabaul between 1968 and 1994 are shown here in relation to five different periods (1–5) and to the progressive but irregular uplift (red line) at the southern end of Matupit Island.

Source: Itikarai (2008, adapted from Figure 2.9 by Johnson et al. 2010, Figure 46).

Several volcanologists, including myself, had from the beginning of the 1994 eruption been taking a special interest in the chemical composition and mineral content of the ashes and larger lava fragments being produced from Tavurvur and Vulcan.[15] A conclusion, which soon became evident, was that the rocks being produced at both volcanoes did not represent a single type of homogeneous magma. Rather, they preserved a record of past 'mixing' or 'mingling' events at different times beneath the volcanoes between magmas of different compositions, and before eruptions took place. The rocks looked liked they carried mixed loads of crystals from different, old, magma batches, like a palimpsest manuscript. This was particularly evident in some Tavurvur samples in which, without the use of a magnifying glass or microscope, clusters of minerals could be identified that normally are found only in basalt. These minerals were much less common in Vulcan rocks. A 'basaltic component' was especially notable in phase 2 samples from Tavurvur. Yet, the overall compositions of the chemically analysed rock samples were those of andesite

---

15 Rabaul Petrology Group (1995), Roggensack et al. (1996) and Patia (2004).

or dacite, magmas which are petrologically different from basalt. Scientific questions that arose included: Where does the basalt come from? Where does the magma mixing and mingling take place? Can magma chambers or reservoirs be identified beneath Rabaul? How do the magmas get to the surface?

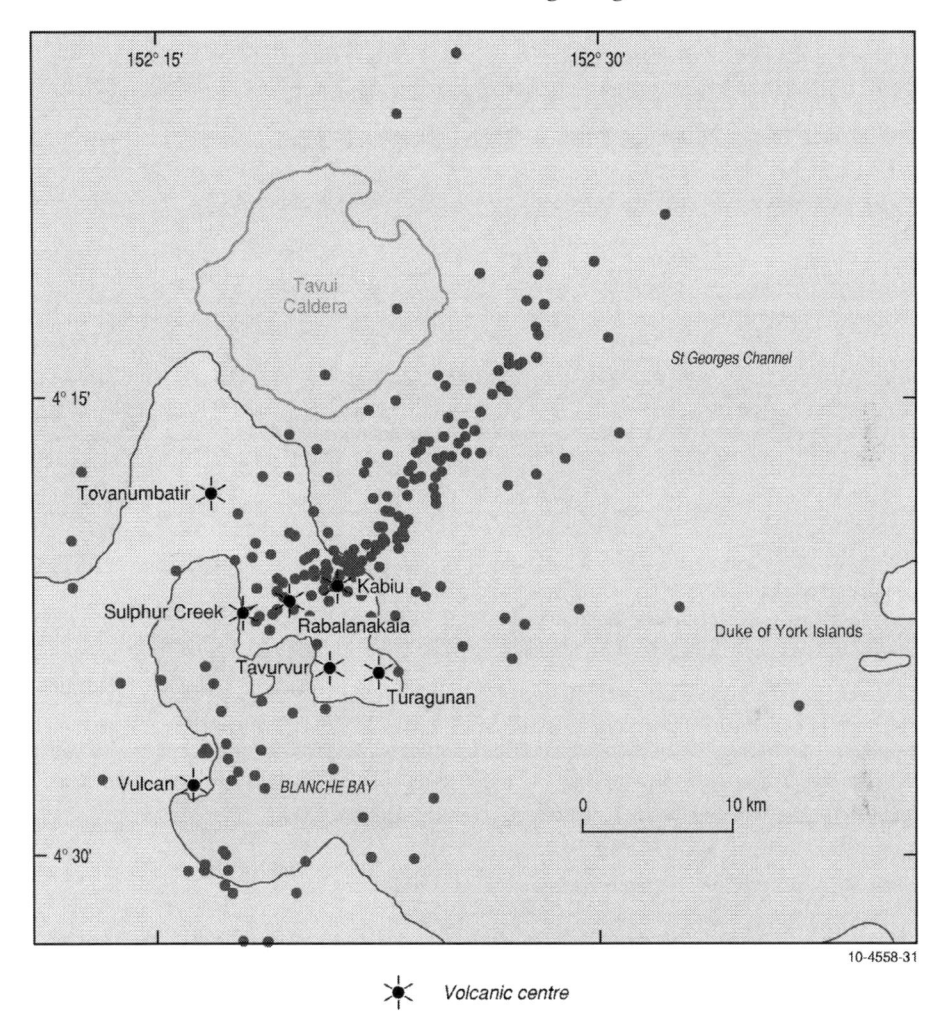

Figure 113. Earthquake epicentres for the period October 1994 to December 1998, showing the zone of earthquakes running north-eastwards from Rabaul into St Georges Channel.

Source: Itikarai (2008, adapted from Figure 5.10a by Johnson et al., 2010. Figure 31).

Possible answers to these questions began to emerge after a major, multi-agency, geophysical survey of the Rabaul Caldera was conducted involving the principles of 'seismic tomography', a technique rather analogous to the CAT-scan method used in investigative medicine. The deep structure of the Rabaul Caldera area

Fire Mountains of the Islands

was determined in three dimensions by mapping out the speeds with which seismic waves from both earthquakes and artificial explosions let off from a ship move through different parts of the Earth's crust down to a depth of about 12 kilometres. This survey was regarded initially as a high-risk venture because scores of earthquake recorders had to be deployed throughout the Rabaul area for several months in 1997–1998. There were fears that the valuable equipment might be stolen or vandalised, bearing in mind the extensive looting that had taken place in 1994. The local communities, however, committed strongly to both the planning and execution of the fieldwork and the geophysical survey was completed successfully without significant incidents.

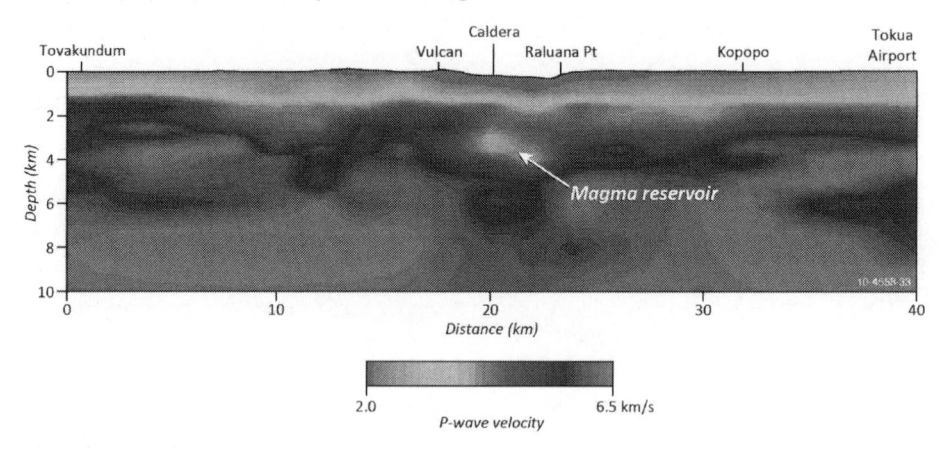

Figure 114. This diagram represents a vertical section through the Earth's crust beneath the Rabaul area. The different colours represent materials that have different 'seismic velocities' — the speeds of earthquake waves as they pass through the crust — and which have been used to define a possible reservoir of magma deep beneath the caldera.

Source: Finlayson et al. (2003, adapted from Figure 9). Reproduced with the permission of Elsevier.

A major result of the survey was the detection of a three-to-six-kilometre-deep magma reservoir.[16] The seismic velocities of the reservoir were sufficiently low that the presence of molten rock could be assumed. This is not to say that the entire reservoir contained only magmatic liquid — like an underground oil tank surrounded by rock. Rather, it was likely to be a mix or 'mush' of magma, crystals, and rock, its spatial limits being ill-defined. There would be further insights in later years when other scientists interpreted the data from the geophysical survey, when petrologists and geochemists obtained further analytical data, and when attempts were made to draw together all the known information on Rabaul volcano into a comprehensive interpretation of how the volcano might 'work'.

---

16   Finlayson et al. (2003).

Restoring the North-eastern Gazelle Peninsula

Shifting the socio-economic functions of Rabaul had been a dominant, but largely inconclusive, theme of discussion during the 1983–1985 seismo-deformational crisis. Construction of the Tokua Airstrip east of Kokopo on newly purchased land is, perhaps, the most outstanding example of post-crisis disaster preparation, undertaken in recognition of the prospect that Lakunai Airstrip, closer to Rabaul town, would not be operational during a volcanic eruption. Furthermore, in the decades before the 1983–1985 crisis, land had been purchased away from Rabaul as a consequence of land in or near Rabaul becoming more scarce for development. Redevelopment accelerated dramatically, however, after the 1994 disaster. Recovering from the 1994 eruption and reorganising the north-eastern Gazelle Peninsula as a whole, and the Kokopo area in particular, are post-disaster processes that continue today.[17]

A Gazelle Restoration Authority (GRA) was established in February 1995 and supported financially by several international development assistance agencies, perhaps most notably the World Bank, AusAID, and the Japan International Cooperation Agency (JICA). GRA was based in Port Moresby, close to central government, but an implementation unit was set up in Kokopo in May 1995 in order to oversee development and rebuilding of the north-eastern Gazelle region. Implementation of the Gazelle Restoration Program, under GRA, was planned in three stages, and the first or Immediate Term Restoration Program was scheduled to run from 1995 to 2000. The aims overall were to establish resettlement estates at Gelegela, Warena, Clifton and Sikut in the Warangoi Valley; to expand Kokopo town and develop nearby Kenebot as a new housing area, as well as Baliora as a satellite town; and to create a light industrial area at Takubar, eventually as part of a development 'corridor' towards Tokua Airport.

A further aim of the program was to upgrade existing transport infrastructure to cater for population shifts away from Rabaul to new areas. New roads were built in and around Kokopo and plantation land was bought by the Provincial Government for village settlements and services at Gelegela, Warena, Clifton and Sikut, which are all well south of the active caldera. They would be linked to the Kokopo area and the coast by an improved road network. Redevelopment in the following years and into the new millennium, however, did not go exactly to plan in these rural areas. Furthermore, not all of the new economic development arrangements were advantageous to the province. Introduction of the *Organic Law on Provincial and Local Level Government* meant that the budget available to the province after 1995 for maintenance of services, such as roads, would be provided through the national government rather than directly from provincial taxes.

---

17   Neumann (1996a), Lentfer & Boyd (2001) and Scales (2010).

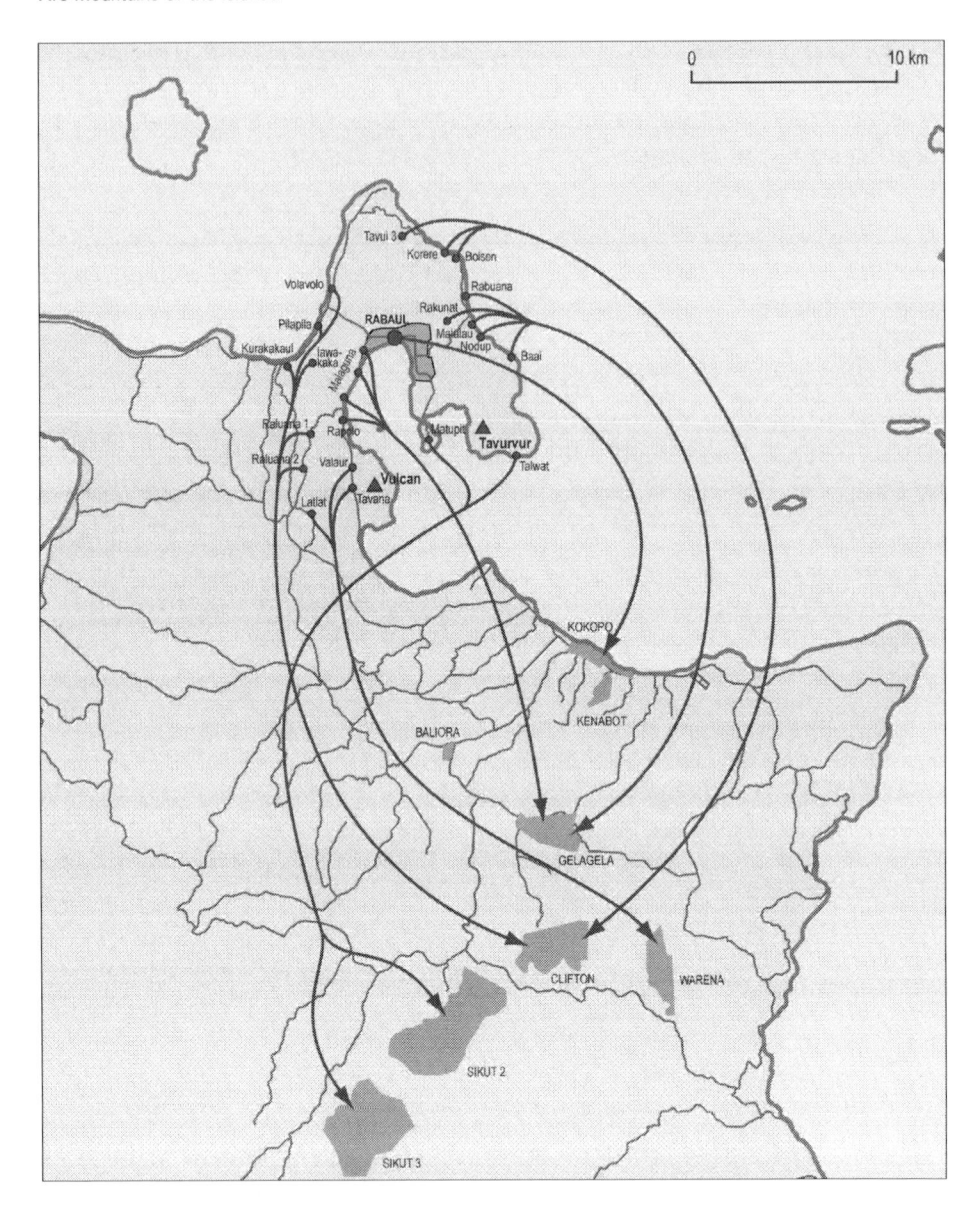

Figure 115. Populations in the Rabaul area after the 1994 eruption initially moved southwards to Kokopo and new settlements in the Warangoi Valley, as shown by the flow lines. Roads are shown in pale grey.

Source: Provided courtesy of I. Scales, who adapted this map from Figure 16 in his account (Scales, 2010).

The concept of urban safety is a relative concept and natural-hazard risk had not been eliminated by the new developments in the Kokopo area. Kokopo is built on the 1,400-year-old Rabaul ignimbrite — a reminder of past volcanic catastrophe and future possibilities. Much of old Kokopo is built on a terrace that sits several metres above sea level at high tide, but parts of the newly developed areas, especially coastal Takubar, are lower and exposed more directly to tsunamis in the Solomon Sea than is Kokopo itself. Furthermore, earthquake hazard at Kokopo was the main reason why N.H. Fisher did not favour the site for redevelopment immediately after the Second World War.[18]

Another early, and key, post-eruption decision was to develop a new land-use zoning scheme for Rabaul town itself. A steering committee was established under the umbrella of the East New Britain Provincial Physical Planning Board and, following extensive consultations with the community and the private sector, it released its development plan in August 1997. The long-term vision of the steering committee was 'for a town more limited in extent than previously'.[19] The declared starting point for the committee was equally pragmatic, direct, and logical: the port and wharf facilities at Rabaul were largely unaffected by the eruption; the harbour was the only deep, all-weather port in the New Guinea Islands region; the core of the province's economy was port–related industry together with commercial ventures in Rabaul town; and, the validity of rebuilding Rabaul using public funds depended solely on the intentions of private firms to reinvest in property and to re-establish business.

The overwhelming response of Rabaul property and business owners at a public meeting held on 31 July 1997, was that they would not be deterred by the ongoing volcanic risk and that they wanted to start rebuilding Rabaul town without further delay, despite problems with obtaining comprehensive insurance cover. Thus, businesses that depended critically on the Rabaul wharfs did not move from the town area — for example, the large and largely undamaged coconut-oil refinery of Coconut Products Limited (CPL), at Toboi Wharf. Initial development of the town, however, would not extend south of Kamerere Street, which ran a few hundred metres south of and parallel to Malaguna Road, although power would be continued to two hotels and the Rabaul Yacht Club at the southern end of Mango Avenue. No development would take place south of Toma Street towards Tavurvur in the greatly devastated southern part of the old town, including Malaytown near Sulphur Creek. The vulnerable and unstable caldera wall area behind Rabaul town was declared a conservation area, which would be off limits to any development, and even to gardening.

The immediate development area would be the western part of the old town, the so-called 'Sector 1' area, running from Malaguna in the west, eastwards on either side of Malaguna Road, to Mango Avenue, thus excluding, for example, the Queen Elizabeth Park recreational area and the old market or *bung*. The market was renowned in the province as a magnet for commerce and social activity and, after the 1994 eruption, it had been re-established informally at Page Park

---

18 Fisher (1946).
19 East New Britain Provincial Planning Board (1997).

at the opposite or western end of Malaguna Road. Support infrastructure was eventually provided for the market at the park, together with pick-up and set-down areas for public motor vehicles (PMV). The new market is near the foot of Tunnel Hill, meaning that people working and shopping there are in a far safer position for any required evacuation from the caldera, compared with the site of the old *bung* in the extreme east.

The 1994 volcanic eruption, therefore, reduced and redefined, but did not eliminate, Rabaul town. There has been, however, a momentous socio-economic shift away from what had been — starting in 1910 — the capital of German New Guinea, then the capital of the Australian-administered Territory of New Guinea between the two world wars and, finally, the capital of East New Britain Province after Papua New Guinea became an independent nation in 1975. Discussions on Rabaul's relocation had started in 1937 when Dr W.G. Woolnough, in particular, recommended unequivocally that the government functions of the Territory capital at Rabaul be moved elsewhere. More specifically, Administrator J.K. Murray's wish in the late 1940s — after the destruction of the town during the Second World War — to relocate the key functions of Rabaul town to Kokopo, had been fulfilled. Thus, Dr Albert Hahl's original decision to move the old German capital from Kokopo to Rabaul in the first place was reversed in both a strategic and physical sense.

The narrative of the 1994 eruption, including its build-up from 1971 onwards as well as its impact on the people of East New Britain, did not by any means end as the twentieth century drew to a close. Other important eruptions and disasters had been taking place in Near Oceania, even though the events at Rabaul from 1994 to 1999 appear to dominate the history of the last decade of the century. There had also been profound technological and attitudinal changes in the world at large, which impacted on the direction taken by volcanic disaster risk reduction strategies in general in the region.

References

AIDAB, 1994. *Rabaul Volcanic Disaster Needs Assessment Mission: Final Report*. Australian International Development Assistance Bureau, Canberra.

Blong, R.J., 2003. 'Building Damage in Rabaul, Papua New Guinea, 1994', *Bulletin of Volcanology*, 65, pp. 43–54.

Blong, R.J. & C.O. McKee, 1995. *The Rabaul Eruption 1994: Destruction of a Town*. Natural Hazards Research Centre, Macquarie University, Sydney.

Davies, H., 1995a. *The 1994 Eruption of Rabaul Volcano — A Case Study in Disaster Management*. Report to United Nations Development Program, University of Papua New Guinea, Port Moresby.

———, 1995b. *The 1994 Rabaul Eruption*. Inaugural Professorial Lecture, University of Papua New Guinea.

Dent, A.W., G. Davies, P. Barrett & P.J.A. de Saint Ours, 1995. 'The 1994 Eruption of the Rabaul Volcano, Papua New Guinea: Injuries Sustained and Medical Response', *Medical Journal of Australia*, 163, pp. 635–39.

Durieux, J. & A. de Wildenberg, 1994. 'Sous nos yeux, l'enfer a mis le feu au Paradis', *Figaro*, 15 October, pp. 48–58.

East New Britain Provincial Planning Board, 1997. 'Rabaul Subject (Zoning) Development Plan'. Rabaul Subject (Zoning) Development Plan Steering Committee, East New Britain Province.

Finlayson, D.M., O. Gudmundsson, I. Itikarai, Y. Nishimura & H. Shimamura, 2003. 'Rabaul Volcano, Papua New Guinea: Seismic Tomographic Imaging of an Active Caldera', *Journal of Volcanology and Geothermal Research*, 124, pp. 153–71.

Finnimore, E.T., B.S. Low, R.J. Martin, P. Karam, I.A. Nairn & B.J. Scott, 1995. *Contingency Planning for and Emergency Management of the 1994 Rabaul Volcanic Eruption, Papua New Guinea*. New Zealand Ministry of Civil Defence, Wellington.

Fisher, N.H., 1946. *Administrative Centre for the Rabaul District*. Department of Supply and Shipping, Mineral Resources Survey Branch Report 1946/32.

IAVCEI Subcommittee for Crisis Protocols, 1999. 'Professional Conduct of Scientists During Volcanic Crises', *Bulletin of Volcanology*, 60, pp. 323–34.

Itikarai, I., 2008. 'The 3-D Structure and Earthquake Locations at Rabaul Caldera, Papua New Guinea'. Master of Philosophy thesis, The Australian National University, Canberra.

Johnson, R.W., I. Itikarai, H. Patia, & C.O. McKee, 2010. *Volcanic Systems of the Northeastern Gazelle Peninsula, Papua New Guinea: Synopsis, Evaluation, and a Model for Rabaul Volcano*. Rabaul Volcano Workshop Report. Papua New Guinea Department of Mineral Policy and Geohazards Management and the Australian Agency for International Development, Port Moresby.

Lauer, S., 1995. *Pumice and Ash: An Account of the 1994 Rabaul Volcanic Eruptions*. CPD Resources, Lismore.

Le Blond, J.S., C.J. Horwell, B.J. Williamson, S. Michnowicz, F. Kelly & P. Delmelle, 2008. *Report on the Mineralogical and Geochemical Characterisation of Rabaul Ash for the Assessment of Respiratory Health Hazard*, Department of Geography, University of Cambridge.

Lentfer, C., & B. Boyd, 2001. *Maunten Paia: Volcanoes, People, and Environment: The 1994 Rabaul Volcanic Eruptions.* Southern Cross University Press, Lismore.

Lindley, I.D., 1995. *The 1994-1995 Rabaul Volcanic Eruptions: Human Aspects*, Bulletin of the Royal Society of New South Wales, 188, pp. 7–9; 190, pp. 4–6.

Lokinap, R.I., 1994. *Rabaul Volcanic Emergency: 'Operation Unity' Report to the National Parliament by the Controller* (Report No 1). Government of Papua New Guinea, Port Moresby.

———, 1995. *Rabaul Volcanic Emergency: 'Operation Unity' Report to the National Parliament by the Controller* (Report No 4). Government of Papua New Guinea, Port Moresby.

McKee, C.O., 1999. 'Lessons from Volcanic Crises and Eruptions at Rabaul'. Geological Survey of Papua New Guinea Report 99/17 (not yet released, assigned number only).

McKee, C.O., RVO Staff & R.W. Johnson, 1994. 'Rabaul', *Global Volcanism Network Bulletin*, 19, no. 9, pp. 4–7.

Nairn, I.A. & B.J. Scott, 1995. *Scientific Management of the 1994 Rabaul Eruption: Lessons for New Zealand.* Institute of Geological and Nuclear Sciences, New Zealand, Report 95/26.

Neumann, K., 1995. *Tavurvur I Puongo!: Students' Accounts of the 1994 Eruptions in East New Britain.* Department of East New Britain, Vunadidir.

———, 1996a. *Rabaul: Yu Swit Moa Yet: Surviving the 1994 Volcanic Eruption.* Oxford University Press.

———, 1996b. *The 1994 Volcanic Disaster in East New Britain and its Aftermath: Comments and Observations.* Emergency Management Australia, Canberra, on behalf of the Australian International Decade for Natural Disaster Reduction Coordination Committee.

———, 1997. 'Nostalgia for Rabaul', *Oceania*, 67, pp. 177–93.

Patia, H., 2004. 'Petrology and Geochemistry of the Recent Eruption History at Rabaul Caldera, Papua New Guinea: Implications for Magmatic Processes and Recurring Volcanic Activity'. Master of Philosophy thesis, The Australian National University, Canberra.

Rabaul Petrology Group, 1995. 'Taking Petrologic Pathways toward Understanding Rabaul's Restless Caldera', *Transactions of the American Geophysical Union*, 76, no. 17, pp. 171, 180.

Roggensack, K., S.N. Williams, S.J. Schaefer, & R.A. Parnell Jr., 1996. 'Volatiles from the 1994 Eruptions of Rabaul: Understanding Large Caldera Systems', *Science*, 273, pp. 490–93.

Rose, W.I., D.J. Delene, D.J. Schneider, G.J.S. Bluth, A.J. Krueger, I. Sprod, C. McKee, H.L. Davies & G.G.J. Ernst, 1995. 'Ice in the 1994 Rabaul Eruption Cloud: Implications for Volcano Hazard and Atmospheric Effects', *Nature*, 375, pp. 477–79.

Scales, I., 2010. *Roads in Gazelle Peninsula Development*. Australian Agency for International Development, Canberra.

Smithsonian Institution, 1994. 'Rabaul', *Global Volcanism Network Bulletin*, 19, no. 8, pp. 2–6.

Tomblin, J., & J. Chung, 1995. *Papua New Guinea Analysis of Lessons Learnt from Rabaul Volcanic Eruption and Programming for Disaster Mitigation Activities in Other Parts of the Country*. United Nations Department of Humanitarian Affairs, Geneva.

Williams, S.N., 1996. 'Double Trouble', *Earth: The Science of Our Planet*, August, pp. 43–49.

World Publishing, 1994. *Volcano*. A Special Project of World Publishing Co Pty Ltd, Publishers of the Times of Papua New Guinea, Wantok, Weekend Sport, and PNG Business, Port Moresby.

14. Eruptions of the Early Twenty-first Century: 1998–2008

> ... it has been a very eventful week ... Tavurvur resumed eruption on Monday evening and Manam produced another paroxysmal eruption on Thursday night ... . Our observation post at Warisi village was wiped out completely by what is described as pyroclastic flow ... . All our equipment was destroyed by the event. There were about 14 people at Warisi at the time of the eruption and all got injured while trying to escape ...

Ima Itikarai, 29 January 2005

International Developments and Modern Near Oceania

The year 2000 marked the 300th anniversary of William Dampier's 'volcanological voyage' through the Near Oceania region, when eruptions and volcanoes were described in greater detail than in any of the known records of the earlier Spanish and Dutch explorers. The three centuries had started during the European Enlightenment with visions of a happier, more rational and more informed world following the Scientific Revolution of the seventeenth century. Those three centuries concluded in the last few decades of the twentieth century with technological advances that would have been unimaginable to the early 'natural philosophers' and secular humanists of the Enlightenment. The world by the end of the twentieth century had also become economically and politically globalised, to an unprecedented degree. Old European empires had disappeared, but inequity remained in the distribution of wealth and political power amongst the 193 countries represented as the United Nations. Smaller 'developing' countries struggled to keep up with massive, global, political and social changes and with the influence of more powerful nations and of multinational companies, such as in the mineral and petroleum industries.

The world's high technology had produced extraordinary advances in electronic communication, and changes to the way in which people measured their place in the world geographically, socially and philosophically. The list of technological achievement is extraordinary: invention of the silicon chip and development of personal computers and laptops; development of the Internet and the World Wide Web and their enormous capacity for linking to vast databases and information sites, and for social networking; electronic mail, mobile and satellite telephones, and the development of a 'smart-phone' industry; GPS positioning, geographic information systems, and spatially referenced digital datasets in general; digitisation and computer-base cataloguing of printed materials; and the ongoing sophistication of the collection of images of a range of earthbound parameters from space.

Earth-observing platforms and satellites were equipped with multispectral scanners, ground temperatures on volcanoes could be measured using infrared detectors, and deformation of volcano surfaces could be measured by space-borne radar interferometry. International aviation was being served by a global network of Volcanic Ash Advisory Centres (VAAC), including one at Darwin in northern Australia, which together formed an International Volcano Watch. Land-based networks were globalised too, including the worldwide monitoring of earthquakes, and the International Monitoring System of the Comprehensive Test Ban Treaty Organisation (CTBTO), for the detection of nuclear explosions whether recorded seismically through the earth, hydroacoustically through the oceans, or infrasonically through the atmosphere — including volcanic explosions. All of these advances showed no signs of slowing as the first decade of the new millennium unfolded, and inevitably they impacted on the way volcanoes were monitored, and volcanological information was obtained, stored, and used, in Near Oceania.

Tracing in narrative form a history of volcano discoveries and volcanic disasters in the near-backwater of 'Near Oceania' inevitably has become focused on only one part of the region — that of the single nation state of Papua New Guinea. No active volcanoes are known in the New Guinea Island part of eastern Indonesia — that is, in western Near Oceania — although Dampier had mistakenly reported an active volcano there in 1700. Furthermore, the Solomon Islands have played, to date, only a marginal part in the volcanological history. This is despite strong, ongoing concerns that volcanoes such as Savo, Simbo and Kavachi cannot be ignored and that they require the maintenance of strong linkages between at-risk communities and government volcanic-risk management programs.[1] The small volcanological agencies in both countries, however, were becoming more fully integrated into the national structures of natural hazard disaster management. Papua New Guinea and the Solomon Islands were also having to work in an enlarged and more complex environment of increased international development assistance, particularly bilaterally with Australia, the European Union, the United States, Japan, and increasingly with China, as well as multilateral agencies such as the United Nations Office for the Coordination of Humanitarian Affairs, the United Nations Development Program, and the World Bank, whose offices are in Port Moresby.

---

1   Cronin et al. (2004).

Ulawun: A Decade Volcano

International volcanology advanced steadily on several fronts into the first decade of the new millennium. Applied volcanology was no longer the 'Cinderella science' of the 1960s and was becoming increasingly multifaceted and multidisciplinary. This meant, for agencies such as the Rabaul Volcanological Observatory (RVO), an increased capacity to cover many aspects of the volcanic risk reduction spectrum. Perhaps most significant have been the international and national efforts made to link with the social sciences and to focus on how communities are structured in both a physical and social sense; the extents to which different societies are vulnerable, or susceptible, to the impact of volcanic disasters; the importance of volcanic-risk awareness-raising campaigns in volcanically vulnerable communities; and, how the volcanological sciences can link with and assist communities in developing strategies for economic development. One example of such efforts internationally was the 'Decade Volcanoes' initiative which ran during the 1990s as part of the United Nations-sponsored International Decade for Natural Disaster Reduction (IDNDR).[2] Sixteen volcanoes worldwide were selected for special attention, including Ulawun volcano in West New Britain Province — the only 'Decade' volcano in Near Oceania.

A major disaster in Papua New Guinea on 17 July 1998 diverted public attention away from questions about Rabaul and volcanic risk in general when a tectonic earthquake in the Bismarck Sea triggered a submarine landslide off the north coast of New Guinea near Aitape, producing a devastating tsunami that killed more than 2,100 people.[3] RVO staff were not involved in the post-disaster scientific assessments of the tsunami and its impacts, as these were undertaken largely by international visitors and staff from the University of Papua New Guinea and Port Moresby Geophysical Observatory. Nevertheless, the vulnerability of shoreline communities in the Bismarck Sea area and knowledge of the disastrous tsunamigenic collapse of Ritter Island in 1888, drew the attention of RVO staff once again to the tsunami-generating capabilities of coastal volcanoes — and to Ulawun.

Ulawun, from most directions, appears as a high, steep, and symmetrical cone. It is, at about 2,350 metres above sea level, the highest volcano in the Bismarck Volcanic Arc — higher than its immediate and equally imposing neighbour Bamus at about 2,250 metres.[4] The vegetation-free slopes of Ulawun near to the summit exceed 35° in places, and the symmetry of the volcano is broken on the southern flank and in views from the east and west, by a long, northward-facing, east-west escarpment against which the upper and younger parts of

---

2 Barberi et al. (1990).
3 See, for example, Davies (2002) and Synolakis et al. (2002).
4 Johnson et al. (1983).

Fire Mountains of the Islands

the active volcano have been built. Volcanoes do not grow indefinitely. Some, like Rabaul, form calderas that reduce the original height of any 'ancestral' mountain that may have existed previously. Others simply become too high and too steep and they collapse gravitationally, like Ritter Island did in 1888. Indeed, the east-west escarpment on Ulawun is similar to the double-cusp shape of the escarpment at present-day Ritter, leading to the idea that Ulawun may have collapsed catastrophically in the distant past towards the north, possibly entering the sea and generating a tsunami.

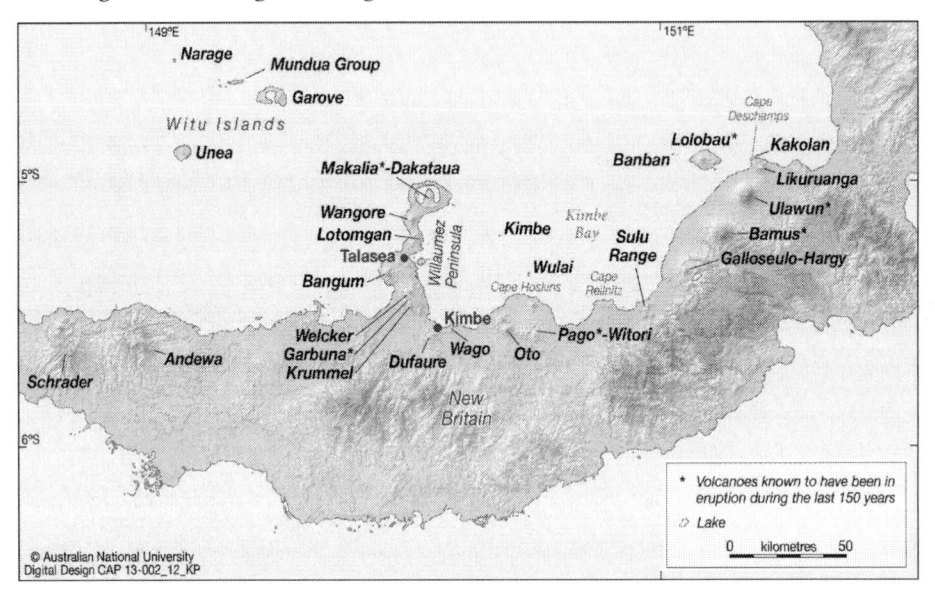

Figure 116. Distribution of main volcanoes in central-north New Britain, including offshore islands. Several eruptive centres, such as at Cape Hoskins, Cape Reilnitz and Cape Deschamps are not labelled here.

Source: Adapted from Johnson (1977, Figure 6).

The evidence in support of a prehistoric cone-collapse at Ulawun is not as strong as one would like. In particular, much of the large debris-avalanche deposit that would form as a result of such a catastrophic collapse may not be well exposed on land because it could have been covered by materials from later eruptions at Ulawun. A debris-avalanche deposit was identified about 15–30 kilometres north of the Ulawun shoreline during a marine-geoscience research cruise in 2004.[5] The deposit, however, is partially buried by sediment and it has not been traced into shallower water nearer the coast, where it may well be concealed by even thicker sediments washed into the sea from the coastal volcanoes. Thus, a direct link between the deposit and Ulawun cannot yet be made. Other origins

---

5   Silver et al. (2009).

for the escarpment are possible, but again, supporting evidence for these may well be concealed beneath the young cone. For example, the escarpment is perhaps a simple geological fault cutting across the volcano and the older part of the volcano to the north conceivably could have dropped down vertically by slower and non-catastrophic movements. Nevertheless, Ulawun does seem too high and too steep and the possibility of future Ritter-like catastrophic collapses cannot be disregarded.

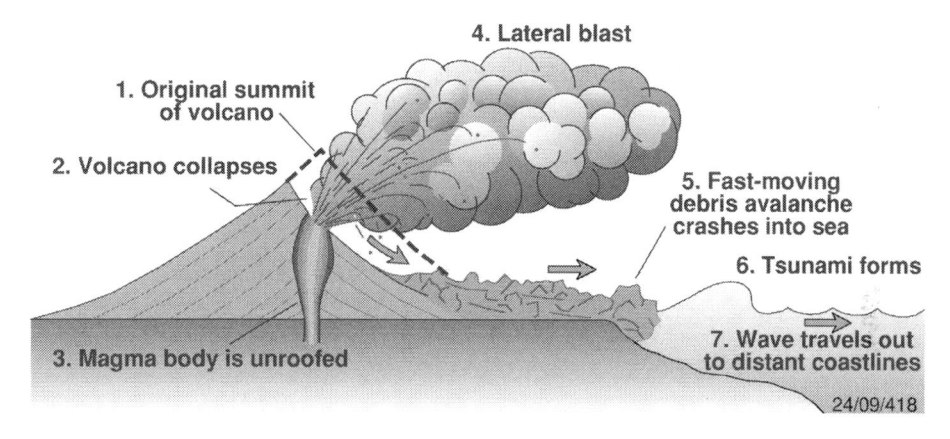

Figure 117. One way in which a high, steep-sided and coastal volcano, such as Ulawun, might collapse and produce a devastating tsunami is illustrated in stages 1–7 of this cartoon. Stages 1–4 are based on what is thought to have happened at Mount St Helens, United States, in 1980.

Source: Ulawun Workshop Report (1998, p. 14).

The tragic Aitape disaster in 1998 was coincidental — to the extent that an IDNDR workshop entitled 'Volcano Cone Collapses and Tsunamis' was already being planned to be held later in 1998 at Walindi in West New Britain Province, using Ulawun volcano as a case study. There remained at this time a still-elevated international interest in volcanic cone collapses as a result of the Mount St Helens eruption of 1980 in the western United States, when flank collapse released pressurised magma and produced a deadly lateral blast. The 40 participants at the Walindi workshop — half of them from overseas, including the United States — produced 25 recommendations. The overall conclusion was that volcanoes as steep and as high as Ulawun could indeed be inherently and gravitationally unstable, and that such volcanoes require special vigilance, particularly those at sea level which might have tsunamigenic potential — assuming that the cone collapses were rapid and therefore catastrophic.[6]

---

6 Ulawun Workshop Report (1989).

Ulawun 2000: A Short-lived, Powerful Eruption

Ulawun was one of five high-risk volcanoes in Papua New Guinea selected in 1995 for an upgrade of instrumental monitoring during the Australian Agency for International Development (AusAID)-funded Papua New Guinea – Australia Volcanological Service Support Project (VSS Project), and subsequent Twinning Program. The other four volcanoes were Rabaul, Manam, Karkar and Lamington. A seismometer and electronic tiltmeter were installed on the slopes of Ulawun, electronic signals were relayed by radio-telemetry to Ulamona Mission at the coast and then by HF-radio to a central 'hub' computer in the recording room at RVO headquarters in Rabaul. Lindsay Miller and Trevor Dalziell, both Geoscience Australia (GA) technicians, designed and developed the hub system, and United States Geological Survey (USGS) Volcanic Disaster Assistance Program (VDAP) software was used to process the received signals, including from the other remote volcanoes of Manam, Karkar and Lamington. Thus, for the first time, a national, instrumental, 'volcano watch' could be made from the RVO recording room of four high-risk volcanoes in the distant parts of Papua New Guinea, which represented a significant technical landmark.

Ulawun had been in eruption in 1985 and in most of the previous years since its spectacular activity in 1978. Some mild explosive activity had also taken place from the summit crater of Ulawun in 1989 and 1993, but for some years the volcano had produced only emissions of vapour — for example, as seen during the Walindi workshop in October 1998. This situation lasted until October 1999 when explosive activity resumed and then, a year later, in September 2000, a new, rapid-onset, but short-lived eruption began. The response to this powerful eruption by affected people in West New Britain and the involvement of RVO represent a significant, but perhaps singular, example of how the results of instrumental monitoring of a Papua New Guinea volcano can be crucial in making decisions to evacuate at-risk communities to safer refuges.

RVO volcanologists in Rabaul began to detect significant changes at Ulawun on the evening of Wednesday 27 September 2000, from their analysis of the earthquake recordings being sent to Rabaul by HF-radio every 20 minutes from Ulamona Mission — a distance of about 130 kilometres.[7] Authorities in West New Britain Province were informed of the changes and alert stages 1 and 2 were declared overnight. RVO staff in Rabaul were in contact with Martina Taumosi, the volcano observer at Ulamona, by radio and together they tracked these and further changes in the earthquake activity at the volcano. A rapid build-up of seismic intensity started during the evening of Thursday 28 September, when constant ground shaking or 'tremor' began to be recorded, rather than single,

---

7 H. Patia (personal communication, 2012) and Smithsonian Institution (1999–2000).

14. Eruptions of the Early Twenty-first Century: 1998–2008

separate earthquakes. Taumosi rang the church bell that evening to inform the local community at nearby Ubili village and at Ulamona Mission on the status of activity of the volcano. The local people, who had gathered at the mission station, began to hear noises from the volcano and to see glow at the summit of Ulawun at about 8.45–9.00 pm. People living near Ulawun had had the opportunity only three weeks earlier, during a public-awareness campaign, to hear RVO staff talk about volcanic hazards at Ulawun, so they were well aware of the potential dangers from the volcano. The villagers were concerned about the current condition of the volcano. The Ulamona observer discussed the situation with RVO staff in Rabaul by radio, and a recommendation was provided by RVO to evacuate. An eruption began at 10.40 pm that night, and a full-scale, incandescent strombolian eruption peaked about four hours later in the otherwise darkened night sky.

Figure 118. Pyroclastic flows were produced at Ulawun during vulcanian activity in 1985. A pyroclastic flow is seen here discharged down the north-western valley of the volcano on 20 November, eventually dumping its load when it came to rest east of Ulamona Sawmill, in the foreground. The hot pyroclastic material at the flow front then rose thermally, giving the false impression of a new eruption taking place low on the flanks of Ulawun.

Source: P.L. Lowenstein, Rabaul Volcanological Observatory.

Many hundreds of people evacuated overnight, mainly by road from settlements west and north-west of the volcano, assisted by road transport provided by Ulamona Sawmill and using other private vehicles. Villagers from vulnerable places to the north and north-east such as Nuau, Voluvolu and Painave moved north-eastwards to Bakada, on the coast well to the north of Ulawun. Other people, from places such as Navo Plantation and from Ubili near Ulamona Mission and Sawmill, moved south-westwards to Kabaia and Soi, two places previously designated as refuge centres. A spectacular, forcefully ejecting ash column, which reached heights of 12–15 kilometres, could be seen after daybreak on the morning of Friday 29 September and throughout that day. Staff at the Volcanic Ash Advisory Centre in Darwin on the same day were able to observe an arc-shaped ash cloud on satellite images, and to issue a notification to aviation authorities. Ash fallout destroyed some gardens downwind of Ulawun, and three pyroclastic flows were produced, but they did not reach settlements. Strong discharges from the summit crater had ceased by early on 30 September. The results of this powerful but short-lived eruption had not been disastrous, but there remained the persistent question of whether similarly or even more intense eruptions might trigger gravitational collapse of the Ulawun cone, as discussed at the Walindi workshop. Evacuations under these circumstances are clearly prudent.

The time-sequence of instrumental early-warning signs, followed by a prompt recommendation to at-risk communities at Ulawun to evacuate, is not always as efficient as in the eruption of late-September 2000. Another explosive eruption took place on 25–30 April 2001 when, again, there were early-warning signs.[8] A Stage 1 alert was declared, but this eruption was relatively minor and only limited evacuations took place. More significantly, there was evidence seen after the 2001 eruption that a fissure eruption had taken place high on the flanks of Ulawun, again raising questions about the structural integrity and gravitational instability of the volcano.

Particularly strong earthquake activity was recorded at Ulawun in 2006 and again in 2008 and 2010, but there were no immediate eruptions. Nevertheless, there were concerns about public safety and the currency of existing evacuation planning, recognising particularly that oil-palm plantations and therefore the number of people, were increasing around the volcano. This led to a meeting in 2010 of representatives from both the East and West New Britain provincial administrations, and from RVO, and the production of a new response plan.[9]

---

8   Smithsonian Institution (2001).
9   East/West New Britain Provincial Governments (2010).

14. Eruptions of the Early Twenty-first Century: 1998–2008

Pago 2002–2003: An Unexpected Eruption

Explosive eruptions at Pago volcano on 3 and 5 August 2002 surprised the staff at RVO in Rabaul. They also surprised the thousands of people in the Cape Hoskins area of West New Britain, including those in the provincial capital at Kimbe, only 40 kilometres west of Pago. Pago had not been active since the eruptive phase of 1911–1933,[10] there was no instrumental monitoring on the volcano, and no reports had been received of any early warning signs of an eruption. Aircraft pilots reported the presence of ash clouds on both 3 and 5 August and, on 6 August, Air Niugini reported a 1.8–2.4-kilometre-thick ash plume extending about 150 kilometres to the west-north-west of the volcano at a height of 7.6 kilometres.[11] Fine ash had spread over a sector of the Cape Hoskins area, affecting a population of about 20,000, triggering evacuations of villages and hamlets, and closing the Hoskins Airport to air traffic. Lava began to flow from vents on a fracture on the north-western side of the Pago.

RVO staff, led by Ima Itikarai, responded promptly by installing a seismograph at Malilimi Plantation, south-west of Pago, and establishing a surveying line from Malilimi towards the volcano in order to measure any ground-surface movements. Aerial inspections were made of the volcano and volcano information bulletins were released. A scientific Japanese Disaster Relief Team had arrived independently to investigate the eruption and its members stayed in the area from 25 August to 3 September, working in partnership with RVO. Foreign assistance had already been sought by RVO through official channels and a request had been directed to the Office of Foreign Disaster Assistance of the US Agency for International Development (USAID), as had happened after the 1994 eruption in Rabaul. A team of three from the USGS-VDAP left the United States on 5 September, and stayed in West New Britain until 13 October. RVO was heavily involved in coordinating this international volcanological effort.

External assistance also arrived in the province to deal with the welfare of 12,000 displaced persons, more than 8,000 of whom were in seven temporary care centres scattered along the coast westwards towards Kimbe, where some evacuees were accommodated in the high school.[12] The Provincial Disaster Committee (PDC) in West New Britain worked closely with a 'Mt Pago Disaster Task Force' established by the National Disaster Management Office and which included representation from national government departments, donor countries, and church groups. Officers from both United Nations Office for the

---
10 Cooke (1981).
11 Smithsonian Institution (2002–2006). Newspaper coverage of the Pago eruption includes feature articles and accompanying photographs in a special issue of the *Post-Courier* (2002). See also, Davies (2003). Much unpublished information is held by RVO, AusAID, and other agencies.
12 Macatol et al. (2002).

Coordination of Humanitarian Affairs (OCHA) and AusAID were also involved with provincial authorities in assessing the situation, making a risk assessment, and addressing key concerns and questions about food and water supplies, health and sanitation. The closure of Hoskins Airport did not help the disaster response, although smaller airstrips were used at both Bialla and at an old airstrip at Talasea, which was reopened. There were also delays in deciding whether evacuees should return to their villages, despite the minimal impacts and ash thicknesses of only few millimetres. Crucial questions to be answered regarded the potential length of the eruption and the possibility of it escalating into a worse incident. There was also consideration to be given to the longer-term impacts on the province's transport links and agricultural industries, especially on the newly established oil-palm plantations. An eruption of the scale and duration of the previous eruptive period in 1911–1933 could have worse effects in 2002 because of the greater number of people who would be affected and the risk to investment.

Figure 119. Dark lava is issuing slowly from vents on the north-western flank of Pago volcano, which is seen in the lower right. This undated view is taken from the south-south-west later during the 2002–2003 eruptive period. The escarpment of Witori Caldera is visible across the upper part of the photograph and the large-volume lava from the most north-westerly vent is seen buttressed and curved against it. The lava has also spilled a short distance towards the south-west.

Source: E. Endo. United States Geological Survey. Originally published in the IAVCEI Calendar for 2005 (December).

Figure 120. The Pago eruption was observed from space. This satellite image is of the contoured thicknesses of the large-volume lava flow erupted in 2002–2003. The thicknesses were obtained by calculations from data obtained from successive overpasses of a satellite carrying a distance-measuring radar instrument.

Source: Image supplied courtesy of C. Wicks (see Wicks et al. 2008).

The explosive eruptions at Pago in August 2002 did not escalate. Pago itself is in an unpopulated area isolated within Witori Caldera and the explosions and ash falls inflicted little damage on settlements outside the caldera. No lives were lost. Explosive activity declined at Pago and, by January 2003 — after the turn of seasons and the arrival of the north-west monsoon — Hoskins Airport was reopened and normal life was resumed in the province. The most striking geological product of the Pago eruption was not the ash deposits but, rather, the multiple extrusions of lava from the vents in a line on the north-western flank of Pago. The lowest of these vents produced a huge, slow-moving lava flow, which gradually moved eastwards around the foot of the caldera escarpment of Witori. This lava probably stopped flowing in about March 2003, thus marking the likely end of the eruption, although emissions of vapour and gas continued for many months. Large, caldera-constrained lava flows were also produced during the early part of the 1911–1933 eruption, particularly in 1914–1918.

An important outcome of the Pago eruption was establishment of permanent instrumental monitoring at Pago and the addition of Pago to the list of high-risk volcanoes for ongoing support by the AusAID-funded Twinning Program. Digital signals from a seismometer and GPS station within the caldera near Pago, which had been donated by USAID and installed by the USGS-VDAP team, were received at a recording station near the governor's residence at Kimbe. There is, however, no direct line of sight between the instruments in Witori Caldera and the Kimbe Volcanological Observatory, so the signals must be relayed through another station on the high summit of the intervening Mount Oto. This monitoring system now has to be maintained and sustained by RVO, but accessing the isolated instruments within the caldera and climbing to the top of Mount Oto for repairs or clearance of vegetation is not easy. Questions of sustainability and identification of a suitable source of ongoing funding, therefore, could be raised, especially for a volcano such as Pago which — like many others in Papua New Guinea and the Solomon Islands — may not erupt again for many years, even decades. However, the Kimbe Volcanological Observatory can also be used to receive signals from other monitored volcanoes in the province and so its ongoing use, despite the difficulties, may be justified.

Threat of a Caldera-Forming Eruption at Pago-Witori

Disaster managers and volcanologists at Pago in 2002–2003 were aware of the possibility of a much larger, caldera-forming, explosive eruption developing in the area. Pago is a small volcano within a prominent caldera atop Witori volcano, the low-angle slopes of which give it a profile rather like that of an inverted saucer. Geologists had demonstrated from earlier field studies that Witori had produced several major eruptions during the last 10,000 years, leading to the deposition of widespread ignimbrites and related air-fall deposits. Such eruptions, were they to recur in 2002, could devastate the economy, lives and livelihoods in this part of New Britain.[13]

The north-central coast of West New Britain, including its hinterland and Willaumez Peninsula, are vulnerable to the effects of a range of eruption types, including large catastrophic ones from several caldera systems, and not from Witori alone. Other active and potentially active caldera systems in the New Britain area as a whole are at Garove, Unea, Dakataua, Hargy and Lolobau, as well as at Rabaul and Tavui. Furthermore, considerable archaeological evidence is available that past eruptions of this large scale affected pre-industrial communities in West New Britain, an area well known and valued for its sources of obsidian for trade and exchange purposes.

Willaumez Peninsula in West New Britain Province is an area that has received considerable research attention over many years from a large, multidisciplinary group of archaeologists, volcanic geologists and historical environmentalists.[14] Indeed, no other area in New Oceania has received the same level of interdisciplinary research attention, and it is clearly the best example for consideration of the ways in which people in prehistory may have coped with and adapted to the threat and impact of volcanic hazards in the region. Volcanological studies in the Willaumez Peninsula area benefitted greatly from this work, especially the improved descriptions of the tephra stratigraphy and the increased number of radiocarbon dates on the tephras and their palaeosols. The area has been called a 'remarkable 40,000 year record of human-volcano interactions'.[15]

Two main areas on Willaumez Peninsula have received intensive archaeological treatment: Garua Island east of Talasea, and the so-called 'isthmus' area at the southern end of Willaumez Peninsula. Correlations have also been made with sequences elsewhere in New Britain, including at Yombon in the central ranges to the south of the peninsula. Evidence of human occupation on Willaumez Peninsula extends back 35,000–45,000 years, to when human colonisation is thought to have first taken place, but the best exposed sequences are from the mid- to late Holocene, a time when the area was subject to the known impact of nine explosive eruptions and four minor episodes of more limited extent.[16] Four of five especially large eruptions of VEI 5–6 scale took place from nearby Witori volcano to the

---

13   Blake & Bleeker (1970), Blake & McDougall (1973), Blake (1976), Machida et al. (1996) and McKee & Kuduon (2005).
14   See, for example, Torrence et al. (2000, 2004, 2009), Boyd et al. (2005), Torrence & Doleman (2007) and Neall et al. (2008). These references are examples of the now extensive literature on the prehistory of Willaumez Peninsula ranging back to the early 1980s (see, for example, Specht, 1981). The more recent papers quoted here include extensive bibliographies of earlier work.
15   Cashman & Giordano (2008, p. 326).
16   Machida et al. (1996), Torrence et al (2004, 2009), Neall et al. (2008) and Petrie & Torrence (2008).

south-east and produced tephras which have been named W–K 1–4, and the fifth from Dakataua volcano —its tephra is labelled 'Dk' — at the northern end of the Peninsula. W–K4 overlies Dk, but there is no intervening palaeosol and the radiocarbon dates are indistinguishable from each other. All five of these eruptions clearly must have had a major impact on the people living nearby, blanketing livable landscapes including reef and forest resources, and creating new land surfaces that would have remained barren and unsettled for years.

Table 7. Dates of Major Eruptions from Witori (W–K) and Dakataua (Dk) Volcanoes

Tephra	Radiocarbon date (years Before Present, BP)	Date range (years BP) from Bayesian analysis[a]
W–K4	1344±38[b]	1310–1170
Dk	1383±28[b]	1350–1270
W–K3	ca 1800[c]	1740–1540
W–K2	ca 3300[c]	3480–3160
W–K1	ca 5600[c]	6160–5740

a. Petrie & Torrence (2008).

b. McKee et al. (2011).

c. Machida et al. (1996).

All listed eruptions are VEI–5 or 5–6.

The artefacts found on Willaumez Peninsula are predominantly of worked and retouched obsidian pieces, together with some 'manuports' — stones believed to have been transported by humans — and a few objects made from crystalline rocks. Lapita-style pottery appears later in the Holocene sequence. Four obsidian sources have been identified in West New Britain: three in the Talasea area — Kutau/Bao, Gulu, and Baki and Hamilton together on Garua Island — and one at Mopir in the south-western part of Witori volcano.[17] These Talasea-Mopir obsidians have been 'finger-printed' geochemically and are, therefore, known to have been traded and exchanged widely, although to different extents, throughout much of Melanesia and into Remote Oceania. The obsidian could, therefore, have been valued and quarried as a commodity in West New Britain, thus encouraging local settlement. Mopir appears to have been a well-exploited source during the Pleistocene, but less so during the Holocene when access may have been made more difficult because of burial by young tephras from nearby Witori. One of the more striking features of the artefact sequence is the presence of finely and skillfully constructed 'stemmed' blades of obsidian. These are pieces of technological craftmanship that may have had cultural values beyond

17 Fullagar et al. (1991) and Summerhayes (2009).

simply the utilitarian, such as the delicate cutting of skin, including perhaps for tattooing.[18] The stemmed blades are not found above the 3,300-year-old W–K 2 tephra, whereas Lapita pottery is not found below it.

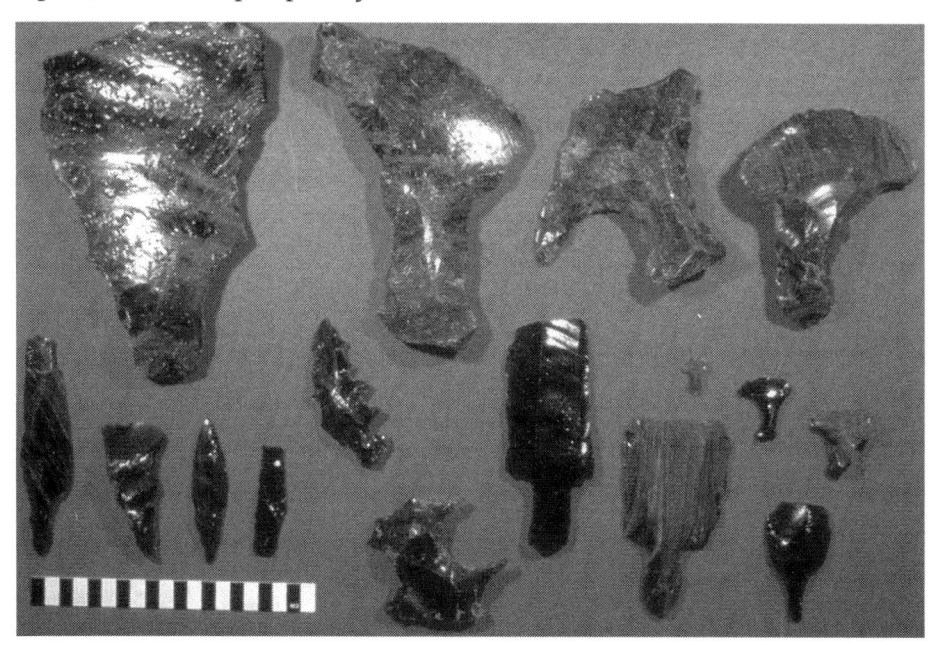

Figure 121. Obsidian artefacts from the Talasea–Mopir area of New Britain have a wide range of size and shape. Some blades include skillfully produced stems or hafts. The graded scale is 20 centimetres long.

Source: R. Torrence and G. Britain, Australian Museum.

These archeological results have been used as a basis for extensive and detailed discussions on cultural evolution and possible human behavior, and especially for hypotheses and speculations about what may have happened in terms of volcanic disaster recovery, resettlement, and cultural adaptation. For example, much of the discussion has focused on the return periods for resettlement after each eruption; on the degrees to which people formed mobile or settled communities; on sites of possible refuge and their links and distance along long-distance trading networks; on the possible urgency of resettlement caused by the need to reclaim obsidian sources; and, on inferred population-density changes. Key questions, however, related to human decision-making concerning volcanic risk reduction, strategies for family safety, and such matters as early warning identification and evacuation are virtually impossible to answer solely on the basis of stone and pottery artefacts and their distributions in space and time, as the answers to the questions are fundamentally social rather than archaeological ones.

---

18 Araho et al. (2002) and Specht (2005).

Manam 2004–2005: Abandoning a Volcanic Island?

Many eruptions continued at Manam after the 1956–1960 volcanic crisis and related evacuation, thus confirming Manam — together with Bagana — as one of the two most frequently active volcanoes in Papua New Guinea. Especially large or 'paroxysmal' eruptions at Manam during the 1990s and up to recent times include several examples where dangerous pyroclastic flows descended and, in some cases, spilled over the sides of the radial valleys on the island. Particularly large eruptions, for example, had taken place in October and November 1992, producing pyroclastic flows as well as lava flows, which in the North East Valley, destroyed bush-material houses, small cocoa fermentaries, and food gardens.[19]

Another, but more deadly, paroxysmal eruption took place from the summit craters on 3 December 1996, following a build-up of smaller eruptions over several weeks.[20] Large pyroclastic flows descended the South East and South West valleys on the afternoon of 3 December, including one that reached the coast, killing 13 people at the site of Budua Old Village in the South West Valley, and flowing into the sea. Neither the start of the 1996 eruption nor its strong phases were anticipated by RVO from the monitoring data from the Tabele Volcanological Observatory, which was established in 1963–1964.

Eruptions at Manam in 2004–2005, starting on 24 October 2004, were even larger and more damaging than were those in 1992 and 1996.[21] Indeed, they could represent the most devastating of any reported eruptions on Manam, as they affected virtually the whole of the island, led to a major evacuation of the Manams to care centres on the mainland, and created social problems that continue to the present day. The full disaster-management story of this major eruption is, therefore, still unfolding and, indeed, a full scientific account of the eruption and disaster has yet to be completed. Manam had been included on the list of high-risk volcanoes for attention during the AusAID-supported RVO Twinning Program, and signals from the monitoring equipment at Tabele Volcanological Observatory could, up to 2001, be sent by HF-radio to Bogia on the nearby mainland and then on to RVO headquarters in Rabaul. This potential benefit of eruption early warning ceased in January 2001, however, because of compensation and access issues raised by the owner of the land at the observatory site, and because of subsequent vandalism. A smaller, makeshift observatory station equipped with a single seismograph, together with capability for radio-

---

19 Smithsonian Institution (1992).
20 Smithsonian Institution (1996).
21 Smithsonian Institution (2004–2005). Other information on the 2004–2005 eruption used here has been taken from a range of unpublished sources including RVO and AusAID correspondence and related reports as well as newspaper articles.

voice reporting from the local observer to the volcanologists in Rabaul, was therefore established at Warisi village, eastern Manam, in May 2001. The Warisi station was limited in its scope and capacity to cope with the emergence and impact of the devastating eruptions of 2004–2005, although RVO staff continued to advise provincial disaster authorities with reports on the state of the volcano and provide advice on alert levels.

Figure 122. A pyroclastic flow descends the South East Valley on Manam volcano during the paroxysmal eruption of 3 December 1996. A second flow can be seen on the left and in the distance moving down the South West Valley.

Source: P. de Saint Ours, Rabaul Volcanological Observatory.

RVO had noted increased earthquake activity on Manam in mid-October 2004 and had declared a low Level 1 alert, but the eruption that began on 24 October and those that followed in November and December were much greater, and early warning of their severity was not possible. A pyroclastic flow ran down the South East Valley to the sea, and much of the eastern side of the island was affected by ash and scoria fallout from the October eruption clouds. The western and north-western sides of the island became affected by fallout too, from Tabele clockwise around the coast to Baliau in the north. Fist-sized rock fragments were reported to have punched through the thatch roofs of houses.

People moved to safer areas on the island, but evacuation of thousands had begun by late November to care centres on the mainland, attracting the start of external financial support for their welfare.

Eruptions on 27–28 January 2005 were even more severe than those in October–December of 2004, adding to the already difficult living conditions on the island.[22] Many people by this time were on the mainland in care centres, although some had returned to Manam. The whole of the island was covered by a layer of pyroclastic materials, up to several tens of centimetres thick in coastal areas and much thicker upslope towards the summit craters. Immediate resumption of living on Manam was untenable and the island was almost deserted. The volcanological facility and equipment at Warisi were destroyed by what seems to have been the cold edge of a pyroclastic flow. This included loss of a satellite telephone, which had been provided by Qantas Airways to provide a direct telecommunications link between Manam and Qantas headquarters in Australia. Those people who were still at Warisi managed to escape by a small boat, which, however, was struck by a falling rock and sank. They were injured and reached the shore. One person who had remained near Warisi was killed. RVO headquarters in Rabaul had been struck by lightning before Christmas 2004, thereby limiting the effectiveness of outside telecommunications and adding to the problem of responding effectively to the major disaster on Manam.

Monitoring the Manam eruption in 2004–2005 from space was as important as it had been during the earlier Pago eruption, and especially from an aviation-safety standpoint. In particular, meteorological staff of the Volcanic Ash Advisory Centre at Darwin in northern Australia, which is part of the International Airways Volcano Watch, paid special attention to the heights of the eruption plumes from Manam. There were several occasions when heights exceeded 10 kilometres — that is, a common flight level for international aircraft overflying the area. The highest detected ash column from Manam was for the eruption of 27–28 January, which reached well into the stratosphere to an estimated height of 21–24 kilometres.[23] Detecting such ash plumes is not straightforward in this region of tropical cloudiness and the reports from in-flight pilots and from volcanological observatories on the ground are important, if not critical. There were in this instance no aircraft/ash-cloud encounters of any significance, at least in part because of these cooperative measures.

News of the January 2005 eruption on Manam was not dealt with as fully by national newspapers as it had been in October–November 2004. A contributing

---

22   Tupper et al. (2007) and Smithsonian Institution (2005a).
23   The major eruption of 27–28 January 2005 appears to represent the climax of the eruptive period at Manam. The eruption climax at Manam in 1958 took place on 25 January — a similar time in the same month as in 2005. This raises again the possibility of a 'seasonal' relationship to earth tides for some eruptions at Manam, as suggested by Taylor (1960).

factor for this may have been the diversion of media attention towards the Indian Ocean or 'Boxing Day' tsunami of 2004 in Indonesia, its destructive impact throughout the northern Indian Ocean, and its huge death of toll of more than 220,000. The exact number of fatalities at Manam in 2004–2005 is unclear, but in contrast to the tsunami death toll, it was very few — almost certainly less than ten, judging by the content of newspaper articles. Appeals for funding support for the evacuated Manams had already started, but those appeals by early 2005 included calls for additional support for in-need people and communities in neighbouring Indonesia. Nevertheless, the national newspapers played a crucial role in the following years in drawing attention to the plight of the thousands of Manam evacuees still in mainland care centres at Potsdam, Asuramba and Mangem.

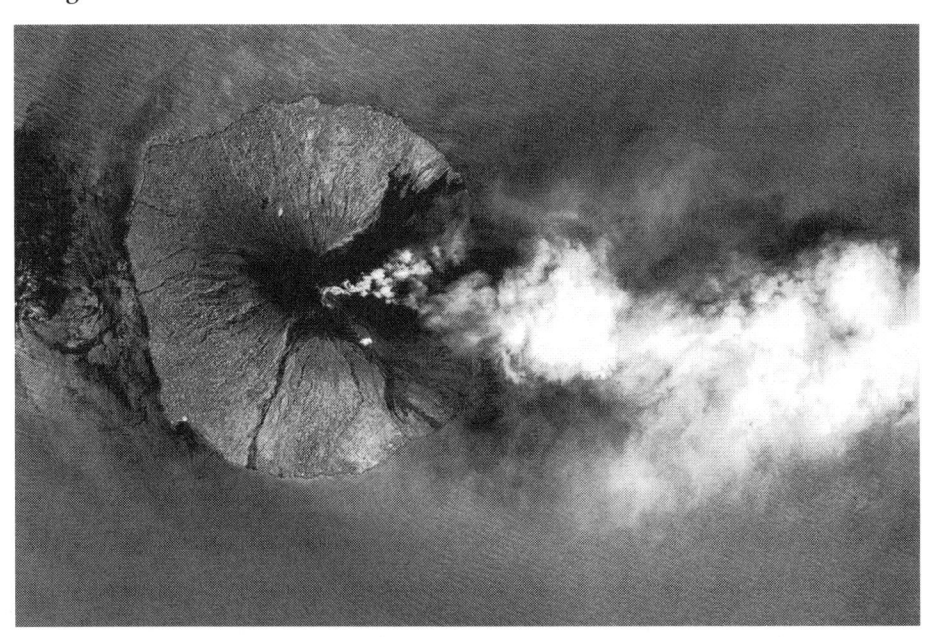

Figure 123. Manam volcano was imaged on 8 November 2004 from a SPOT (Système Pour l'Observation de la Terre) satellite. Key features are the fan-like area of devastation on the eastern flank of the volcano, another area of pyroclastic fallout on the north-western flank reaching to the coast, and vapour emission from the summit. The east-west width of the island through the summit is about 10 kilometres.

Source: Image supplied by P. L. Shearman, University of Papua New Guinea. Remote Sensing Centre.

Tensions between mainland landowners and the Manams reached crisis point in the face of growing pressure on land for gardening for essential foods, and as serious issues of law and order developed. Eruptions continued on Manam, including another powerful one in February 2006, and four people died on 13

March 2008 when mudflows engulfed their homes at Dangala village on the north-north-eastern coast of the island. Permanent resettlement of the Manams seemed to be the most sensible option, if land could be identified on the mainland and the necessary funding found. A Manam Restoration Authority, which was rather similar in concept to the Gazelle Restoration Authority in Rabaul, was established. Furthermore, a draft plan by a Manam Resettlement Taskforce, set up specifically to explore the relocation options, was being developed by the PNG National Disaster Centre in late 2010.[24] Hundreds of Manams have returned to their respective villages on the island where, despite ongoing volcanic activity, including pyroclastic flows, they feel more secure and independent compared with the stressful condition of their lives as displaced people on the mainland.[25] The population on Manam Island today changes daily because villagers travel frequently across the strait between the island and the mainland care centres.

How permanent instrumental monitoring can be continued on Manam volcano remains problematic and is unresolved. The site of the Warisi Observatory, which was destroyed by the January 2005 eruption, had to be abandoned. Furthermore, there has been no resolution to the land-access and compensation issues at the inland Tabele Observatory, which was built at considerable cost in 1964, together with a cellar for instruments. The buildings at the old observatory have, in any case, disappeared as a result of scavenging for building materials. Some instrumental monitoring is achieved today through a single seismograph installed in a room at the mission and school at Tabele on the coast and by using the radiotelemetry link to Bogia on the mainland and then to Rabaul. This situation is not satisfactory for such a high-risk volcano as Manam, which remains a threat to people on the island and to international aviation.

Rabaul 2006 and other Volcanic Crises in 1999–2007

An explosive eruption at Tavurvur on 7 October 2006 was the largest at Rabaul since the 1994 eruption.[26] It produced ash and dust that obscured visibility around the entire Rabaul Harbour area, destroyed vegetation that had been attempting to recover since 1994, thus causing increased run-off and siltation, which affected areas — such as at Ranguna where there had been clearing for new settlements — and covering roads with silt. This 2006 eruption also gave rise to a dramatic umbrella-shaped plume that rose about 18 kilometres into the

---

24 Manam Resettlement Taskforce (2010).
25 Mulina et al. (2011).
26 Smithsonian Institution (2006b).

atmosphere. The wide extent of the plume, as seen from space, was reminiscent of the Vulcan plume in 1994, and there were other similarities, too — including the production of Vulcan-like pumice which fell hot near to Tavurvur, forming a lightly cemented deposit or 'agglutinate'. Another significant period of both strombolian activity and explosive eruptions was in May to December 2008, when the cumulative ash thicknesses in Rabaul grew to be substantial enough to significantly impede day-to-day life in town.[27]

The eruptions and social impacts at Ulawun, Pago, Manam and Rabaul volcanoes were serious enough during 2000–2008, demanding the attention of RVO, provincial and national disaster managers, and international development assistance agencies. There were, however, other threatening situations at about the same time that added to the disaster management work load. The early to middle part of the first decade of the twenty-first century was, in fact, crowded with volcanic disaster management challenges for RVO, and these were having to be dealt with in a governmental context far more complex than had existed during, say, the eruption time-cluster periods of 1951–1957 and 1972–1975. A list can be compiled of seven of these other crises and false alarms.

Kasu, Enga Province, 1999. The crater wall of Kasu, a small tephra cone and crater lake, collapsed into the lake on 20 April 1999, producing a 15-metre-high wave which broke over the crater rim killing a boy, injuring 11 others, destroying two houses, and flattening vegetation.[28]

Lamington, Oro Province, 2002. Rumours circulated in Popondetta town during the first three weeks of April 2002 of renewed volcanic activity at Lamington volcano, including reports of fire, smoke, dead vegetation, noises from the volcano, and people becoming dizzy and dogs fainting in the summit area. Some schools had closed for fear of an eruption. Subsequent investigations by RVO could not confirm any renewed activity at the volcano.[29]

Garbuna, West New Britain Province, 2005. An unexpected eruption and related earthquake activity took place at Garbuna on 17–18 October 2005 within the extensive geothermal area at the summit of the volcano. Mud-like ash was distributed around two vents and a 3–4-kilometre-high ash cloud spread to the north-west. The eruption was the first recorded historical eruption from the volcano. There was concern that the eruption, which seems to have been a geothermal blowout or phreatic eruption, could be a precursor to a larger eruption of new magma, and a major deployment of monitoring equipment was undertaken by RVO in partnership with a USGS-VDAP team. There were, however, no confirmed magmatic eruptions, although vapour emissions and

---

27 Smithsonian Institution (2008b).
28 Wagner et al. (2003).
29 Itikarai & Bosco (2002) and Smithsonian Institution (2003).

earthquake activity continued for some months and suspicion remained that new magma may indeed have been emplaced beneath Garbuna, but not actually erupted in any significant quantities.[30]

Sulu Range, West New Britain Province, 2006. An intense swarm of felt volcanic-type earthquakes began near the volcanoes of the Sulu Range on 6 July 2006. People in nearby villages began self-evacuating, supported soon after by provincial authorities who established temporary care centres for more than 1,000 evacuees. Geyser activity at a large and nearby geothermal field also became pronounced. No eruption took place, however. Magma is thought to have intruded into an active tectonic fault, but did not reach the surface — another example of a volcanic false alarm or so-called 'failed eruption'.[31]

Near Likuruanga, West New Britain Province, 2006. A boy died by asphyxiation on 21 September 2006 at the coastal logging camp of Bakada north of Likuruanga, an extinct volcano. The boy had entered a deep hole — left empty by the logging company, which had been constructing latrines — in order to retrieve his dog and was overcome by inhalation of carbon-dioxide gas of uncertain origin. There were no other signs of potential volcanic unrest and the gas could have been of biogenic origin.[32]

Ritter, Morobe Province, 2006–2007. Small earthquakes followed by vapour emissions and diffuse ash clouds were reported for 17 October 2006. Occasional rocks slides were seen from the avalanche amphitheatre wall at Ritter Island, and an ash plume was observed from space. Further activity was reported for 19 May 2007 when sea surges destroyed a boat and four houses on nearby Umboi Island following an eruption, and 1,500–2,000 people from two coastal villages moved to higher ground.[33]

Karkar, Madang Province, 2007–2008. Signs of unrest were reported at Bagiai cone within the caldera at Karkar Island between early November 2007 and early March 2008, when increased vapour emissions and vegetation dieback were observed on the cone. Earthquakes were recorded using portable seismometers on several visits by RVO staff to the island during this time.[34] No eruption has since taken place.

Are any of these eruptions at Rabaul and at other volcanoes in the Bismarck Volcanic Arc during the 2000–2009 decade — including the so-called 'failed

---

30   Smithsonian Institution (2005b).
31   Smithsonian Institution (2006a), Taranu et al. (2007) and Wicks et al. (2007).
32   Smithsonian Institution (2006c).
33   Smithsonian Institution (2007). An eruption probably also took place at Ritter in 1997 (Saunders & Kuduon, 2009).
34   Smithsonian Institution (2008a). Karkar had been selected as one of the high-risk volcanoes for permanent monitoring during the AusAID VSS Project but the equipment had been vandalised repeatedly, evidently because of local land disputes, and so was not available for use during this period of increased unrest.

14. Eruptions of the Early Twenty-first Century: 1998–2008

eruption' at Sulu Range and the mysterious gas emission near Likuruanga — part of an eruption 'time cluster'? Significant eruptions and periods of volcano restlessness in fact took place at as many as eight, possibly nine of these volcanoes — Likuruanga is the possible exception — within only four years between August 2002 and October 2006 and, accordingly, there is a hint at some sort of time-clustering. The period expands to seven years, and the number of volcanoes to 11, if the Ulawun eruptions of 2000–2001 and the unrest at Karkar in 2007–2008 are added to the list.

Figure 124. Tavurvur volcano, Rabaul, is seen here in explosive eruption at 9.57 am on 7 October 2006. The photograph was taken from the south-east at Takubar, near Kokopo. The nearly vertical western side of the eruption column can be seen clearly on the left, together with heavy ash falls over the sea to the right.

Source: J. McLean, Port Moresby.

The proposed 2002–2006 time cluster is reminiscent of the period, 30 years previously, when six volcanoes in the Bismarck Volcanic Arc were in eruption between 1972 and 1975.[35] An intriguing difference, however, is that all of the six volcanoes, except Ulawun, in 1972–1975 are in the *western* sector of the Bismarck Volcanic Arc, whereas six of the nine restless volcanoes in 2002–2006 are in the *eastern* sector of the arc. Each sector of the Bismarck Volcanic Arc corresponds to a different plate boundary, which raises the possibility that

---

35 Cooke et al. (1976).

different tectonic characteristics control not only differences in the types of magma being formed and in the patterns of volcanic centres,[36] but also the timing of multiple eruptions in each sector.

Table 8. Nine Bismarck Volcanic Arc Volcanoes in Eruption or Restless between August 2002 and October 2006

Pago, 2002-3	First eruptive period since 1914-33
Manam, 2004-5 (and 2006)	Possibly the largest recorded eruption in historical time.
Garbuna, October 2005	First recorded eruption (phreatic) in historical time.
Sulu Range, July 2006	Inferred magma-intrusion event but no eruption.
Likuruanga, September 2006	Emission of carbon dioxide of uncertain volcanic origin.
Rabaul, October 2006	Largest explosive eruption at Tavurvur since 1994 activity.
Ritter, October 2006	First reported eruptions since 1997
Langila, ongoing	Eruptions of its long-lived period of activity continued.
Ulawun, ongoing	Some mild ash eruptions following the large eruption in 2000.

The eruption time clusters of 1951–1957 and 1972–1975 may be related to series of major tectonic earthquakes that preceded them. So too, apparently, was the 2002–2006 eruption cluster. A remarkable sequence of tectonic earthquakes took place in the New Britain and New Ireland area in late 2000.[37] A magnitude 6.8 earthquake along the New Britain submarine trench on 29 October 2000 started the sequence. This was followed on 16 November by a magnitude 8.2 earthquake along the Weitin Fault in southern New Ireland, producing large ground-surface displacements and tsunamis, and followed in turn on 16 and 17 November by two more large earthquakes along the New Britain trench. The earthquake pattern was like a seismic chain reaction of related releases of tectonic stress.

---

36 Johnson (1977).
37 Anton & McKee (2005) and Park & Mori (2007).

References

Anton, L. & C.O. McKee, 2005. *The Great Earthquake of 16 November 2000 and Associated Seismo-Tectonic Events Near the Pacific–Solomon–South Bismarck Triple Junction in Papua New Guinea*. Geological Survey of Papua New Guinea Report 2005/1.

Araho, N., R. Torrence & J.P. White, 2002. 'Valuable and Useful: Mid-Holocene Stemmed Obsidian Artefacts from West New Britain, Papua New Guinea', *Proceedings of the Prehistoric Society*, 68, pp. 61–81.

Barberi, F., R. Blong, S. de la Cruz Reyna, M. Hall, K. Kamo, P. Mothes, C. Newhall, D. Peterson, R. Punongbayan, G. Sigvaldason & N. Zana, 1990. 'Reducing Volcanic Disasters in the 1990s', IAVCEI Task Group for the International Decade for Natural Disaster Reduction (IDNDR), *Bulletin of the Volcanological Society of Japan*, 35, no. 1, pp. 80–95.

Blake, D.H., 1976. 'Pumiceous Pyroclastic Deposits of Witori Volcano, New Britain, Papua New Guinea', in R.W. Johnson (ed.), *Volcanism in Australasia*. Elsevier, Amsterdam, pp. 191–200.

Blake, D.H. & P. Bleeker, 1970. 'Volcanoes of the Cape Hoskins Area, New Britain, Territory of Papua and New Guinea', *Bulletin Volcanologique*, 34, pp. 385–405.

Blake, D.H. & I. McDougall, 1973. 'Ages of the Cape Hoskins Volcanoes, New Britain, Papua New Guinea', *Journal of the Geological Society of Australia*, 20, pp. 199–204.

Boyd, W.E., C.J. Lentfer & J. Parr, 2005. 'Interactions between Human Activity, Volcanic Eruptions and Vegetation during the Holocene at Garua and Numundo, West New Britain, PNG', *Quaternary Research*, 64, pp. 384–98.

Cashman, K.V. & G. Giordana (eds), 2008. 'Volcanoes and Human History', *Journal of Volcanology and Geothermal Resources*, 176, no. 3 (special edn), pp. 325–437.

Cooke, R.J.S., 1981. 'Eruptions at Pago Volcano, 1911–1933', in R.W. Johnson (ed.), *Cooke-Ravian Volume of Volcanological Papers*. Geological Survey of Papua New Guinea Memoir, 10, pp. 135–46.

Cooke, R.J.S., C.O. McKee, V.F. Dent & D.A. Wallace, 1976. 'Striking Sequence of Volcanic Eruptions in the Bismark Volcanic Arc, Papua New Guinea, in 1972–75', in R.W. Johnson (ed.), *Volcanism in Australasia*. Elsevier, Amsterdam, pp. 141–72.

Cronin, S.J., M.G. Petterson, P.W. Taylor & R. Biliki, 2004. 'Maximising Multi-stakeholder Participation in Government and Community Volcanic Hazard Management Programs; A Case Study from Savo, Solomon Islands', *Natural Hazards*, 33, pp. 105–36.

Davies, H., 2002. 'Tsunamis and the Coastal Communities of Papua New Guinea', in R. Torrence & J. Grattan (eds), *Natural Disasters and Cultural Change*. Routledge, London, pp. 28–42.

——, 2003. 'Mt Pago Awakes From its Slumber', Natural Disasters 2002, *Papua New Guinea Yearbook 2003*, pp. 12–14.

East/West New Britain Provincial Governments, 2010. 'Mt. Ulawun — Integrated Emergency Response Plan'. Unpublished report, East and West New Britain Provincial Governments.

Fullagar, R., G. Summerhayes, B. Ivuyo & J. Specht, 1991. 'Obsidian Sources at Mopir, West New Britain Province, Papua New Guinea', *Archaeology in Oceania*, 26, pp. 110–14.

Itikarai, I., 2005. Travelling to Manam. Letter to S. Nancarrow, 29 January 2005, Rabaul Volcanological Observatory, File A8002(OAR)[A]/A1008.

Itikarai, I. & J. Bosco, 2002. 'Investigations at Lamington Volcano, Popondetta 21st – 25th April 2002'. Unpublished report, Rabaul Volcanological Observatory.

Johnson, R.W., 1977. 'Distribution and Major-Element Chemistry of Late Cainozoic Volcanoes at the Southern Margin of the Bismarck Sea, Papua New Guinea', Bureau of Mineral Resources, Canberra, Report 188.

Johnson, R.W., R.P. Macnab, R.J. Arculus, R.J. Ryburn, R.J.S. Cooke & B. W. Chappell, 1983. 'Bamus Volcano, Papua New Guinea: Dormant Neighbour of Ulawun, and Magnesian-Andesite Locality', *Geologische Rundschau*, 72, pp. 207–37.

Macatol, I., V. Golpak & K. Jimbade, 2002. 'Needs Assessment Report 1: Mount Pago Volcanic Eruption, West New Britain Province, Papua New Guinea'. Unpublished report, West New Britain Provincial Administration.

Machida, H., R.J. Blong, J. Specht, H. Moriwaki, R. Torrence, Y. Hayakawa, B. Talai, D. Lolok & C.F. Pain, 1996. 'Holocene Explosive Eruptions of Witori and Dakataua Caldera Volcanoes in West New Britain, Papua New Guinea', *Quaternary International*, pp. 34–36, 65–78.

Manam Resettlement Taskforce, 2010. *Manam Resettlement Program Plan*. National Disaster Centre, Port Moresby.

McKee, C.O. & J. Kuduon, 2005. *Assessment of Volcanic Hazards at Pago-Witori Volcano, West New Britain, Papua New Guinea*. Papua New Guinea Department of Mineral Policy and Geohazards Management, Report 2005/3.

McKee, C.O., V.E. Neall & R. Torrence, 2011. 'A Remarkable Pulse of Large-scale Volcanism on New Britain Island, Papua New Guinea', *Bulletin of Volcanology*, 73, pp. 27–37.

Mulina, K., J. Sukua & H. Tibong, 2011. *Madang Volcanic Hazards Awareness 06th September — 29th September 2011*. Department of Mineral Policy and Geohazards Management Open File Report 2011/01.

Neall, V.E., R.C. Wallace & R. Torrence, 2008. 'The Volcanic Environment for 40,000 Years of Human Occupation on the Willaumez Isthmus, West New Britain, Papua New Guinea', *Journal of Volcanology and Geothermal Research*, 176, no. 3 (special edn), pp. 330–43.

Park, S.-C. & J. Mori, 2007. 'Triggering of Earthquakes during the 2000 Papua New Guinea Earthquake Sequence', *Journal of Geophysical Research*, 112, B03302, doi: 10.1029/2006JB004480.

Petrie, C.A. & R. Torrence, 2008. 'Assessing the Effects of Volcanic Disasters on Human Settlement in the Willaumez Peninsula, Papua New Guinea: A Beysian Approach to Radiocarbon Calibration', *The Holocene*, 18, pp. 729–44.

Post-Courier, 2002. 'West New Britain's Mt Pago in Focus: A *Post-Courier* Supplement', Thursday 31 October 2002, pp. 13–20.

Saunders, S.J. & J. Kuduon, 2009. *The June 2009 Investigation of Ritter Volcano, with a Brief Discussion on its Current Nature*. Papua New Guinea Department of Mineral Policy and Geohazards Management, Open File Report 003/2009.

Silver, E., S. Day, S. Ward, G. Hoffmann, P. Llanes, N. Driscoll, B. Applegate & S. Saunders, 2009. 'Volcano Collapse and Tsunami Generation in the Bismarck Volcanic Arc, Papua New Guinea', *Journal of Volcanology and Geothermal Research*, 186, pp. 210–22.

Smithsonian Institution, 1992. 'Manam', *Bulletin of the Global Volcanism Network*, 17 (10, 11).

——, 1996. 'Manam', *Bulletin of the Global Volcanism Network*, 21 (12).

——, 1999–2000. 'Ulawun', *Bulletin of the Global Volcanism Network*, 24 (10, 12); 25 (3, 7, 8, 11).

——, 2001. 'Ulawun', *Bulletin of the Global Volcanism Network*, 26 (5, 6).

———, 2002–2006. 'Pago', *Bulletin of the Global Volcanism Network*, 27 (7, 8, 9, 12); 28 (1, 3, 9, 12); 29 (2, 4, 7); 30 (7, 9); 31 (2).

———, 2003. 'Lamington', *Bulletin of the Global Volcanism Network*, 28 (1).

———, 2004–2005. 'Manam', *Bulletin of the Global Volcanism Network*, 29 (10, 11), 30 (2).

———, 2005a. 'Manam', *Bulletin of the Global Volcanism Network*, 30 (2).

———, 2005b. 'Garbuna', *Bulletin of the Global Volcanism Network*, 30 (11).

———, 2006a. 'Sulu Range', *Bulletin of the Global Volcanism Network*, 31 (7, 9).

———, 2006b. 'Rabaul', *Bulletin of the Global Volcanism Network*, 31 (9).

———, 2006c. 'Likuruanga', *Bulletin of the Global Volcanism Network*, 31 (10).

———, 2007. 'Ritter', *Bulletin of the Global Volcanism Network*, 32 (3,5).

———, 2008a. 'Karkar', *Bulletin of the Global Volcanism Network*, 33 (3).

———, 2008b. 'Rabaul', *Bulletin of the Global Volcanism Network*, 33 (11).

Specht. J., 1981. 'Obsidian Sources at Talasea, West New Britain, Papua New Guinea', *Journal of the Polynesian Society*, 90, pp. 337–56.

———, 2005. 'Obsidian Stemmed Tools in New Britain: Aspects of their Role and Value in Mid-Holocene Papua New Guinea', in I. Macfarlane, R. Paton & M. Mountain (eds), *Many Exchanges: Archaeology, History, Community and the Work of Isabel McBryde*. The Australian National University, Aboriginal History Monograph, 1, pp. 357–72.

Summerhayes, G.R., 2009. 'Obsidian Network Patterns in Melanesia — Sources, Characterisation and Distribution', *Bulletin of the Indo-Pacific Prehistory Association*, 29, pp. 109–23.

Synolakis, C.E., J.-P. Bardet, J.C. Borrero, H.L Davies, E.A. Okal, E.A. Silver, S. Sweet & D.R. Tappin, 2002. 'The Slump Origin of the 1998 Papua New Guinea Tsunami', *Proceedings of the Royal Society of London*, A458, pp. 763–89.

Taranu, F., C. Collins, L. Miller, K. Mulina, R. White, A. Lockhart & P. Cummins, 2007. 'The 2006 Earthquake Swarm in the Sulu Range, Central New Britain, Papua New Guinea', poster presented at the 2007 Australian Earthquake Engineering Conference, Sydney.

Taylor, G.A., 1960. *An Experiment in Volcanic Prediction*, Bureau of Mineral Resources, Canberra, Record 1960/74.

Torrence, R. & T. Doelman, 2007. 'Chaos and Selection in Catastrophic Environments: Willaumez Peninsula, Papua New Guinea', in J. Gratton & R. Torrence (eds), *Living Under the Shadow: The Cultural Impacts of Volcanic Eruptions*. Left Coast Press, Walnut Creek, California, pp. 42–66.

Torrence, R., V. Neall, T. Doelman, E. Rhodes, C. McKee, H. Davies, R. Bonetti, A. Guglielmetti, A. Manzoni, M. Oddone, J. Parr & C. Wallace, 2004. 'Pleistocene Colonisation of the Bismarck Archipelago: New Evidence from West New Britain', *Archaeology in Oceania*, 39, pp. 101–30.

Torrence, R., V. Neall & W.E. Boyd, 2009. 'Volcanism and Historical Ecology on the Willaumez Peninsula, Papua New Guinea', *Pacific Science*, 63, pp. 507–35.

Torrence, R., C. Pavlides, P. Jackson & J. Webb, 2000. 'Volcanic Disasters and Cultural Discontinuities in Holocene Time, in West New Britain, Papua New Guinea', in W.G. McGuire, D.R. Griffiths, P.L. Hancock & I.S. Stewart (eds), *The Archaeology of Geological Catastrophes*. Geological Society of London, Special Publications, pp. 171, 225–44.

Tupper, A., I. Itikarai, M. Richards, F. Prata, S. Carn & D. Rosenfield, 2007. 'Facing the Challenges of the International Airways Volcano Watch: The 2004/05 Eruptions of Manam, Papua New Guinea', *Weather and Forecasting*, American Meteorological Society, 22, pp. 175–91.

Ulawun Workshop Report, 1998. *Volcanic Cone Collapses and Tsunamis: Issues for Emergency Management in the Southwest Pacific Region*. IAVCEI Workshop on Ulawun Decade Volcano, Papua New Guinea. International Decade for Natural Disaster Reduction, Emergency Management Australia, Canberra.

Wagner, T.P., C.O. McKee, J. Kuduon & R. Kombua, 2003. 'Landslide-Induced Wave in a Small Volcanic Lake: Kasu Tephra Cone, Papua New Guinea', *International Journal of Earth Sciences*, 92, pp. 405–06.

Wicks, C, R. White, H. Patia, C. Collins, W. Johnson, S. Saunders & H. Yarai, 2007. 'Surface Deformation from ALOS Interferometry Related to the July 2006 Seismic Crisis and Dike Intrusion on Central New Britain Island, Papua New Guinea', *Transactions of the American Geophysical Union*, 88, no. 5, AGU Fall Meeting, Supplement, Abstract G51C-0621.

——, 2008. 'ALOS/PALSAR Interferometry of Surface Deformation in the Sulu Volcanic Range and Lava Thickness at Pago Volcano in Central New Britain Island, Papua New Guinea', *Transactions of the American Geophysical Union*, 89, no. 23, AGU Western Pacific Geophysics Meeting, Supplement, Abstract V44A-06.

15. Reassessing Volcanic Risk in the North-eastern Gazelle Peninsula: 2000–2012

> There is nothing more certain in the disaster management business than the fact that once a disaster starts to unfold, it is too late to start looking for the information needed to manage it.
>
> K.J. Granger (2000)

Restructuring a Society

Rabaul town celebrated its centenary in 2010. The 100 years that have passed since German colonists established the town on Simpson Harbour in 1910, after draining the mangrove swamps, is brief in the context of archaeological and geological time frames. Yet the story is intimately interwoven with, and is a recurrent theme of, the larger history of volcanic disasters in Near Oceania, because of the drama of its high-risk location. The extraordinary history of Rabaul town is, of course, punctuated not only by the relatively small eruptions of 1937–1943 and the ongoing eruptive activity that started in 1994, but by military invasions by Australia in 1914 during the First World War and by Japan in 1942 during the Second World War. Thus, navies, too, have appreciated the strategic benefits of Simpson Harbour. Nevertheless, the volcanoes, their historical eruptions, and the threat of larger eruptions and even the catastrophic formation of a new caldera, form the major part of contemporary risk perceptions. The town's brief history, accordingly, is significant for the present day because of the tension that has been recorded between, on the one hand, the needs of economic development based on fertile volcanic soils and ideal rainfall, coupled with the export wharfs bordering the superb breached-caldera harbour at Blanche Bay and, on the other hand, an ongoing requirement for community safety and the need to reduce disaster risk from devastating volcanic eruptions in the future. A modern history of Rabaul and its environs is, therefore, a narrative of volcanic risk management, and the dramatic tension is an ongoing question of what level of risk is acceptable.

The Rabaul–Kokopo area, by the year 2000, had already undergone great socio-economic changes since the 1994 eruption. This included, especially, the withdrawal of many people from the Blanche Bay area to new settlements outside of the active caldera area, despite the nostalgia for and strong sentimental attachments to 'old Rabaul'. Urban development of the Kokopo, Baliora, Kenebot and Takubar areas was undertaken successfully. Furthermore, there has been, and continues to be, a determination by modern-day leaders in East New Britain Province to reduce volcanic risk in the north-eastern Gazelle Peninsula through planned resettlement in parallel with sensible policies for economic development. This ongoing commitment was reflected in the 2003 Provincial Development Plan and its 'growth centre' policy of encouraging development at four main centres in the north-eastern Gazelle — Kokopo, Kerevat, Vunakanau and Kurakakaul — but not at Rabaul or, indeed, anywhere within the caldera area of Blanche Bay, including Matupit Island. The ongoing eruptions at Tavurvur assisted this process of social change as they were, and are, a constant reminder that the decision to focus development away from Rabaul is a sensible one. Nevertheless, Rabaul itself clearly remains a volcanically vulnerable place. Furthermore, the challenges of economic growth in East New Britain Province are set against a difficult national backdrop of population growth, poor employment opportunities, and a wide range of developmental issues affecting the whole of Papua New Guinea — such as poverty, inadequate health services, law and order, education and governance issues. These challenges have continued up to the present day in East New Britain, even as the Gazelle Restoration Authority (GRA) closes down at the end of the formal, post-disaster, restoration and relocation program.

The major socio-economic restructuring in the rural areas of the north-eastern Gazelle has not all gone to plan, in part because of the absence of a thorough disaster-preparedness plan of land acquisition supported by rigorous socio-economic modelling in the years immediately after the 1983–1985 seismo-deformational crises.[1] Ian Scales, for example, in his report on roads in the post-eruption economic landscape of the province concluded that 'The bulk of resettlement has ultimately been unofficial, unplanned and unfacilitated … [and the] formal resettlement areas, their roads and other infrastructure are carrying much lower population than planned'.[2] Problems were compounded by some delays in post-disaster settlement construction and by the reluctance of some communities to resettle in the prescribed areas. Furthermore, upkeep of the existing road networks has not been possible, and the stretch of the economically important Kokopo–Rabaul trunk road, west of Vulcan, still remains unsealed. Economic development in recent years has also been affected adversely by much reduced cocoa harvests that were caused by a pod-borer infestation.

Many people after the 1994 eruption made their own resettlement arrangements by moving informally to the peri-urban area of Kokopo, or buying smallholdings from Baining people and other landowners in rural areas, thus contributing to the development of a settlement belt which covers the outermost parts of

---

[1] Neumann (1996) and Scales (2010).
[2] Scales (2010), p. 65.

the best agricultural land in the Rabaul area. Furthermore, some people in the early years, for any one of several reasons, moved back from the new, still poorly serviced, inland settlement areas to their original villages near Rabaul — including at Matupit Island, for example, where several hundred villagers still prefer to live on their traditional coastal land adjacent to Tavurvur, enduring living conditions that are substandard compared to other parts of the province.

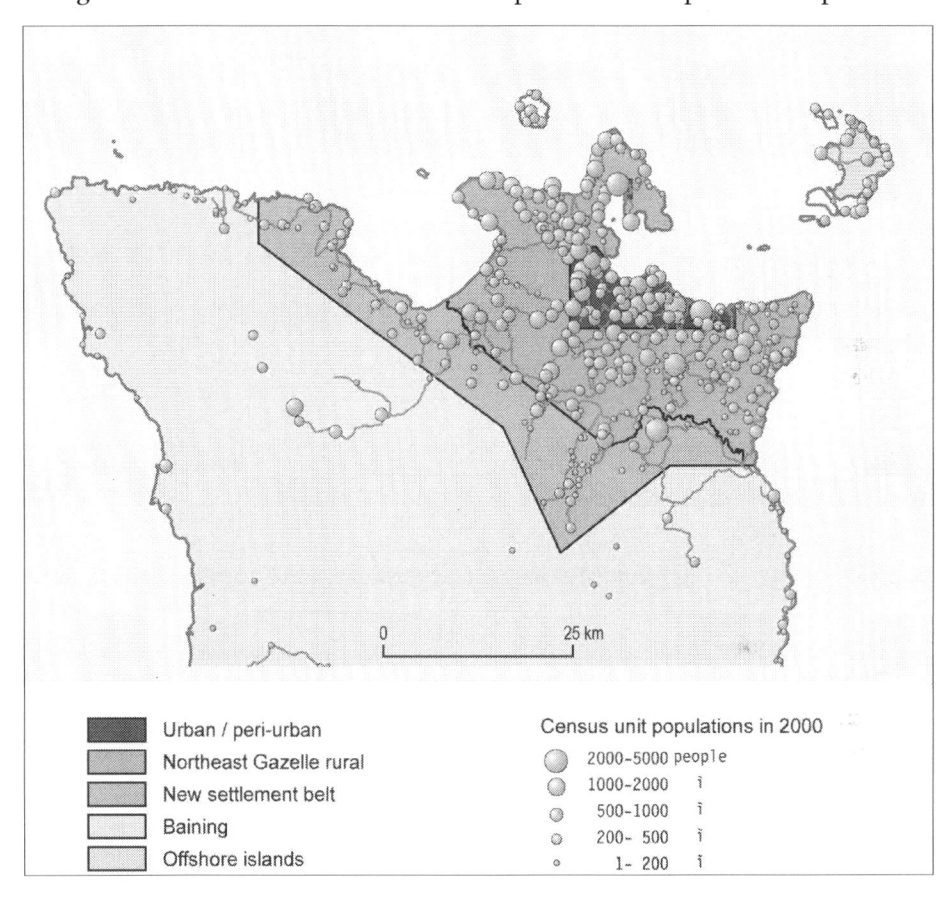

Figure 125. Economic zones, including the new settlement belt together with population distribution in 2000, are here mapped for the northern part of East New Britain Province.

Source: Provided courtesy of Ian Scales who adapted Figure 4 in his original account (Scales, 2010).

The 1994 eruption triggered a demographic shift to places away from the dangerous, volcanically active caldera area of Blanche Bay, but what is the current level of risk from natural hazards for the redistributed population? Are people today, together with their settlements and infrastructure, any safer? Sector 1 of Rabaul town, its economically crucial wharf on Simpson Harbour, and the Kokopo trunk road have survived, but what is the ongoing risk to their viability within the active caldera area? Answering these questions means that the 'risk formula' from the 1980s has had to be kept in mind — perhaps even more so than it was before the 1994 eruption. Formulaic 'risk' is a product of the following three factors and is 'zero' if any one factor is zero:

- The hazards (e.g. magnitudes, frequencies, early warning precursors)
- Elements of a society exposed to risk (e.g., people, homes, lifelines, etc.)
- The total of the separate vulnerabilities of each element.

This formula serves as a framework for important questions for the future. What is the range of the volcanic hazards that can now be identified and which threaten the communities? Is there an improved understanding of how the volcanic systems at Rabaul 'work' in a geophysical sense? Can volcanic risk be reduced through improved early warnings by the Rabaul Volcanological Observatory (RVO) of impending volcanic eruptions? Can community vulnerability be assessed more rigorously? And, overall, what is the current understanding of natural hazard risk, including volcanic risk, in the north-eastern Gazelle Peninsula, and can it be quantified?

Volcanic Hazards

A good deal is now known about the wide range of volcanic hazards in the north-eastern Gazelle Peninsula, based on the experiences of the 1878, 1937–1943, and ongoing 1994 eruptions at Vulcan and Tavurvur.[3] These eruptions are, however, generally regarded as only 'small' to 'moderately large' in volume and scale — that is, mainly 1–3 on the Volcanic Explosivity Index (VEI) scale, although the current eruptive period at Rabaul eventually may well classify up to VEI 4.[4] Different volcanic hazards, irrespective of the size of an eruption, affect the vulnerable Rabaul-Kokopo area differently, and therefore need to be treated separately where considering risk.

The main hazards from the small, intra-caldera volcanoes in Blanche Bay are from explosive eruptions. The main threat to life is within one or two kilometres of the active vents, from small pyroclastic flows and surges and from the impact of ballistic lava blocks and bombs flung out on parabolic trajectories. The most damaging hazard, even from small–moderate explosive eruptions, is from the widespread fall of ash from clouds driven over settlements and agricultural lands

---

3   McKee et al. (1985), Blong & Aislabie (1988), Blong & McKee (1995) and McKee et al. (2012).
4   Siebert et al. (2010) tentatively classified the combined Vulcan and Tavurvur eruptions of both 1937 and 1994 as VEI '4?' (note the question mark) or 'large'. There is some debate amongst volcanologists about the appropriateness of adjectives such as 'large' for VEI 4 eruptions. These are considerably smaller than those having VEIs of 6–8, corresponding to tephra volumes with orders of magnitude smaller than for VEI 6–8 eruptions.

by prevailing winds. Ash clouds may also produce deadly lightning strikes, such as are believed to have caused one death at Ralalar village, south of Vulcan, in 1994. Clouds and lightning can be a threat to aviation and airport operations, including Tokua Airport, during north-west monsoon seasons, but even at great distances, to which ash may drift from the volcano. Ash falling on buildings may produce roof collapses that can kill any entrapped victims. A combination of heavy ash fall and pyroclastic flows and surges was probably responsible for the 500 deaths at Vulcan in 1937. Explosive eruptions also produce atmospheric shock waves that may rattle or break window glass.

Figure 126. Ash from Tavurvur falls on Rabaul and clogs the air on 6 September 2008. Piles of ash from previous eruptions, and which have been removed from the road, are seen in front of the Cathedral of St Francis Xavier on the left, but winds re-suspend the dry ash.

Source: J. McLean, Port Moresby.

Mudflows and floods contemporaneous with eruptive periods, and caused by heavy rain, can have serious consequences to property and, potentially, to life. Some Blanche Bay eruptions can entrain seawater and contribute to the severe run-off and resultant problems of rapid surface erosion. Other run-off and erosional problems are caused after eruptions, when normal rainfall degrades areas of vegetation stripped earlier during the eruptions, or through clearing by people for new, post-eruption, settlement areas outside of the caldera area. Mudflows and floods are still generated on the steep walls of the calderas in Blanche Bay, especially during the north-western monsoon. They cause problems in Rabaul town and periodically cut the Rabaul–Kokopo trunk road. In contrast, lava flows from Tavurvur during the ongoing 1994 eruption are localised, slow-moving, and are not nearly as much a threat to property as they are, for example, in the radial valleys on Manam Island.

A range of volcanic gases can be hazardous, even during non-eruption times when a volcano may be quietly degassing. Some gases are a severe irritant to eyes and the respiratory tract. Carbon dioxide is an asphyxiant in low-lying areas, such as at Tavurvur in 1990, when six people suffocated in gas-filled depressions. Sulphur dioxide and halogen gases can form aerosols and acid rain, which attack exposed metals and kill vegetation. Long-term exposure to the ingestion of fine ash particles may produce respiratory problems. Significant ingestion of silica minerals, such as crystobalite, could lead, in the longer term, to silicosis, although this has not been confirmed as a significant problem during and since the 1994 eruption. The long-term persistence of fine ash in the atmosphere, including ash re-suspended by winds or by traffic on busy roads, can also be psychologically wearing on the spirit of a community. Dust finds its way into homes, businesses, gutters, vehicles and the fabric of clothes. It enters ears, nostrils and eyes. Wearing goggles and wide-brimmed hats can be advantageous at such times.

Other volcanic hazards include tsunamis, which are formed where pyroclastic flows crash onto the waters of Blanche Bay, as in 1994, for example, or where part of a volcano collapses into water generating a debris avalanche and then a tsunami, such as happened on the northern flank of Tavurvur in 2006 when the rock avalanche ploughed into the waters of Greet Harbour. Pumice floating on water as rafts can be dangerous if the pumice surface is thought by pedestrians to be firm; or damaging to ships, if they are navigated by captains who do not recognise the abrasive properties of pumice on painted hulls; or to boats, when their water-cooling systems ingest pumice.

Evidence for past eruptions at Rabaul that were larger than those witnessed since 1878, has been found as a result of extensive geological fieldwork and study of the old volcanic deposits around Blanche Bay.[5] Such fieldwork is not without its challenges as exposure of the deposits and rocks is generally inferior because of the cover of tropical vegetation, and making stratigraphic correlations over large distances between one part of the area and another can be extremely difficult. Nevertheless, a general tephra-stratigraphy has been established, supported by isotopic dating on a range of samples. One important result is the identification of material from prehistoric explosive eruptions of

---

5   Nairn et al. (1995). C.O. McKee and RVO staff in recent years have undertaken extensive geological fieldwork and their largely unpublished results supplement and extend those provided by Nairn et al. (1995).

'intermediate' scale, corresponding to VEI values of 4 or 5.[6] These may not have been large enough to have been accompanied by formation of a caldera in every case, or at all, but nevertheless are thought to have been sourced within Blanche Bay and to be of a scale that, were they to recur today, would affect large tracts of landscape outside of the Blanche Bay area, as well as within, and likely would cause hazards and disasters much greater than anything experienced historically.

Evidence for even larger, catastrophic, VEI–6 eruptions is also found at Rabaul. The youngest and best-exposed deposits of these large-volume eruptions are the so-called 'Rabaul Pyroclastics', dated by the radiocarbon method and correlated to a calendar date of AD 720–750 ±20 years.[7] The Rabaul Pyroclastics have an estimated volume of 11 cubic kilometres and consist of thick, pumiceous, airfall deposits together with overlying ignimbrites laid down by pyroclastic flows and which can be traced up to 50 kilometres from their source in Blanche Bay, including over water to Watom Island. Kokopo is built on the deposits of the eruption, which most probably devastated the whole of the north-eastern Gazelle Peninsula and presumably made the area uninhabitable for quite some time. The eruption is thought to have been accompanied by a caldera-forming event, probably together with tsunamis. Whether this habitation 'vacuum' in the Rabaul area presented a subsequent opportunity for the ancestors of the present-day Tolai, who are thought to have come from New Ireland, to reoccupy a recovering and fertile landscape is unknown.

A key challenge concerning future volcanic risk in the north-eastern Gazelle Peninsula is determination of the frequency of occurrence of eruptions of different sizes at Rabaul. A general rule in applied volcanology is that larger eruptions are separated by longer time intervals than are smaller ones — or, in other words, eruption frequency decreases as eruption size increases.[8] This relationship could apply to Rabaul. The historical, intra-caldera, VEI 1–3 eruptive periods at Rabaul are separated roughly by decades of inactivity, whereas there may have been, according to one estimate, at least five and possibly nine 'significant ignimbrite eruptions' — say, VEI 6 — possibly together with accompanying caldera-forming events, over the last 20,000 years or so, thus giving a rough recurrence rate of between 2,000 and 3,600 years.[9] Intermediate-size and possibly non-caldera-forming eruptions presumably are somewhere between these two limits.

---

[6] McKee & De Saint Ours (1998). Latter & Hurst (1987) suggested that intermediate-scale eruptions may pose a greater 'annually apportioned risk' in the Rabaul–Kokopo area than do larger eruptions. This risk value is obtained by first estimating how many people are at risk for each size category of eruption, VEI 2–6, and then dividing these population numbers by the interval between eruptions of the same size category. Latter & Hurst considered that the larger eruptions at Rabaul were too infrequent to provide higher annually apportioned risk values, even though the field evidence for these frequencies was unknown at the time.

[7] Walker et al. (1981), Nairn et al. (1995) and C.O. McKee (personal communication, 2011).

[8] See, for example, Siebert et al. (2010).

[9] Nairn et al. (1995).

The question that arises from such considerations of hazard types and eruption sizes and frequencies at Rabaul is not only whether eruptions can be forecast effectively, but whether their sizes and durations can be forecast too. In other words, can risk be reduced by effective early warnings, which are sufficiently accurate and early enough to permit effective evacuations of an appropriate scale? A starting point in attempting to answer such questions is to consider what is now known about the sub-surface geophysical structure at Rabaul, the location of magma reservoirs, how magmas reach the surface, and what kind of useful signals they provide as eruption precursors.

Early Warnings and a New Model for Rabaul Volcano

Eruptions at Tavurvur have continued up to 2013 — that is, 19 years after the start of the volcanic activity at Rabaul in 1994. Serious attempts over this time had been made by RVO scientists and their colleagues to reach a general consensus on how Rabaul volcano may 'work' in a geophysical sense. Professional volcanological work at Rabaul began in 1939 with publication of the report on the 1937 eruption by N.H. Fisher, while Ima Itikarai completed a seismological study of Rabaul for a postgraduate thesis in 2008. The years between have seen major contributions by dozens of other geoscientists based both in Papua New Guinea and abroad.

A workshop was held at the Rabaul Hotel in November 2009 in order to discuss the results of these 70 years of information gathering, and a report was published in 2010.[10] An overall conclusion of the workshop was that the feature known as 'Rabaul volcano' was best understood not as a single volcano but rather by the concept of *three* volcanic systems working together. These three are: (1) the Blanche Bay area, where there is evidence for the presence of several calderas of different ages; (2) the line of stratovolcanoes from Watom Island in the north-west to Turagunan volcano in the south-east; and (3) the largely submarine Tavui Caldera to the north-east. Thus, the old idea of 'Rabaul volcano' once being a towering ancestral mountain having smaller stratovolcanoes as satellites on its flanks, and which disappeared to form a single caldera at Blanche Bay, has been abandoned.

Another important aspect of the favoured model is the existence of two adjacent magma reservoirs about 4–5 kilometres deep beneath the Rabaul area. These reservoirs were identified as a result of the mapping of low seismic velocities, known as 'low-velocity anomalies' (LVA), during the RELACS geophysical

---

10 Johnson et al. (2010).

survey in 1996–1997. One reservoir is the Rabuana LVA, which is thought to underlie the north-eastern coast on St Georges Channel. Rising basalt magmas erupt from this magma reservoir, which together with its precursors in more distant times, have formed the volcanoes of the Watom–Turagunan Zone. Some of the basalt in the present-day reservoir is thought to have been injected south-westwards into a second magma reservoir beneath Rabaul Caldera, its lateral passage probably marked by the occurrence of the north-east earthquakes. This second reservoir corresponds to the Harbour LVA where the basaltic magma mixes and mingles with resident dacitic magma beneath Rabaul Harbour. The 'mixed' magmas then inject the lower parts of ring faults beneath Blanche Bay. This causes earthquakes which form the seismic annulus, as well as uplift at places such as Matupit Island. Such caldera-wide geophysical unrest is followed eventually by near-synchronous eruptions at Tavurvur and Vulcan.

This process is thought be a long-lived one, such that the reservoir beneath Blanche Bay has been replenished repeatedly with new magma over recent geological time. Furthermore, a key aspect of the sequence of events is that there are opportunities for RVO to track changes instrumentally at different stages and to assign new levels of volcanic alert. Thus, the north-east earthquakes can be detected and mapped by seismometers as the first stage of the progression of events. Formation of the seismic annulus, or geodetically measured uplift at Matupit and Vulcan, or both, can be detected, mapped, and interpreted as magma 'on the move' after mixing beneath the Greet Harbour area. Similarly, any measured increases in temperatures and changes to the chemistry of fumaroles in the Greet Harbour area, including at Rabalanakaia and Tavurvur, can be taken as a trigger for declaring a higher level of alert following magma-mixing beneath the area. But can provision of alerts be accomplished any better than they were in 1994?

The experience of the build-up to the 1994 eruption is that magma mixing, accompanied by magma intrusion up the ring faults of the seismic annulus, may have taken place over many years without leading to an immediate eruption. The magma system underlying Rabaul can remain inflated and unstable for a long time — somewhat analogous to a filled balloon or bladder waiting to burst. Thus, the 1994 eruption at Rabaul seems to have had 13 years of build-up time between 1971 and 1984, and then a further ten years of relative quiescence before an eruption finally took place 23 years later, after only 27 hours of immediate warning. RVO may have to affirm, and Rabaul people may have to accept, that 27 hours is the only time they may have in the future if eruptions similar to those in 1937 and 1994 happen again. Rabaul people will have to accept, too, that RVO will continue to struggle in understanding the workings of Rabaul volcano. RVO deals with scientific uncertainty, caldera complexity, and only generalised eruption forecasting, and not with precise prediction of geophysical events.

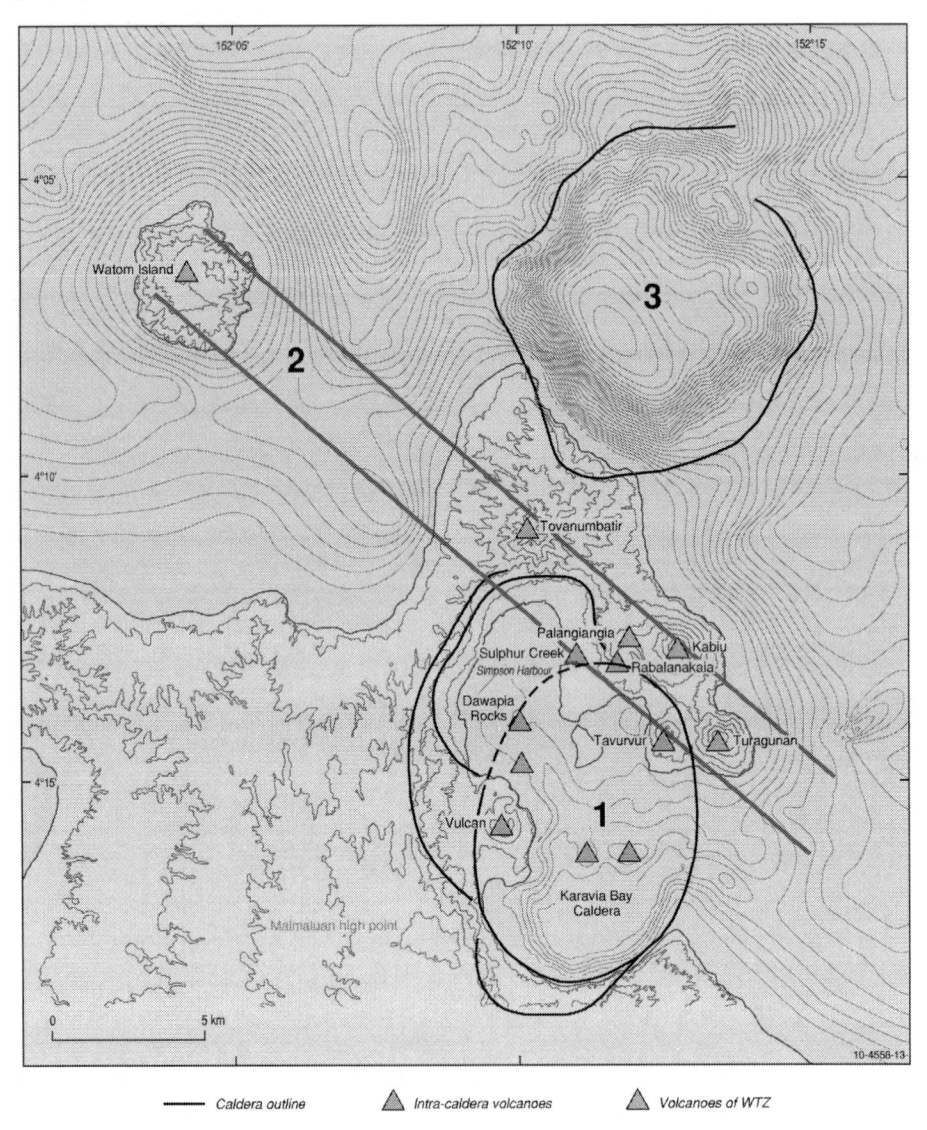

Figure 127. The three volcanic systems of Rabaul volcano are (1) the cluster of 'nested' caldera escarpments in the south, (2) the Watom-Turagunan Zone in the middle, and (3) Tavui Caldera in the north-east.

Source: Johnson et al. (2010, Figure 13).

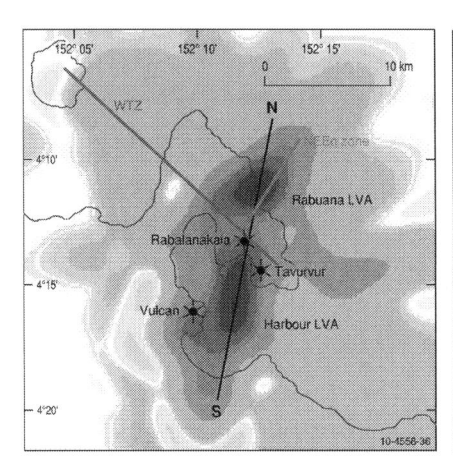

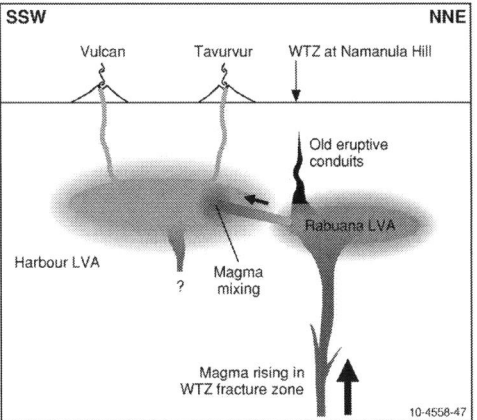

Figure 128. Two large reservoirs that are thought to contain magma are shown in the left-hand map of a horizontal slice, about five kilometres deep, beneath the Rabaul area. The different colours from black outwards to light green refer to rocks having increasing seismic velocities. Rabalanakaia volcano is near where the two 'low-velocity anomalies' or magma reservoirs connect, and where the trend of the north-east-earthquake zone or NEEq intersects the Watom to Turagunan Zone, or WTZ, of stratovolcanoes. The line N–S refers to the vertical cross section shown in the right-hand cartoon, where new magma from the more northerly reservoir, Rabuana, is injected into older magma of the Harbour reservoir, causing mixing or mingling of the different magma types. The magma then moves up the fractured seismic annulus and erupts from Vulcan and Tavurvur volcanoes.

Source: Itikarai (2008, adapted from Figure 6.3 by Johnson et al., 2010, Figures 36 and 47).

One remaining question is whether the 1994 eruption is typical of Rabaul eruptions. Eruptions are continuing at Tavurvur 19 years after the 1994 activity, whereas eruptive activity following the 1937 eruption was restricted only to 1941–1943. The current pattern is different, and so extreme care must be taken in assuming that the lessons learnt from 1994, and indeed from 1878 and 1937–1943, will necessarily apply in the future. This point, too, will have to be taken into consideration by those who continue to live in the Rabaul area and who may still retain the desire of restoring the town to what it was before 1994.

Determining Risk Today

More attention since the year 2000 has been focused both internationally and in the Rabaul–Kokopo area on defining the exposure and vulnerability factors of the 'risk formula' — especially the susceptibility to hazard impacts of populations, settlements, housing, commercial buildings, power supplies, critical infrastructure and transport facilities, including roads, ports and airports. Furthermore, since 2000, work that is aimed at reducing risks arising from natural hazards has involved a general trend towards undertaking more formal risk assessments — a trend that, in large part, was stimulated by the risk-reduction focus of the United Nation's International Decade for Natural Disaster Reduction (IDNDR) in the 1990s. The direct result of such assessments is evidence-based identification of the highest risk areas, which should lead in turn to key decisions about new economic development strategies and community safety. This may, in the case of East New Britain Province, result in further migration of people and investment to safer areas, but ultimately will require a determination of what risks are acceptable and why. East New Britain, like any at-risk society, will determine its own 'risk equilibrium' based on the best possible information and analysis.

Risk assessment of natural hazards, for many years before 1994, had been a general part of the work of the insurance and reinsurance industries in determining the costs of premiums, but now there is wider interest in the topic amongst members of the disaster management community, both locally and internationally. Two of the risk studies that were undertaken at Rabaul during the 1980s were commissioned by companies of the Insurance Underwriters' Association of Papua New Guinea,[11] thus providing the industry with some basis for reassessment of insurance premiums. The extent to which these results, and those of the other hazard and risk studies of the 1980s, had a direct impact on strategies for governmental policies on economic development and natural-hazard risk-reduction is not clear, but there is a possibility that results were largely forgotten, or at least put to one side, in the ten years after the 1983–1985 crisis when the expectation of an eruption declined.

The first attempt to address natural-hazard risk in East New Britain Province after the 1994 eruption was in 2001–2002, when the provincial administration, the GRA, and the Australian Agency for International Development (AusAID) collaborated in a multi-hazard risk assessment of the province. A trained disaster risk assessment adviser, Isolde Macatol, was contracted to join a team led by Levi Mano, advisor for the administration's Planning and Research Division. Macatol compiled and edited a final report, which included benchmark papers on individual hazards by RVO staff: volcanic hazards (Jonathan Kuduon), earthquakes (Itikarai), landslides (Felix Taranu), and tsunamis (Kila Mulina). Other natural hazards considered were active faults and floods and, to a lesser extent, land subsidence, sea level rise, coastal erosion and drought. The report also included development by Macatol of a 'hazard and vulnerability index' for

---

11 Latter & Hurst (1987) and Blong & Aislabie (1988).

the province, as well as a range of risk-reduction recommendations.[12] In addition, discussion papers on different aspects of risk — economic, social, political and cultural — arising from extreme natural hazard events were prepared by officers from the provincial administration and GRA.[13] A third report on community 'risk perceptions' was planned, but was not developed to production stage.

Figure 129. The artificial lights in this night-time photograph of strombolian activity at Tavurvur volcano, as seen from RVO on 16 April 2008, serve as a proxy to illustrate the closeness of settlements to the active volcanic centres of the Blanche Bay area. Rabaul town and the wharf area extend along the length of the foreground. The Rabaul Yacht Club and its marina, together with the Travelodge Hotel, are in the middle distance in front of Tavurvur. Kokopo town is in the far distance to the right of Tavurvur and outside Blanche Bay.

Source: S. V. Hohl, Freie Universitat Berlin.

The 'hazard and vulnerability index' for East New Britain Province consisted of 119 computer-drafted maps for a range of natural hazards that were classified

---
12 Macatol (2002).
13 Macatol (no date). A photocopy of this second report was kindly provided by Levi Mano, East New Britain Province Administration.

into groups of different severities, and shown in relation to settlement locations. These maps were supplemented by 24 tables of village population data based on the 2000 census, together with related statistics. Risk assessment was, therefore, focused largely on estimating the numbers of people living the different hazard zones. A significant conclusion was that settlements in the province as a whole were threatened by mainly geological hazards: volcanic eruptions, earthquakes, tsunamis, landslides and related flooding, and rapid shoreline loss. One other principal conclusion was that 85 per cent of people still living in the Rabaul district in 2000 — including Rabaul town and Matupit Island — occupied zones of high to very high risk. All of this population-based risk work contributed substantially to the information base required for ongoing risk-reduction and development strategies in East New Britain.[14]

Risk perceptions in any multicultural society, such as that found in the north-eastern Gazelle Peninsula, are wide-ranging, commonly subjective, even conflicting, and not always based on the best factual evidence. A major, but complex, challenge in disaster management, then, is to provide an objective assessment of risk based on the integration and analysis of observable, but multifaceted, facts about hazard, exposure and vulnerability. An important historical development in this regard, which was evident in Australia and elsewhere for some years up to 2000, was the use of spatially referenced, or 'georeferenced', digital information in computer-based geographic information systems (GIS). The origin of GIS can be traced back to the 1960s and 1970s, when computers began to be used for cartographic purposes, replacing pen and ink, but the full power of GIS began to be recognised when spatial data from field land-resource surveys, geophysics and remote sensing from aircraft and satellites, were married with traditional computer-assisted cartography.[15]

GIS, in the context of disaster management, represents an attempt to use the spatial 'overlays' as digital maps to quantify and then analyse multiple natural hazard risks. K.J. Granger undertook GIS research work for crisis-management purposes at Rabaul immediately after the 1983–1985 seismo-deformational crisis[16] and subsequently, with others, he was at the forefront of this type of research in Australia. This included promotion of the concept of national information infrastructures for disaster-management purposes in Pacific islands countries.[17] Granger, in 1997, introduced the expression 'Risk-GIS' as a way of focusing attention on the benefits of GIS as an essential tool for disaster risk-reduction analysis.[18] Modern field collection of spatially referenced information for GIS usage is facilitated by handheld computers that are pre-programmed

---

14 L. Mano (personal communication, 2011).
15 See, for example, Burrough (1986).
16 Granger (1988).
17 Granger (2000).
18 See, for example, Granger (1997).

for exposure and damage assessments, and equipped with GPS, camera, and preloaded imagery and building addresses. Similar equipment can be mounted on vehicles for data collection throughout urban areas at large-city scale.

Georeferenced spatial data sets for land-use, population, and natural resource purposes, including from East New Britain, had already been captured in a digital inventory called the Papua New Guinea Resource Information System (PNGRIS) as a result of mapping methodologies pioneered by CSIRO in the 1970s for land-use and agricultural purposes.[19] New spatial-data sets for East New Britain have also been obtained more recently as a result of, for example, a provincial GIS-based transport study.[20] Furthermore, and importantly, spatial data for the province has been compiled by the University of Papua New Guinea in a digital 'Geobook' for the province, supported by the National Economic and Fiscal Commission.[21] Similarly, the World Bank and Asia Development Bank recently funded a Pacific Disaster Risk Assessment of 15 countries, including Papua New Guinea, which included the collection of some georeferenced data for the urban areas of Rabaul and Kokopo.[22]

East New Britain is now in the fortunate position of having a comprehensive collection of such spatial sets for use in disaster management and economic development. These are being exploited currently by a new, but low budget, risk-assessment project that is supported by the provincial administration and AusAID.[23] The results are potentially important because they may well demonstrate the applicability of such a GIS-based methodology as a means of reducing volcanic and other sudden-impact, natural hazard risks in other parts of Papua New Guinea and the Solomon Islands. Data management and analysis is at least as important for volcanic risk reduction as is the development of effective, instrumentally based, warning systems.

References

Blong, R. & C. Aislabie, 1988. *The Impact of Volcanic Hazards at Rabaul, Papua New Guinea*. Institute of National Affairs Discussion Paper 33, Port Moresby.

Blong, R.J. & C.O. McKee, 1995. *The Rabaul Eruption 1994: Destruction of a Town*. Natural Hazards Research Centre, Macquarie University, Sydney.

---

19 See, for example, Bryan & Shearman (2008).
20 Scales (2010).
21 University of Papua New Guinea Remote Sensing Unit (2010).
22 See, for example, SOPAC (2010).
23 This new activity is is entitled 'Strengthening Natural Hazard Risk Assessment Capacity in Papua New Guinea'. It is scheduled to be completed in 2013–2014.

Bryan, J.E. & P.L. Shearman (comps), 2008. *Papua New Guinea Resource Information System Handbook*. 3rd edn. University of Papua New Guinea, Port Moresby.

Burrough, P.A., 1986. *Principles of Geographical Information Systems for Land Resources Assessment*. Oxford University Press, New York.

Granger, K.J., 1988. 'The Rabaul Volcanoes: An Application of Geographic Information Systems to Crisis Management'. Master of Arts thesis, The Australian National University, Canberra.

——, 1997. 'Risk-GIS: More than Just a Tool for Disaster Impact Mitigation', in L. Searle & W. Sayers (eds), *Northern Australia Remote Sensing and Geographic Information System Forum (NARGIS 97)*, no pagination. Paper presented in Forum Proceedings, Cairns.

——, 2000. 'An Information Infrastructure for Disaster Management in Pacific Island Countries', *Australian Journal of Emergency Management*, Autumn, pp. 20–32.

Itikarai, I., 2008. 'The 3-D Structure and Earthquake Locations at Rabaul Caldera, Papua New Guinea'. Master of Philosophy thesis, The Australian National University, Canberra.

Johnson, R.W., I. Itikarai, H. Patia & C.O. McKee, 2010. *Volcanic Systems of the Northeastern Gazelle Peninsula, Papua New Guinea: Synopsis, Evaluation, and a Model for Rabaul Volcano*. Rabaul Volcano Workshop Report. Papua New Guinea Department of Mineral Policy and Geohazards Management and the Australian Agency for International Development, Port Moresby.

Latter, J.H. & A.W. Hurst, 1987. *An Assessment of Volcanic, Seismic, and Tsunami Hazard at Rabaul and Neighbouring Areas of New Britain, Papua New Guinea*. New Zealand Department of Scientific and Industrial Research, Wellington, Contract Report 25.

Macatol, I.C. (ed.), 2002. *Hazard Assessment. Disaster Risk Assessment Report 1*. Gazelle Restoration Authority, East New Britain Provincial Administration, and AusAID PNG Advisory Support Facility Project.

——, (ed.), no date (about 2002). 'Vulnerability Assessment Report for East New Britain Province'. Gazelle Restoration Authority, East New Britain Provincial Administration, and AusAID.

McKee, C.O., & P. De Saint-Ours, 1998. 'How Dangerous is Rabaul Volcano?'. Proceedings of the Institution of Engineers of Papua New Guinea Conference, Rabaul, East New Britain Province, 24–27 September.

McKee, C.O., J. Kuduon & H. Patia, 2012. 'Recent Eruption History at Rabaul: A Report Prepared for the Gazelle Restoration Authority'. Rabaul Volcanological Observatory.

McKee, C.O., R.W. Johnson, P.L. Lowenstein, S.J. Riley, R.J. Blong, P. de Saint Ours & B. Talai, 1985. 'Rabaul Caldera, Papua New Guinea: Volcanic Hazards, Surveillance, and Eruption Contingency Planning', *Journal of Volcanology and Geothermal Research*, 23, pp. 195–237.

Nairn, I.A., C.O. McKee, B. Talai & C.P. Wood, 1995. 'Geology and Eruptive History of the Rabaul Caldera Area', *Journal of Volcanology and Geothermal Research*, 69, pp. 255–84.

Neumann, K., 1996. *Rabaul Yu Swit Moa Yet: Surviving the 1994 Volcanic Eruption*. Oxford University Press.

Scales, I., 2010. *Roads in Gazelle Peninsula Development: Impact of Roads in the Post-eruption Economic Landscape of East New Britain*. Australian Agency for International Development, Canberra.

Siebert, L., T. Simkin & P. Kimberley, 2010. *Volcanoes of the World: A Regional Directory, Gazetteer, and Chronology of Volcanism During the Last 10,000 Years*. Smithsonian Institution, Washington D.C., and University of California Press, Berkeley.

SOPAC, 2010. *Pacific Exposure Database: Asia Development Bank Regional Partnerships for Climate Change Adaptation and Disaster Preparedness*. SOPAC Pacific Islands Applied Geoscience Commission, Joint Contribution Report 21, in collaboration with New Zealand Geological and Nuclear Sciences (GNS), the World Bank, and the Pacific Disaster Center.

University of Papua New Guinea Remote Sensing Unit, 2010. *East New Britain Geobook*. In association with the National and Economic Fiscal Commission, Port Moresby.

Walker, G.P.L., R.F. Heming, T.J. Sprod & H.R. Walker, 1981. 'Latest Major Eruptions of Rabaul Volcano', in R.W. Johnson (ed.), *Cooke Ravian Volume of Volcanological Papers*. Geological Survey of Papua New Guinea Memoir, 10, pp. 181–93.

16. Historical Analysis and Volcanic Disaster-Risk Reduction

Science does what it must; humans comprehend what they can.

J. Byrne (2012)

Patterns in the Historical Record

Documented histories are complex time series, punctuated by major events that commonly define turning points. This history of volcanic eruptions and disaster management is no different. Historical trends also have impetus and trajectories, which define future challenges and even, on the basis of lessons learnt from history, ways to deal with them.

Slicing seamless narratives into time sectors hardly acknowledges the continuities of histories and the long-lived interdependencies of the many factors defining them, but here there is value in drawing together some key threads, first, in recognising five periods or phases which, spliced end-to-end, define a framework of recorded volcanic crises and disasters in Near Oceania.

The first, longest, and least complete phase is the 330 years between 1545, when Spanish voyagers probably noticed volcanoes in Near Oceania, through to about 1875, when Europeans began to settle permanently in volcanically active areas, most notably in St Georges Channel and in the Rabaul area. This phase encompasses the European Scientific Revolution, the Enlightenment of the 'long eighteenth century', and the Industrial Revolution. Near Oceania was still remote to Europeans and few written records of extended observations of volcanoes or their eruptive activity were produced. One notable exception is the account left by William Dampier of his voyage through the islands north and north-east of New Guinea Island in 1700 AD. An especially large volcanic eruption at Long Island during this phase created a 'time of darkness' across much of mainland New Guinea, the memory of which is still retained in traditional stories, but the eruption was not recorded in written history by actual eyewitnesses.

German, British and Australian colonialism dominated the second phase between 1876 and 1941. Colonial headquarters were established by the Germans at Rabaul in 1910 on the foreshores of Simpson Harbour in Blanche Bay, where damaging eruptions at both Vulcan and Tavurvur had taken place at Rabaul in 1878. These eruptions did not present a particular concern for the German, or later, Australian authorities, bearing in mind their common recognition of the considerable economic advantages of the superb natural harbour formed by the sea-breached calderas of Blanche Bay. A repeat of the 'double eruption' phenomena at Tavurvur and Vulcan in 1937, however, changed the attitudes of the Australian authorities to the dangers of volcanic eruptions when about 500 people were killed and ash fell heavily on Rabaul, causing little damage but leading to temporary evacuation of the town. The future of Rabaul as a government administrative centre was reviewed after this unexpected eruption. Volcanologist C.E. Stehn wrote that a transfer to a safer place was unnecessary if a well-equipped volcanological observatory were established in order to provide science-based warnings of eruptions. In sharp contrast, and in the same report, however, W.G. Woolnough stated the opposite — that, bearing in mind the uncertainties involved in eruption prediction '... the provision of elaborate warning systems should not be entertained.'[1] Stehn's view prevailed, however, and a volcanological observatory became operational in 1940, thus marking the start of an expectation or even promise of early warnings for eruptions at Rabaul. The 'promise' arguably represents the most important scientific turning point in this volcanological history and one that would be tested in future years, particularly in 1983–1985 and 1994.

The next ten years of the third phase, from 1942 to 1951 inclusive, were dramatic both volcanologically and in terms of world events. Australian operations at the new volcanological observatory were interrupted by the Second World War when, following earlier military actions against China and the United States, the Japanese invaded Rabaul in January 1942. The Japanese military established a volcanological observatory at Sulphur Creek to assist its naval operations in Rabaul Harbour. The building was destroyed by Allied bombing, however, and not until 1950 did Australia restart volcanological operations at Rabaul by appointing Australian volcanologist G.A.M. 'Tony' Taylor to the task of running the re-established facility on Observatory Ridge. Taylor soon became heavily involved in investigative work at Lamington volcano, Papua, following the large explosive eruption there in January 1951 when almost 3,000 people were killed. This is still the largest known death toll from the impact of a natural geological hazard recorded anywhere in Near Oceania. The Lamington tragedy secured the future of the volcanological observatory at Rabaul as a required service that, from this time on, covered active volcanism throughout the whole of the then Territory of Papua and New Guinea — a task much larger than the one defined for Rabaul volcano alone.

More than 30 years of the fourth phase between 1952 and 1985 represent a period of general growth for the observatory at Rabaul, especially with regard to the installation and operation of improved instrumental systems for volcano monitoring at Rabaul itself. Observatories with instrumental cellars were also

---

1   Stehn & Woolnough (1937), p. 157.

constructed at both Manam and Esa'ala in 1964, and were part of a local-observer network that also included Ulawun and Langila volcanoes. Authorities in the 1950s continued to be influenced by the Lamington disaster. They initiated or supported evacuations at Long Island in 1953, Bam in 1954, and Manam in 1957, and a government-supported evacuation also took place at Esa'ala in 1969. Papua New Guinea became an independent nation in 1975 and the Solomon Islands in 1978, thus taking over the responsibility for volcano monitoring from Australia and Britain, respectively. A major seismo-deformational crisis at Rabaul in 1983–1985 led to forecasts of an imminent eruption from intra-caldera vents, but none had taken place by the end of the crisis period.

Reduced general concern for the situation at Rabaul during the first eight years of the final period from 1986 to the present day, translated into a decline in instrumental monitoring capacity for the national volcanological service of Papua New Guinea centred on the Rabaul Volcanological Observatory (RVO). Adequate early warnings were not issued in the hours preceding the disastrous eruptions at Tavurvur and Vulcan, Rabaul, in September 1994, and evacuations there were initiated largely by the affected communities themselves. The governments of Australia and the United States after 1994–1995 began providing support to the Government of Papua New Guinea for the strengthening of the RVO through a series of international development-assistance projects.

*

Four volcanoes in New Oceania have been the most dangerous since the mid-1870s, based on the historical evidence of dated volcanic disasters. All of them are in Papua New Guinea: Rabaul in 1878, 1937 and 1994; Ritter in 1888; Lamington in 1951; and Manam on several occasions, but especially in 2004–2005. Other historically or potentially active volcanoes have caused justifiable concern for the safety and livelihoods of local communities. These include Karkar, Pago, Ulawun and, to a lesser but still significant extent, Kadovar, Bam, Long, Langila, Garbuna, Sulu Range, Tuluman, Bagana, Kavachi, Simbo, Goropu, Victory, Koranga and the volcanoes of the Dawson Strait area.

The precise number of active or potentially active volcanoes in New Oceania is unknown, and the total of 57 subaerial volcanoes mentioned at the beginning of this book and listed in the Smithsonian Institution database — whether the 'Holocene' attribution is certain or just possible — includes probable examples of volcanoes that will never be in eruption again. Nevertheless, many eruptive centres in Near Oceania are a threat, not only from the general type of eruptions seen at the historically active volcanoes of the region but also from larger-scale eruptions that have not been observed and recorded. Savo in the Solomon Islands, Bamus and Makalaia-Dakataua in New Britain, and Victory in Oro Province are just four examples of 'sleeper' volcanoes. All four are capable of producing eruptions larger than when each was last active in the nineteenth century. Geothermally active Yelia, in the highlands of mainland New Guinea, is an example of a type of 'sleeper' volcano that has not been active historically, but which may well break out in eruption again. Furthermore, the stories of a 'time of darkness' in the highlands of mainland New Guinea, caused by a VEI (volcanic explosivity index) 6 eruption at Long Island, apparently during the mid-seventeenth century, are a reminder that the mild eruptions witnessed historically at Long are not representative of much larger possible eruptions from such caldera-related volcanoes. Other, similar, caldera systems include Rabaul-Tavui, Lolobau, Galloseulo-Hargy, Pago-Witori, Makalaia-Dakataua, Garove, Unea, Umboi, Karkar and Loloru. Then there are high coastal volcanoes such as Ulawun, Bamus, Manam and Victory, which could produce major gravitational collapses and widespread tsunamis in the future, or else calderas, together with large accompanying eruptions.

Following, then, is an important conclusion. The many eruptions recorded since the mid-1870s in Near Oceania are all small — that is, VEIs of no more than 4 — compared to some recorded geologically but of a scale not yet witnessed in recent times. The threat from these much larger eruptions in the future is real. Furthermore, volcanic risk in Near Oceania is higher today than at any time since people first entered the region more than 40,000 years ago because of the much larger, sedentary populations, and the permanent agricultural lands and built environments on which they depend.

Another striking feature of the historical record is the eruption 'time clusters' of 1951–1957, 1972–1975, and 2002–2006. There are also hints of clusters in 1875 to 1878 and 1884 to 1899, but these are less clear and conceivably both could represent two parts of one longer period of enhanced eruptive activity. The starts and ends of all of these periods are not well defined, and there exists the possibility that these are simply statistical anomalies, rather than the clusters having any real geophysical significance. Nevertheless, volcanologists who were involved in the assessment of the 1951–1957 and 1972–1975 eruption time clusters proposed that these periods may have begun as a result of preceding tectonic-earthquake activity or 'regional geophysical unrest'. Furthermore, the remarkable earthquake sequence in the northern Solomon Sea in 2000 conceivably could have triggered the eruption cluster of 2002–2006. Perhaps the most practical lesson to be learnt from these eruption clusters is that they require extra work from volcanologists and that consideration should be given to requests for external, temporary assistance to offset the excessive workloads at these times.[2]

---

2   There is an intriguing corollary to these comments on eruption time clusters, which can be made with reference to the example of one particular volcano. Ulawun produced severe eruptions, first in January 1970 after a long period of relatively minor activity, and again in September 2000. Adding both of these

Evacuations, Early Warnings and False Alarms

This history contains 13 documented examples of at-risk communities evacuating either before or after the initial outbreak of a nearby volcanic eruption or because of the immediate threat of one. All 13 examples date from 1937. Evacuations of people from active volcanic areas undoubtedly took place before 1937 — for example, at Victory in the 1880s — but little if anything is known about them. Furthermore, 1937 is the year when a new, more scientific and instrumental approach to monitoring volcanoes was adopted in the hope of providing early warnings of eruptions. Each of the 13 documented examples has its own unique combination of factors, but the examples as a whole can be used to assess the degree of success of authorities, including volcanologists, in declaring evacuations that saved lives.

Table 9. Thirteen Evacuations in Papua New Guinea

Group A	Group B	Group C
Rabaul 1937	Esa'ala 1969	Long 1953
Goropu 1943	Rabaul 1984	Bam 1954
Lamington 1951	Rabaul 1994	Manam 1957
Pago 2002	Sulu Range 2006	Ulawun 2000
Manam 2004		

Nine of the 13 documented examples, shown loosely as Groups A and B in the table, are cases where evacuations took place without them first being declared by authorities, whether they were assisted by volcanologists or not. Group A evacuations are instances where the actual outbreak of an unexpected volcanic eruption triggered the movement of people away from a volcano to places of greater safety, in most cases immediately. Authorities in all five cases played no part in disaster-mitigation efforts before the outbreak of eruptions, but inevitably took a lead in post-disaster relief and recovery work amongst the displaced communities. The Lamington eruption of 21 January 1951 is a

---

eruptions to the respective eruption time clusters that followed them soon after in 1972–1975 and 2002–2006 respectively, expands each of the clusters by two years to 1970–1975 and 2000–2006, in which case Ulawun was the first of the volcanoes of each cluster to begin eruptive activity. Note, however, that both the January 1970 and September 2000 eruptions at Ulawun *preceded* the respective series of tectonic earthquakes that have been proposed as 'triggers' for the two eruption time clusters. Thus, either the timings of these two particular Ulawun eruptions are coincidental — that is, they are not linked at all to the time clusters and are therefore irrelevant to them — or else both eruptions are a hint that, in some cases, eruptions of a cluster can precede large tectonic-earthquake series. In other words, the eruptions and tectonic earthquakes are both simply expressions of the same period of 'regional geophysical unrest', irrespective of which comes first. Furthermore, some eruptions at Ulawun perhaps can be regarded as precursor volcanic signals for subsequent major, tectonic earthquake series. This last conclusion, given on the basis of current but limited evidence, is speculative, but the idea could be tested by future eruption/earthquake events.

standout example in this group of five. The deadly eruption there developed so quickly and savagely that few people within the devastated area could escape, although some apprehensive villagers did manage to self-evacuate before the catastrophe on that Sunday morning. The number of dead, almost 3,000, greatly exceeded the number of survivors who were from the destroyed settlements, who happened to be away from home at the time of the catastrophe, and who, with others from villages outside of the devastated area, moved further away to areas of refuge.

The phase of eruptions that started at Manam on 24 October 2004, also classified in Group A, eventually led in late November and December of that year to a slow evacuation from the island to care centres on the mainland. Limited data from a single seismograph at the makeshift observatory at Warisi on Manam in 2004–2005 was used by RVO to assess the general seismic condition of the volcano. The extent to which these data and the reports provided by RVO from Rabaul were strongly influential in authorities arranging for the eventual evacuation of the islanders is unclear, however, given the low-level nature of the available scientific information and the unreliability of the communication links between Rabaul and decision-makers in Madang and Port Moresby. The initial eruption of 24 October 2004 and the major eruption of 27–28 January 2005 were both unexpected.

The four examples in Group B are examples of at-risk communities evacuating because they were themselves sufficiently concerned about warning signs of an imminent eruption that they were not prepared to wait for authorities to announce an evacuation. Government authorities in these cases soon gave their support in one way or another to these 'self-evacuating' communities, even though Rabaul in 1994 is the only one of the four where an eruption actually broke out soon after the evacuation. The other three were in effect volcanic 'false alarms'.

The remaining four examples shown in Group C are of evacuations taking place as the result of early warnings and declarations being made by authorities with or without the recommendations of volcanologists. Evacuation at Long was undertaken without any prior input from volcanologists, whereas the successful evacuation of Manam Island in 1957 was a case where Taylor's recommendation to the authorities was based in large part on his forecast of major eruptions at Manam being expected at times of strong earth tides and after general tectonic unrest in the region, and based only to a limited extent, if any, on instrumental results.

Perhaps the most telling of the three Group C examples during the 1950s was the tragic case of Bam Island in 1954 when volcanologists recommended the permanent evacuation of the island because of the perceived high risk from future major eruptions. The Bam islanders were taken to care centres on the mainland in November 1954, and 24 of them died there, apparently mainly through disease. Mild explosive eruptions had been taking place from the summit crater of Bam, but no one in authority seems to have considered seriously the self-sufficiency of the Bams themselves in coping with the dangers of these eruptions.

The three volcanic alarms raised by the authorities and volcanologists in the 1950s were almost certainly an overreaction influenced by the shadow of the Lamington tragedy in 1951, and by the fear of similar tragedies taking place at Long, Bam and Manam, for which the authorities and volcanologists might be held responsible. Lamington itself was also a case where more lives might have been saved had the opinions of two Europeans in authority not been heeded by so many local people. There is, therefore, in all three cases a strong element of misplaced colonial paternalism — that white men had a greater understanding of such things compared to people in traditional societies. This seems to have been a common misperception at that time, and not just in Near Oceania, although later European research into natural disasters, social structures, and change in traditional communities soon demonstrated otherwise.[3]

Ulawun in 2000 is only one example in the total of 13 where instrumental results were important in providing timely early warnings of an impending volcanic eruption, thus contributing significantly to decisions being made regarding evacuation. Seismic activity recorded at Ulawun in September 2000 was used by RVO as justification for recommending an immediate evacuation. The final recommendation had to be made promptly by RVO and directly to the affected communities as the final seismic build-up took place rapidly over only a few hours, including during the hours of darkness immediately after sunset. Evacuations began on the night of 28–29 September, only a few hours before Ulawun broke out into full eruption. Other, equally important factors, however, that contributed to this success were: (1) a highly capable local volcano observer at Ulamona Mission and Sawmill, Martina Taumosi, with whom RVO volcanologists in Rabaul discussed by radio the instrumental results being received by her at Ulamona; (2) the high level of hazard awareness in the local community caused by an RVO-led awareness campaign that was undertaken only three weeks previously; (3) people in the community seeing the beginning of the eruption and so being motivated to evacuate promptly; and, (4) the willingness of local businesses to quickly provide road transport. Thus, instrumental early warning was not the only factor in ensuring a successful evacuation. The short-lived eruption at Ulawun in 2000 did not have any disastrous impacts, other than destruction of some gardens, but this does not mean that the evacuation was unnecessary. On the contrary, powerful eruptions like those in 2000 could

---

3   See, for example, Torry (1978).

easily be the trigger for future large-scale cone collapse on Ulawun, as well as even larger and more disastrous eruptions and possible tsunami impact over a wide area of the Bismarck Sea.

There are numerous examples in Near Oceania, including some not mentioned in this history, of what may loosely be called volcanic 'false alarms' — that is, cases where early warning signs on active volcanoes were *not* followed by damaging volcanic eruptions. Some 'early warning' signs on volcanoes have nothing to do with an impending eruption. Water vapour trails streaming off the summits of high conical volcanoes can be misinterpreted by aircraft pilots and others as being plumes of volcanic origin, whereas in fact they are a meteorological effect of strong upper winds on normal weather clouds. Landslides on volcanoes may be the result of heavy rainfall, or shaking from distant tectonic earthquakes, or both, and have no connection with the movements of magma beneath a volcano. Changes in the extent and behaviour of geothermal areas on volcanoes can be part of the normal evolution of shallow geothermal systems as hot geothermal fluids and mineralisation excavate new channels, rather than being indicative of emplacement and impending eruption of magma. One example of this is the local reports in February 1984 of the increased activity in the thermal areas bordering Dawson Strait area, Milne Bay Province, including coral-reef die off at Dobu Island. Intense swarms of felt, local earthquakes in volcanic areas may also cause concern, but these may not be of volcanic origin and will not necessarily signify subsequent eruptions.

The 'false alarm' at Sulu Range in 2006 can be highlighted because the scientific evidence pointed strongly to intrusion of magma beneath the volcanic area. People evacuated because of the threat of an eruption, but none has taken place — at least to date. Magmatic intrusion without eruption, such as at Sulu, can indeed be regarded as a *volcanic* false alarm, but there is a need for at-risk communities and supporting authorities to accept that such 'intrusive' events are not unusual beneath volcanoes and that they will not necessarily lead to a volcanic eruption. Another case in this context is Kadovar in 1976, when a newly enlarged thermal area gave cause for concern. Did this represent normal, non-threatening, geothermal activity finding a new way to the surface, or was it triggered by a magmatic intrusion beneath the volcano? The answer remains unknown.

Two unusual examples of false alarms are at Koranga volcano in 1967, when volcanic-like events in the crater probably represented the burning of reactive iron-pyrites exposed to the atmosphere; and, at Simbo volcano in 1993, when threatening black emission clouds turned out to be combustible sulphur set alight by a bushfire that had been apparently lit by tourists. Plans for evacuations were made in both of these instances, but the plans did not need to be implemented at the time. A false alarm at Lamington volcano in 2002 was caused by escalating rumours following reports of smoke and noises from the volcano, and people becoming dizzy and dogs fainting in the summit area, none of which could be confirmed by later investigations. Rumours can be founded on inconsequential or poorly understood phenomena, but may gain in strength unnecessarily through the powerful and rapid transmission of events by word of mouth. There is also the example of the 1983–1985 seismo-deformational crisis at Rabaul being recognised by some people for several years afterwards as a 'false alarm', whereas it can now be regarded as the precursor to damaging eruptions ten years later in 1994. The alarm in this case proved not to be 'false'.

This brief review of evacuations and false alarms since 1937 can be concluded by making two remarks. First, the management of events suspected by volcanologists to be of little consequence and which are expected to lead to false alarms, or just minor benign eruptions, is just as important as managing the lead-up to volcanic eruptions and disasters considered by volcanologists to be more likely and of larger impact. Both are times when open dialogue with affected communities is equally essential. Second, the self-sufficiency and resilience of at-risk communities on volcanoes in Near Oceania are important factors to identify in asking the question: what are the best approaches and methodologies that should be adopted for reducing volcanic risk in such areas? Strong, reactive communities in one sense challenge authorities to do better in deciding when to declare official evacuations, including the volcanologists who monitor volcanoes instrumentally and who aim to provide the best possible scientific advice to authorities. Strong communities will not necessarily wait for official advice from volcanologists who, understandably, may be cautious and restrained in issuing an alert. Rather, communities may respond spontaneously, promptly and confidently when the conditions warrant such action. The evacuations at Rabaul in September 1994 are an excellent example of this where people from Matupit Island said that their old people, who in 1937 had experienced the precursory signs of the eruptions 57 years earlier, urged the community to leave — and the community did so in timely fashion.

Artefacts and Oral Traditions

Adequate volcanic-disaster information from the historical record of observations in New Oceania covers less than 150 years, but are there other potential sources of information on volcanic disasters and on the ways that people have dealt with the threat of volcanic hazards and coped with their impacts? One such source is traditional archaeology, coupled with studies of historical ecology and changes in vegetation and foods that supported people. A strong partnership between archaeology and volcanology has existed ever since the nineteenth century excavations at Pompeii of the volcanic disaster in 79 AD.

This collaboration has included even the debated role of volcanic eruptions as the cause of the demise of some ancient civilisations. One well-known example is the caldera-forming, possibly VEI 7 eruption at Thera Island, Greece, in about 1610 BC, which resulted in the destruction of the Bronze Age city of Akrotiri on present-day Santorini and, conjecturally, determined the eventual end by volcanic-tsunami impact of the Minoan culture on Crete. Another example is the proposed elimination of some early Mayan cities, in about 450 AD, by a VEI 6+ eruption at Ilopango volcano, El Salvador.[4]

The principal benefits for archaeologists working in volcanic areas are the widespread sequences of airfall tephras that can be dated by geochronological methods. The broad areal extents of the tephras represent individual time horizons for spatial correlation and comparisons of contemporaneous cultural artefacts, as well as of edible plant remains such as nuts and coconut shells, together with peat, pollen, phytoliths and starch grains. The tephras may build up long-duration sequences at any one site of archaeological and palaeo-environmental interest, although the number of such sites that can be studied across any one time horizon may, for different reasons, be limited. This applies to the results of the extensive archaeological work on Willaumez Peninsula in West New Britain Province, where the evidence has been used in discussions of whether the volcanic eruptions are the cause of subsequent cultural change, and thus possibly even to adaptations to the ongoing threat of volcanic hazards. These discussions, however, are fraught with uncertainty, especially where the physical evidence consists mainly of obsidian tools and chips, and where the important discussions on risk and strategy amongst the affected peoples themselves remain archaeologically invisible. Furthermore, the relative dating of and connections between events commonly remains controversial and inconclusive in the absence of precise geochronolgial data. Which came first, volcanic eruption or social change? And, if definitely the former, is that necessarily conclusive that cultural adaptation to the volcanic threat necessarily followed the eruption?

What people on Willaumez Peninsula thought and how they made decisions about volcanic threats and evacuation are not recorded archaeologically. Neither are their deductions about volcanic early warning signs nor how they coped with false alarms. There is, consequently, a temptation to apply conclusions from present-day view points or through application of results from pre-history studies in other parts of the world. Syncretism and imaginative interpretation of minimal evidence, however, must be and have been guarded against. There are, after all, no known equivalents of Pompeii in Near Oceania, or reported evidence yet for people being killed by the VEI 5–6 eruptions at Witori and Dakataua

---

4 The radiocarbon ages and VEI values quoted here for Santorini and Ilopango are from Siebert et al. (2010). See also Harris (2000).

— for example, tephra-buried human bones or destroyed settlements of any significant size or number. The only certainty is the rapid burial of landscapes and therefore habitable surfaces by airfall tephras and pyroclastic flows, and the new surfaces so formed remaining uninhabitable for people, presumably until gardens could be re-established, or forest and reefs returned to provide their required food resources. Nevertheless, there is some appeal in the general conclusion that the people on Willaumez Peninsula who were affected by the Witori-Dakataua eruptions may well have been resilient in facing the effects of volcanic eruptions, bearing in mind that they probably had a low dependency on built environments, a high degree of mobility including movements by both land and sea, a minimal reliance on intensive agriculture and domesticated food stuffs, and presumably low population densities.[5] Risk, therefore, was probably much lower compared to that for the more populous and sedentary communities of present-day West New Britain Province.

Navigating the written records for this narrative has involved, as in many other historical efforts, the search for the Holy Grail of factual objectivity. All historical narratives, however, are shaped by the many tellers, writers, listeners and readers of real events and so, inevitably, are subjective, although verification can be used more favourably to check factual veracity in the histories of more modern times. Oral histories are arguably no different and so, in the absence of written records, are potentially useful sources of facts, or of what originally may have been facts.[6] Myths and legends are an integral part of all histories, including scientific ones, and they have coloured the histories of volcanic eruptions and disasters over the millennia. But can oral traditions carrying points of volcanological interest such as eruptions, disasters, and coping strategies, be used in isolation as scientifically valuable testimonies for past events? Interrogation of volcano-related stories transmitted through oral traditions has been another popular, multidisciplinary, research activity in recent years yet, like its partner field of volcanic archaeology, the results are generally characterised by uncertainty if not inconclusiveness.[7] Nevertheless, some verification is possible where the stories are consistent with local volcanic geology and where they overlap with eras of written histories, and where both the written and orally transmitted histories are consistent with each other.

Pre-industrial societies in Near Oceania are known through their stories, and in European anthropological accounts of them, to have accepted the existence and

---

5 Highly mobile communities may be the norm for many societies in circum-Pacific regions before European contact. Sheets (2007), for example, described examples from eruption-affected parts of Central America, which are similar in this respect to the societies inferred for the Willaumez Peninsula area.
6 See, for example, the general review and discussion on oral history and oral traditions by Neumann (1992) in his book on the Tolai people of Rabaul.
7 See, for example, recent compilations of conference papers on volcanic oral traditions edited by Grattan & Torrence (2007) and Cashman & Giordano (2008). These compilations include specific papers on oral traditions in volcanic areas of the south-west Pacific by Cronin & Cashman (2007) and Cashman & Cronin (2008).

recognised the importance of spirit worlds involving deities, ancestors, totems, as well as fiends who exist alongside humans in the physical universe. Powerful spirit figures in volcanic areas were feared, or at least recognised as requiring respect or some sort of propitiation through ritual and ceremony. The many features of the volcanic landscape at Rabaul are an example where different kinds of spirits or *kaia* inhabit different landscape features and where the word 'kaia' itself in Tolai society was in some instances synonymous with active volcanoes. Tolai oral tradition includes the story of the eruption at Sulphur Creek in about 1850, which is told allegorically and dramatically as a fight between a large crab and a snake from the spirit world.

The names of some spirits in parts of Near Oceania are those of active volcanoes themselves. Ulawun and neighboring Bamus are examples, both deriving from the story of the 'smoking' Ulawun who is also known as the Father. The spirit Ulawun is said to have introduced tobacco to the local Nakanai people. He married a Nakanai woman, and Bamus is one of their sons — hence 'South Son' as a synonym for Bamus. In addition, Karkar volcano is named after a founding ancestor who is now present in the spirit world. The summit caldera area of Karkar, including Kanagioi peak, is where people live after death, together with deities and ancestors, as well as being the site of the Christian 'heaven' and a place where 'cargo' is manufactured. Perhaps the most dramatic volcano-related personality in Near Oceania is Zaria, the terrifying female spirit who lives in the active crater of Manam Island and who features in stories of the origin of fire and whose unfortunate husband is seen frozen in geological form as Yabu Rock on the south-western side of the island. The crags at the summit of Lamington volcano in Oro Province also held geo-legendary significance for the local Orokaiva who regarded the summit as the centre of their universe. The ancestor Sumbiri was the first Orokaivan to die. He became a spirit master, living inside the mountain along with those Orokaivan people who died after him.

All of these stories — at Ulawun, Bamus, Karkar, Manam and Lamington — refer to the summits or active craters of high volcanic mountains inhabited by personalities from the spirit world, but avoided by humans. Non-settlement by people of these volcanically dangerous summit areas is a sensible precaution, of course, and the stories do raise the possibility that the existence of the spirits worked as a good reason for people to stay away. The spirits in practice may, therefore, have been effective in reducing volcanic risk.

People living on or near three other volcanoes in Near Oceania have stories relating to eruptions that were not witnessed by Europeans, but which, to differing degrees, are potentially valuable as sources of volcanological information. The villagers of Savo Island in the Solomon Islands have a rich oral history of large prehistoric eruptions for the volcano that today is regarded, on the basis of its volcanic geology and historical activity, to be still dangerous. Strong oral traditions also underpin the existence of Yomba volcano, south-east of Long Island, where Hankow Reef now exists, and whose strange disappearance — according to some of the stories — is said to have been catastrophic. Hankow Reef may well be built on a volcanic foundation, but side-scan surveying of the surrounding sea floor has not revealed any further evidence relating to the past geological history of this legendary volcanic island. The third example is the comprehensive collection of stories collected from mainland New Guinea, and from the highlands region in particular, of the ash falls that produced the Tibito and Olgaboli tephras exposed, for example, at the Kuk archaeological site. The younger Tibito Tephra appears to correlate with the Matapun Beds on Long Island, which were deposited at about the time of the latest, caldera-forming, VEI 6 eruption at Long. The oral traditions in the Huli area of the highlands concerning Tibito Tephra include the well-documented and ash-fall-related phenomena of *mbingi*, which includes the preparations that must be made traditionally for similar events in the future — an example of prehistoric disaster mitigation.

All three of these examples of volcanic oral tradition have ongoing issues concerning the accurate dating of the events or observations being reported, and relating their timing to the modern calendar. Some attention has been given to this challenge by counting the number of reporting generations, of a particular length, back to the time of the original event in a narrative of oral history. Such counting is difficult because the narratives cannot be regarded necessarily as being chronologically precise. Generations may be omitted or added for reasons related to, for example, claims on land, or tribal competition for political and social dominance, or simply by human error. The narratives also cannot be regarded as being designed and promulgated primarily for the transmission of precise historical facts, but rather may have cultural, even artistic and poetic value that has been used for ritual or performance purposes. Furthermore, verification of the claimed generational sequences from more than one authoritative source may be impossible to obtain. Radiocarbon dating of materials hosted by tephras or their palaesols may help in some cases, but accurate correlation of isotopic values for very young materials to just a few years on the modern calendar is still difficult, despite modern advances in radiocarbon dating. Large, global, volcanic events can be dated on an annual basis from suitable ice-core and tree-ring records. Identifying the source volcano, however, involves a process of eliminating other volcanoes and eruptions, which depends on having good radiocarbon dates in the first place.

International Disaster-Risk Reduction

More important than interpretations from archaeology and the oral traditions of former times are the present-day risk perceptions of communities in volcanically active Papua New Guinea and the Solomon Islands. Contemporary risk perceptions cannot be assumed to be those of yesterday. Oral traditions can and should be studied, but modern at-risk communities in Near Oceania now live in a different world, one that is globalised, technologically linked, and governed largely by the forces of secular rationalism rather than by traditional belief systems. The modern world's strengths and opportunities need to be used in the ongoing work of volcanic disaster risk reduction. This world is, however, a global network of interaction and competition, a hierarchical patchwork of national sovereignties in which global wealth and human capacity are unevenly and unfairly distributed. Global climate change, population growth, stresses on essential resources, losses of biodiversity, and the threats of nuclear catastrophe between warring nations, are dominant global threats and risks, rather than volcanic ones.

Financial support for international disaster management efforts is directed largely towards relief and recovery following major disasters, such as the 2004 Indian Ocean tsunami, the 2010 Haiti earthquake, and the 2011 Honshu earthquake and related tsunami and Fukushima nuclear-power disaster. These disaster responses represent episodes of altruistic, humanitarian and material outpouring, which is assisted by the voluntary financial contributions made by concerned citizens from many countries. The needs of the affected countries are promoted dramatically by global media outlets through news stories and high-impact film footage and social media. Funding for disaster recovery may also, in some instances, lead to economic and social benefits that did not exist in the affected country before the disaster. There are also many political and diplomatic rewards and advantages to be gained on the international stage by countries who respond generously and effectively at such times of tragedy. In contrast, however, funding for disaster risk reduction — that is, the mitigation or prevention of and preparation for disasters — tends to be less forthcoming, despite work by the United Nations through its International Strategy for Disaster Reduction, or the World Bank through its Global Facility for Disaster Risk Reduction. Furthermore, successful application of disaster risk reduction efforts does not make for exciting stories and thus it receives little attention from the world media. Disaster risk reduction today is, nevertheless, a well-known theme in the international field of disaster management. It even has its own acronym, DRR, which has become part of disaster management language of government policy-making and academic research.

There are many ways in which DRR can be achieved nationally and locally. Hazard and risk-assessment and mapping are essential as ways of identifying community vulnerabilities, and a clear determination of community risk perceptions is fundamental. Hazard awareness-raising in communities is important, including through education and formal inclusion in school curricula. Legislation to prevent development in areas vulnerable to natural hazards can be productive but difficult to impose. Effective building codes ensuring the construction of safe housing during, say, earthquake-induced ground shaking is another way to achieve DRR, if the resources are available to build safer buildings. Scientists, engineers, technicians and town planners also can consider DRR issues when designing the placement and construction of roads and bearing in mind evacuation routes to predetermined refuges, as well as early warning systems and critical infrastructure such as hospitals, power plants, airports and harbours. The rapid transmission of hazard alerts is critical and ideally must include partnerships with the news media. At-risk communities can practice prescribed evacuations.

DRR is also an element of international development assistance for countries, such as Papua New Guinea. Natural hazard risk reduction concepts are included in the development assistance policies of many national and international donor agencies. Development assistance is well known as an effective instrument of political influence, but in some countries, international disaster management can be a crowded 'market place', requiring recipients of such assistance to manage the diplomatic complexities of implementation that are caused by the attention of multiple, possibly competing donors. Some donor agencies are more concerned than others about the effectiveness of their investments and whether resulting improvements are sustainable once the assistance ends.

This, then, is the general context for the more specific issues of volcanic risk reduction in Near Oceania, including what role volcanological observatories should play.

Observatories and Volcanic Disaster-Risk Reduction

The European concept of the 'volcanological observatory' was imported to the New Oceania region from the Netherlands Indies in 1937, during Australian colonial times, and the first 'volcanological observatories' — built in Italy in the second half of the nineteenth century — were based on the model of astronomical observatories equipped with telescopes for observation and study of the planets and stars. Telescopes were, and still are, used for such direct 'observing' purposes in many volcanological observatories, which may be situated on isolated, underpopulated parts of mountains providing good visibility of summit areas and active craters. This physical isolation of observatories, however, commonly gives rise to perceptions of a monastic culture of volcanologists who focus on narrow scientific endeavours that are separate from the needs of the real world. Furthermore, the observatory culture has developed its own set of customs, mores, and traditions that have been built up over many decades in different parts of the world.

Volcanological services in Third World countries, such as those of the Circum-Pacific region, are run largely by governments, which means that volcanologists and technicians are public servants who are accountable to the government in power. Furthermore, national volcanological services commonly operate from a single headquarters, which may be identified as the country's volcanological centre or 'observatory'. There is a high public expectation that such taxpayer supported volcanological centres will be successful in providing a national eruption early warning service based on the data recorded from instruments installed on threatening volcanoes. Evacuations can be recommended, thus reducing the risk of potentially damaging effects on society. An ideal situation is where such instrumentation on any one active volcano is permanent and can be operated 24 hours a day, seven days a week. This may not be possible, however, if there are many dangerous volcanoes scattered throughout the country and national funding support is inadequate for capital and maintenance costs for instrumentation at all of them. Theft, vandalism, scavenging, and the climatic deterioration of equipment deployed in isolated areas, as well as the destruction of instruments by volcanic eruptions, are additional problems. Indeed the question should be asked: Is the instrument-based volcanological observatory 'model' still relevant to the challenges of volcanic DRR work in modern and independent countries, such as Papua New Guinea? The question is important in recalling that only once during the last 75 years has instrumental data been significant in declaring the need for evacuation anywhere in Near Oceania — at Ulawun in 2000. Alternative strategies may be needed in order to reduce volcanic risk, especially during times of crisis.

Interpreting the results of instrumental monitoring on volcanoes is not straightforward, whether in a wealthy or less privileged country. This is because ground shaking measured by seismographs, or temperatures measured in thermal areas, or ground deformation measured by tiltmeters, GPS equipment, and so forth, are surface phenomena. They are related, but only indirectly, to the activity of deep-seated magma bodies, the important and primary characteristics of which — such as gas content, viscosity, temperature, pressure, and volume of magma likely to be erupted — cannot be measured by these secondary 'surface' techniques. Neither can the robustness of blocked eruption conduits nor the strength of the crustal rocks that confine the pre-eruption magma reservoir, be measured by these methods.

Both Papua New Guinea and the Solomon Islands are young, sovereign, nation states which, like other volcanically active countries in the Circum-Pacific and in the Third World in general, occupy challenging places in a rapidly changing world. Thousands of people in Near Oceania live on or near volcanoes, many of which are capable of producing scales of eruption far greater than any recorded there in historical time. The Solomon Islands and Papua New Guinea, however, have contrasting opportunities and challenges in the field of volcanic DRR. Only seven of the 57 Holocene volcanoes of Near Oceania are in the Solomon Islands chain, and only Kavachi has been repeatedly active historically, so far without significant hazard impact. Historically active Savo, on the other hand, being so close to the capital Honiara, is a constant and visible reminder of future threats from this potentially dangerous volcano. There is no national volcanological service in the Solomon Islands, as there is in Papua New Guinea, and all matters pertaining to geological hazards are covered by a small number of general geoscientific staff in the Ministry of Mines, Energy and Rural Electrification in Honiara.

The population of the Solomons is probably around only 550,000, spread mostly amongst the small islands of the western Solomon Islands chain. Neighbouring Papua New Guinea, in contrast, has a population approaching seven million, and has an abundance of natural resources including minerals and natural gas — particularly on New Guinea Island which is by far the largest land mass in Near Oceania. Papua New Guinea, therefore, has considerable wealth potential, which if realised, could be invested in the strengthening of public service agencies, such as the national volcanological service. RVO in principle, therefore, could become self-sufficient in the years ahead, and ideally may not require international development assistance for undertaking its work.[8]

Determining the type and extent of instrumental monitoring on the many active volcanoes of Near Oceania thus remains a difficult challenge. The posed question then reduces to one of cost-benefit: Are instrumental systems on active volcanoes the best and most cost effective way of reducing volcanic risk through early warning of eruptions?

---

[8] There is value in this context of Papua New Guinea and the Solomon Islands joining with Vanuatu in supporting the creation of a Melanesian Volcanological Network, and collaborating in the exchange and sharing of the volcanological resources in each of the three countries. A proposal for such a network was completed in 2008 by the Applied Geoscience Commission SOPAC, which is now a division within the Secretariat of the Pacific Community, SPC. The proposal has yet to attract financial support.

Strengthening At-Risk Communities in Near Oceania

A striking conclusion from the analysis of evacuations in Near Oceania since 1937 is that communities at risk from volcanic eruptions in Papua New Guinea are commonly self-reliant when decisions have to be made about their own safety and the need for possible evacuations. Such community decisions typically can be made promptly, responses can be rapid, and the evacuating communities can be remarkably mobile, as long as authorities do not interfere negatively with the spontaneity of the process, as in the case of Lamington in 1951. 'Spontaneous' evacuations such as these can, in many circumstances, be a more effective way of volcanic-threat avoidance, compared with waiting for instrument-based early warnings of eruptions from cautious, distant authorities. Well-informed, resilient, and self-sufficient communities may be the first to recognise the earliest signs of volcano change, meaning that professional volcanological assistance can be requested. Volcanologists can then respond rapidly and deploy mobile arrays of instruments, if necessary, for further assessment of the potential threat. This may be a more practical option for countries such as Papua New Guinea where funding support may not exist for expensive, permanent, and sustainable instrumental networks on volcanoes.

A critical success factor for any national volcanological observatory is the way in which it collects, compiles, stores and analyses information about the nation's volcanoes. Making observations of eruptions and collecting instrumental data are important, but there is also a fundamental requirement to ensure that the different types of scientific information obtained by one generation of volcanologists are made available as a legacy for the next. This is where compiling data, writing reports, building bibliographic reference collections, managing geographic information systems, and using national spatial-data infrastructures are crucial, not just for coping with contemporary volcanic crises, but for future ones too. Knowledge is, indeed, crucial and much still needs to be learnt about the dangerous volcanoes of Near Oceania — their geological histories, scales of past eruptions, internal structure, and how threatening magmas come to be emplaced and then evolve beneath the volcano. Additional investment in investigative field surveys and related research must continue, leading to a greater understanding of how threatening volcanoes 'work'.

These themes as a whole give rise to the requirement for an even stronger program of building volcanically resilient at-risk communities in which instrumental monitoring plays a more balanced, cost-effective, and sustainable part. A government volcanological agency such as RVO is required to advise provincial authorities and its host department in Port Moresby regarding volcano status, eruption alerts, and evacuation recommendations, as it has done since its establishment in 1940, but public awareness campaigns are now at the core of RVO's work. Papua New Guinean staff from RVO make repeated visits to volcanically at-risk communities, giving presentations at open meetings, answering questions, and using audio-visual equipment. They visit schools and, in East New Britain, encourage visitors to observatory headquarters in Rabaul to see the volcano-monitoring instrumentation in operation, where necessary providing transport from outlying areas.[9] Volcanology is an international science that progresses resolutely on many fronts and 'does what it must', but there are no reasons why at-risk communities in Near Oceania cannot continue to increase their resilience by taking greater advantage of the transferable knowledge that the science provides.

Figure 130. RVO officer Jonathan Kuduon, on the left, addresses school children and other villagers during a volcano-awareness campaign at Bokure village on Manam volcano in 2000.

Source: D. Okole. Rabaul Volcanological Observatory.

---

9   A different approach is the Participatory Rural Appraisal or PRA methodology used at a workshop on Savo Island in 1999. The PRA methodology was developed overseas and the main drivers for the workshop were three volcanologists from New Zealand and Australia 'who tried to combine the roles of facilitators and educators, and to involve the input of all stakeholders (from community to national government) in the process of volcanic risk management' (Cronin et al. 2004b, p. 105; see also Petterson et al. 2008). The same methodology was used on Ambae Island, Vanuatu, where a special attempt was made to incorporate traditional knowledge of eruptions with scientific concepts (Cronin et al. 2004a).

References

Byrne, J., 2012. 'Noted', *The Monthly*, February, p. 64.

Cashman, K.V. & S.J. Cronin, 2008. 'Welcoming a Monster to the World: Myths, Oral Tradition, and Modern Societal Response to Disasters', in K.V. Cashman & G. Giordano (eds), 'Volcanoes and Human History', *Journal of Volcanology and Geothermal Research*, 176, no. 3 (special edn), pp. 407–418.

Cashman, K.V. & G. Giordano (eds), 2008. 'Volcanoes and Human History', *Journal of Volcanology and Geothermal Resources*, 176, no. 3 (special edn), pp. 325–437.

Cronin, S.J. & K.V. Cashman, 2007. 'Volcanic Oral Traditions in Hazard Assessment and Mitigation', in J. Grattan & R. Torrence (eds), *Living Under the Shadow: Cultural Impacts of Volcanic Eruptions*. Left Coast Press, Walnut Creek, California, pp. 175–202.

Cronin, S.J., D.R. Gaylord, D. Charley, B.V. Alloway, S. Wallez & J.W. Esau, 2004a. 'Participatory Methods of Incorporating Scientific and Traditional Knowledge for Volcanic Hazard Management on Ambae Island, Vanuatu', *Bulletin of Volcanology*, 66, pp. 652–68.

Cronin, S.J., M.G. Petterson, P.W. Taylor & R. Biliki, 2004b. 'Maximising Multi-stakeholder Participation in Government and Community Volcanic Hazard Management Programs; A Case Study from Savo, Solomon Islands', *Natural Hazards*, 33, pp. 105–36.

Grattan, J. & R. Torrence (eds), 2007. *Living Under the Shadow: The Cultural Impacts of Volcanic Eruptions*. Left Coast Press, Walnut Creek, California.

Harris, S.L., 2000. 'Archaeology and Volcanism', in H. Sigurdsson (ed.), *Encyclopedia of Volcanoes*. Academic Press, San Diego, pp. 1301–314.

Neumann, K., 1992. *Not the Way It Really Was: Constructing the Tolai Past*. University of Hawaii Press, Honolulu.

Petterson, M.G., D. Tolai, S.J. Cronin & R. Addison, 2008. 'Communicating Geosciences to Indigenous People: Examples from the Solomon Islands', in D.G.E. Liverman, C.P.G. Pereira & B. Marker (eds), *Communicating Environmental Geoscience*. Geological Society, London, Special Publications, 305, pp. 141–61.

Sheets, P., 2007. 'People and Volcanoes in the Zapotitan Valley, El Salvador', in J. Grattan & R. Torrence (eds), *Living Under the Shadow: Cultural Impacts of Volcanic Eruptions*. Left Coast Press, Walnut Creek, California, pp. 67–89.

Siebert, L., T. Simkin & P. Kimberley, 2010. *Volcanoes of the World*. 3rd edn. Smithsonian Institution, Washington D.C., University of California, Berkeley.

Stehn, Ch.E. & W.G. Woolnough, 1937. 'Report on Vulcanological and Seismological Investigations at Rabaul', *Commonwealth of Australia Parliamentary Paper 84 of 1937*, pp. 149–58.

Torry, W.I., 1978. 'Natural Disasters, Social Structure and Change in Traditional Societies', *Journal of Asian and African Studies*, 13, pp. 167–83.

An Epilogue

Melanesian tradition and scientific accuracy are combined comfortably in this artwork by the late Cecil King Wungi, depicting the internal architecture of a volcano. The image was published originally in 1976 on the cover of a handbook for the Geology Department, University of Papua New Guinea. Wungi worked as a laboratory technician in the department during the 1970s.

Source: C. K. Wungi. Used with permission of the Wungi family.

Appendix: Acronyms and Glossaries

Acronyms

ANGAU: Australian New Guinea Administrative Unit

AusAID (AIDAB): Australian Agency for International Development (formerly Australian International Development Assistance Bureau)

BMR: Bureau of Mineral Resources — now Geoscience Australia (GA)

CPL: Coconut Products Limited

CSIRO: Commonwealth Scientific and Industrial Research Organisation

CTBTO: Comprehensive Test Ban Treaty Organisation

DRR: disaster risk reduction

DSIR: New Zealand Department of Scientific and Industrial Research

GIS: geographic information system

GRA: Gazelle Restoration Authority

IAVCEI: International Association of Volcanology and Chemistry of the Earth's Interior

IDNDR: International Decade for Natural Disaster Reduction (United Nations)

JICA: Japan International Cooperation Agency

NDES: National Disaster Emergency Services

OCHA: Office for the Coordination of Humanitarian Affairs (United Nations)

PDC: Provincial Disaster Committee

PMGO: Port Moresby Geophysical Observatory

PMV: public motor vehicles

PNGRIS: Papua New Guinea Resource Information System

RVO: Rabaul Volcanological Observatory

USGS: United States Geological Survey

VAAC: Volcanic Ash Advisory Centre (International Volcano Watch)

VDAP (VCAT): Volcanic Disaster Assistance Program, USGS (formerly Volcanic Crisis Assistance Team)

VEI: volcanic explosivity index

VSS Project: Papua New Guinea – Australia Volcanological Service Support Project

Glossary 1: Types of Explosive Eruptions

hydrovolcanic: Explosions caused by magma encountering subsurface water and producing intense fracturing and expulsion of fragmented rocks.

peléean: The name derives from the explosive eruptions at Mount Pelée in 1902 when nuées ardentes were first described and named. Both terms are no longer volcanologically fashionable.

plinian: Paroxysmal ejections of large volumes of pyroclastic materials, at times accompanied by caldera formation. The eruptions form high-rising eruption columns and clouds. Large-volume pyroclastic flows may develop when the column collapses, including ignimbrites if pumice is abundant in the flows.

strombolian: Weak to violent ejection of pasty blebs of fluid lava, accompanied by spherical to fusiform 'bombs', cinders, and ash. The activity can be spectacularly incandescent at night-time. Lava flows may be formed.

surtseyan: Distinctive black bursts of pyroclastic materials in cocks' tail or cypressoid-like patterns, and mixed with contrasting white water vapour where a subaqueous eruption breaks through the surface of the sea or lake. Base surges may explode radially outwards from the foot of the sub-aerial eruption cloud.

vulcanian: Violent ejections of solid or viscous hot volcanic fragments, at times in cauliflower- or mushroom-shaped clouds. Pyroclastic flows and lava flows are typically absent.

Glossary 2: Volcanological Terms

andesite: A generally grey volcanic rock containing 53 per cent, up to less than 62 per cent, silica (SiO_2) and forming a chemical series with basalt, dacite and rhyolite.

Appendix: Acronyms and Glossaries

ash: Pieces of generally glassy volcanic rock less than 4 millimetres in diameter.

avalanche amphitheatre: A large, arcuate, or U-shaped escarpment where the flanks of a volcano have collapsed gravitationally producing debris avalanches.

basalt: A dark volcanic rock containing less than 53 per cent silica (SiO_2) and forming a chemical series with andesite, dacite and rhyolite.

base surge: Laterally propelled eruption clouds caused typically by the interaction of magma with lake water, or in the shallow water of coastal areas. They flow across the water surface away from the vent and the base of the main eruption cloud or column.

caldera: Large, generally elliptical or sub-circular surface depressions at least one–two kilometres wide, formed by the collapse of the roofs of magma reservoirs.

dacite: A generally light-greyish volcanic rock containing 62 per cent, up to less than 70 per cent, silica (SiO_2) and forming a chemical series with basalt, andesite and rhyolite.

debris avalanche: Highly mobile flows of broken rock and entrained air formed where the side of a volcano collapses and producing an avalanche amphitheatre.

Holocene: The geological epoch that began at the end of the Pleistocene, about 11,700 years ago, and which continues to the present day. Its start is commonly rounded up to about 12,000 years ago and thus to 10,000 years BC.

ignimbrite: A pumice-rich deposit or rock commonly of dacitic or rhyolitic composition produced by deposition from a pyroclastic flow.

intrusion: Rocks representing magmas that have filled and solidified in fissures, cracks, faults and other spaces beneath volcanoes, but which have not erupted from them.

lava flow: A ground-hugging stream of erupted magma and rock, commonly referred to simply as 'lava' and clearly distinguishable from pyroclastic flows.

maar: Bowl-shaped craters caused by the explosive interaction of magma with groundwater (see hydrovolcanic explosions). They typically cut deeply into the pre-eruption ground surface and may be surrounded by low-angle ramparts of pyroclastic materials.

magma: Sub-surface rock that has become molten and which either erupts from volcanoes or forms intrusions beneath them. Most magmas are technically alumina-silicate liquid and most contain at least some pre-eruption crystals.

nuée ardente: The term means 'glowing cloud' and was introduced for the block-and-ash type pyroclastic flows observed at Mont Pelée on 8 May 1902.

obsidian: A volcanic rock consisting almost entirely of natural glass.

Pleistocene: The geological epoch spanning the world's recent period of glaciations and most recently dated from about 2.6 million to about 11,700 years ago (see also Holocene).

pumice: Pieces of highly frothed volcanic rock, commonly of dacitic or rhyolitic composition, and which can float on water.

pyroclastic: Literally meaning 'fire broken' and applied to fresh magma or hot volcanic rocks that have been broken up during volcanic explosions.

pyroclastic flow: Fast-flowing, hot emulsions or 'avalanches' of pumice, ash, dust, blocks, volcanic ash, and entrained air, which tend to follow the floors of valleys during their emplacement.

pyroclastic surge: The turbulent, lower density part of a pyroclastic flow which is not so constrained by topography as is the main and denser part of the flow.

rhyolite: A volcanic rock containing 70 per cent silica (SiO_2) or more and forming a chemical series with basalt, andesite and dacite. Rhyolite ranges from black, shiny, glassy obsidian to a pale grey or white rock when crystallised.

tephra: A commonly used synonym for pyroclastic materials or pyroclastic rocks in general. The term was used originally by Aristotle.

volcanic field: Volcanic areas characterised by numerous small cones and craters rather than just one major volcano. Many of these small cones are formed by just one volcanic eruption and so are commonly called 'monogenetic'.

Index

An index is unnecessary for a book such as this one being published online and where terms can be searched electronically. Here, however, for the benefit of readers of the hard-copy version, is a two-part index of text references to: (1) the names of the main volcanoes and volcanic areas in Near Oceania; and (2) the names of the local participants in this history, including volcanologists, volcanic-risk specialists, other researchers, decision makers, and other key reporters on volcanoes, eruptions, disasters, and volcanic products in Near Oceania.

Volcanoes and Volcanic Areas:

Aird Hills 93
Bagana 30, 33, 58, 64, 74-75, 77-78, 140, 180, 212, 232, 234, 326, 361
Balbi 30, 75
Bam Island 7-8, 45, 58, 77, 180, 183, 187-194, 202, 232, 361, 363-365
Bamus 55-56, 58, 77-78, 313, 361-362, 370
Blup Blup Island 187
Bosavi 97-98
'Cook' xxiii, 245- 246
'Cornwallis' 29
Crown Island 6, 240
Dakataua: see Makalaia-Dakataua
Dawson Strait area (including Esa'ala) 42, 90, 183, 210, 217-224, 262, 361, 363, 366
Dobu Island 42, 90, 183, 217-221, 262, 366
Doma Peaks 97, 212-213
Du Faur 55-56
Esa'ala: see Dawson Strait area

Favenc 93, 98
Fergusson Island 12, 42, 183, 217-220, 222
Galloseulo-Hargy 107, 122, 276, 323, 362
Garbuna 331-332, 334, 361
Garove Island 323, 362
Giluwe 95, 98-99
Goropu (Waiowa) 136-141, 153, 183, 219, 249, 361, 363
Hagen 95-96, 99, 231, 241
Hahie 200
Hargy: see Galloseulo-Hargy
Hydrographers Range 93, 151
Ialibu 95, 98-99
Kadovar Island 7-9, 33, 187, 232-233, 361, 366
Karimui 98
Karkar Island 7, 11, 22, 25, 33, 76-78, 180, 224, 226, 232, 234-235, 246-250, 271, 276, 316, 332-333, 361-362, 370
Kasu 331
Kavachi 144, 180, 232, 234, 312, 361, 375
Kerewa 97
Koranga 94, 214-217, 223, 361, 366
Kururi 140
Lamington xxiii, 1, 12, 93, 98-99, 118, 133, 139, 143-144, 149-175, 179-180, 182-183, 186, 191, 194, 212-213, 219, 233, 268, 271, 292, 316, 331, 360-361, 363-366, 370, 376
Langila 57-58, 77-78, 180, 210, 232, 234, 262, 334, 361
Lihir Island 2, 22
Likuruanga 32, 55-56, 332-334
Lolobau Island 32, 55, 58, 77, 107, 276, 323, 362
Loloru 362
Long Island 6-7, 112, 140, 180, 183-188, 196, 208, 232, 234, 239-245, 267-268, 276, 359, 361-365, 371
Lou Island 12, 46, 180, 199-202
Makalaia-Dakataua 77-78, 107, 112, 323-324, 361-362, 364, 368-369
Managalase Plateau 139-140

387

Manam Island 7, 9, 13, 24-26, 32-33, 45, 58, 65, 77-78, 106, 108, 140, 179-180, 182-183, 187-188, 192-198, 210, 225-227, 232, 234-235, 262, 268, 271, 276, 311, 316, 326-331, 334, 361-365, 370
Mopir 12, 324
Murray 97
Musa River xxiii
Pago-Witori 76-77, 107, 112, 122, 319-325, 328, 331, 334, 361-363, 368-369
Palangiagia: see Rabalanakaia/Palangiagia
Pam Islands 12, 200
Rabalanakaia/Palangiagia (Rabaul) 27, 34, 41, 76, 256, 349
Rabaul area (including Rabaul town and the Rabaul Volcanological Observatory) xviii, 3, 27, 33-34, 39-42, 46-55, 58, 60, 63, 65, 73, 75-76, 79, 94, 99, 103-106, 109, 112-124, 129-137, 140, 142-144, 149, 154, 179, 181, 183-184, 210-211, 223-225, 227, 231-232, 244-245, 249, 255-277, 283-306, 313-314, 316-317, 319, 322-323, 326-328, 330-332, 334, 341-355, 359-365, 367, 369-370, 375-377
Ritter Island 4-7, 26, 30-33, 57-58, 65-70, 74, 77-79, 232, 234, 239, 240, 245, 313-315, 332, 334, 361
Savo Island 1, 23, 26, 33, 35, 64, 133, 144, 180, 312, 361-362, 370, 375, 377
Simbo Island 277-278, 312, 361, 366
Sisa 97-98
St Andrew Strait area 199, 202, 222
Suaru 98
Sulphur Creek (including the Sulphur Creek Observatory, Rabaul) 34-35, 103, 131-133, 135, 305, 360, 370
Sulu Range 332-334, 361, 363, 366
Talasea area 12, 222, 320, 323-324
Tavui 134-135, 323, 348, 362
Tavurvur (Rabaul) 27, 29, 34, 41, 47, 50-52, 54, 104, 117-118, 122-123, 129-134, 140, 182, 256, 264, 275-276, 283-285, 287, 289, 296, 298-300, 305, 311, 330-331, 334, 342-344, 346, 348-349, 351, 359-361

Tuluman Island 45, 78, 180, 199-202, 361
Ulawun 3, 6, 9, 33, 55-56, 58, 77-78, 140, 210, 232-239, 269, 271, 276, 313-318, 331, 333-334, 361-363, 365-366, 370, 374
Umboi Island 32, 68-69, 184, 240, 332, 362
'Umsini' xxiii, 86-87
Unea Island 323, 362
Victory 42, 77-78, 88-90, 92-93, 139, 361-363
Vulcan (Rabaul) 50-54, 75, 113-119, 122-123, 129, 135-136, 143, 264, 274, 283-286, 289-290, 292, 296, 300, 331, 342, 344-345, 349, 359-361
Waiowa: see Goropu
Wangore 107
Willaumez Peninsula (see also Talasea area) 32, 56, 66, 107, 323-324, 368-369
Witori: see Pago-Witori
Yelia 212, 362
'Yomba' xxiii, 245-246, 371

Participants:

Adamson, C.T.J. 97-99
Addison, R. 277
Aislabie, C. 271
Alexander, B. 286
Anderson, H. 220
Anderson, L. 284, 287
Archibald, M.J. 258
Ashton, D.N. 214
Ball, E.E. 70, 185
Banks, N.G. 267
Bates, C.D. 184
Beasley, J. 220
Behrmann, W. 71
Belshaw, C.S. 164, 170-171
Bemmelen, R.W. van 109, 152
Bensted, F.E. 188
Best, J.G. 151, 154, 179-181, 184-186, 188, 192-193, 200
Bibra, M. de 149-151, 165
Blake, D.H. xviii, xxi, 96
Blaikie, R.W. 157, 162

Index

Blong, R.J. xxi, 241, 243-244, 271, 289, 292
Boas 76
Boegershauser, G. 34, 103
Bougainville, L.-A. de 27
Branch, C.D. xviii, 211-212,
Britain, G. 325
Brown, C. 106
Brown, G. 47, 50-52
Carne, J.E. 91
Carey, S.W. 98-99
Carteret, P. 26-27, 29, 64
Champion, C. 159, 186-187
Champion, I.F. 97-99, 157-159
Chan, J. 284, 290
Chignell, A.K. 85, 88
Cilento, R. 103, 119-120
Cleland, D.M. 172, 187, 192
Clout, L.E. 123
Conroy, W.L. 196
Cooke, H. 233
Cooke, R.J.S. xxi, 7, 67, 231, 234-235, 238, 248-250, 255-256, 258, 263
Couppé, L. 63, 76
Cowley, A. ('Mrs') 153
Cowley, C.E. 153-154, 159, 164
Cowley, E. 159
Crellin, W. 213
D'Addario, G.W. 211, 220-221
Dalziell, T. 316
Dampier, W. 1-16, 26, 31-32, 59, 74, 86, 243-244, 311-312, 359
Davies, H.L. 91, 262, 289
Davies, R. 116
Denehey, M. 219
D'Entrecasteaux, A.-R.-J. Bruny- 26, 29-30, 32, 55-56, 66, 107
Dorsey, G.A. 74
Dumont D'Urville, J. S. C. 9, 30, 32, 244
Durdin, P. 155, 173
Dwyer, M. 94-95
Earl, A.J. 153, 159
Earl, P. 153
Edwards, J. 220
Elliot-Smith, S. 164

Endo, E. 320
Ereman, D. 283
Finsch, O. 65, 71
Fisher, N.H. xviii, xxi-xxii, 120-123, 129-131, 140, 142-144, 151, 180-182, 186, 210, 214, 226, 236, 305, 348
Friederici, G. 73-74
Gallego, H. 23
Gibb Maitland, A. 90, 92
Golson, J. 243
Granger, K.J. xxi, 272-274, 341, 354
Greet, W.F.A. 39-42, 47
Grover, J.G. 144, 180, 246
Gunther, J.T. 157, 160, 167, 172, 191
Guppy, H.B. 64-65
Hahl, A. 71, 73-77, 306
Hand, D. 166-167, 170-171
Hasluck, P.M.C. 172
Hastings, P. 255
Hawnt, E.A. 130
Heming, R.F. 213-214, 236
Hernsheim, E. 47, 49-51, 54, 59, 63
Hiari, M. 152
Hicks, W. 47, 50-52
Hides, J.G. 97
Hilder, B. 117
Hogan, Captain 29
Hohl, S.V. 353
Horne, J.R. 89, 154
Horne, R.G. 214
Hughes, I.M., 185
Humphries, W.R. 170
Hunter, J. 26, 29
Hurley, F. 94
Ischler, Father 107
Itikarai, I. xxi, 258, 297-299, 311, 319, 348, 352
Jacobson, A. 155
Johnston, G.R. 105
Kaad, F.P.C. 164
Kaivovo, E. 288, 293
Keesing, F.M. 164, 170-171
Kendall, R. 174-175
Kennedy, J. 22
Kizawa, T. 131-136, 143

Knight, C.L. 123, 131
Ko 231
Kohl, L. 75
Kuduon J. 352
Labillardiere, J.J.H. de 31-32
Langron, W.J. 161
Latter, J.H. 211
Lauer, N. 285
Lauterbach, C. 71, 96
Lawson, J.A. 85-86
Leahy, M.J. 94-95, 97
Löffler, E. 96
Lokinap, R.I. 288
Lonergen, S.A. 172-173
Lorenz, V. 73
Lowenstein, P. 255, 258, 271, 274, 276, 317
Macatol, I. 352
MacGillivray, J. 90
Maclean, 'Rusty' 157
Macnab, P.R. xix
Maire, J. le 3, 24, 26
Mano, L. 352-354
Marsh, D.R. 138
Martin, J.D. 149, 157-158, 162
Massey, C.H. 104-105
McCarthy, J.K., 115-116
McGrade, S. 286
McGregor, W. 88-90
McKee C.O. xxi, 231, 245, 249-250, 258, 263, 266, 274, 276, 289-290, 292, 296, 347
McLean, J. 333, 345
McNicoll, W.R. 112, 115, 118, 131
Mendaña, A. 1, 21, 23, 26
Meneses, J. de 9, 22
Miklouho-Maclay, N. 43-46, 58, 183, 188
Miller, L. 316
Morell Jr, B. 32
Moresby, J. 42, 87
Mori, J. 258, 263
Morioga, R. 220-221
Mulina, K. 352
Muller, W.G. 208
Murray, H. 91

Murray, J.K. 142-143, 153, 156-157, 172, 306
Nancarrow, S.N. xxi
Nettleton, J. 94
Neumann, K. 273, 293-295
Niall, H.L.R. 157
Nishimura, Y. 133
Noakes, L.C. 123
Nollen, Father 103, 114
Noser, A.A. 191
Okole, D. 377
O'Malley, L.J. 97
Pain, C.F. 241
Palfreyman, W.D. 211
Papabatu, A.K. 278
Park, W. ('Sharkeye') 94, 214
Parker Wilson, D.: see Wilson, D. Parker
Patia, H. 217, 220-221, 316
Paulius, N. 258
Pearce, G. 118
Petterson, M.G. 277
Phillips, F.B. 112, 115, 117-118, 153-154, 169
Phillips, M. 286
Piper, L. 217, 220
Plant, H.T. 164
Polach, H. 244
Porter, R.G. 155
Powell, M. 221
Powell, W. 39, 49-52, 54-58, 66
Ravian, E. 231, 248-250
Reay, M. 157, 174
Retes, I.O. de 21-23
Reynolds, M.A. 179-181, 188, 190, 200, 219
Riddell, P. 220
Rudofsky, S. 220
Russell, M. 217, 220
Ruxton, B.P. 140
Saint Ours, P. de 258, 327
Sapper, K. 63, 71, 73-75, 77, 79, 86, 216-217
Saunders, S.J. 297, 299
Savaadra, A. de Cerón 21-23
Scales, I. 304, 342-343

Schleinitz, G.E.G. 47, 65-68, 71
Schouten, W.C., 2-3, 24, 26
Scott, B.J. 262
Searle, C.E. 156
Shearman, P.L. 329
Simpson, C.H. 39-42, 47
Sinclair, J. 207, 214-215
Skinner, I. 214, 220
Somare, M. 260
Speer, A. 174
Spender, P. 169
Stamm, J. 235-236
Stanley, E.R. 90-93, 106-109, 121
Stanley, O. 90
Stehn, C. (or Ch.) E. 109, 120-122, 142-143, 236, 360
Strong, N.W. 166
Strong, W.M. 88
Sverklys, M. 157
Talai, B.P. 232, 249, 258, 271, 274, 296-297
Taranu, F. 352
Tasman, A.J. 24-25, 245
Taumosi, M. 316-317, 365
Taylor, D.J. 129, 153, 157, 165
Taylor, G.A.M. xviii, xxi, 59, 143-144, 151, 154, 156-157, 161, 163, 167-170, 172, 179-183, 185-186, 188, 193-198, 207, 210-215, 219, 224-227, 231, 235, 240, 243, 360, 364
Taylor, J. 94
Threlfall, N.A. xxi, 259
To Maran 50-51
To Mulue 34
Toba, T. 35
Topue, L. 151, 172, 231, 274, 296
Torrence, R. 325
Trégance, L. 86
Tuohy, A. 214
Wangiga, N. 193, 195, 198
Wanliss, D.S. 115
White, N.H. 165
Wicks, C. 321
Williams, S.N. 289
Wilson, D. Parker 21, 26, 30, 74

Woolnough, W.G. 120-122, 142, 306, 360
Wungi, W.K. 381
Yali 196

Printed in Great Britain
by Amazon